Spare No One

Spare No One

Mass Violence in Roman Warfare

Gabriel Baker

ROWMAN & LITTLEFIELD
Lanham • Boulder • New York • London

Published by Rowman & Littlefield
An imprint of The Rowman & Littlefield Publishing Group, Inc.
4501 Forbes Boulevard, Suite 200, Lanham, Maryland 20706
www.rowman.com

6 Tinworth Street, London SE11 5AL, United Kingdom

Copyright © 2021 by The Rowman & Littlefield Publishing Group, Inc.

All rights reserved. No part of this book may be reproduced in any form or by any electronic or mechanical means, including information storage and retrieval systems, without written permission from the publisher, except by a reviewer who may quote passages in a review.

British Library Cataloguing in Publication Information Available

Library of Congress Cataloging-in-Publication Data

Names: Baker, Gabriel David, author.
Title: Spare no one : mass violence in Roman warfare / Gabriel Baker.
Other titles: Mass violence in Roman warfare
Description: Lanham : Rowman & Littlefield, [2021] | Series: War and society | Includes bibliographical references and index.
Identifiers: LCCN 2020033393 (print) | LCCN 2020033394 (ebook) | ISBN 9781538112205 (cloth) | ISBN 9781538112212 (paperback) | ISBN 9781538112229 (epub)
Subjects: LCSH: Military art and science—Rome—History. | Rome—History, Military. | War and society—Rome.
Classification: LCC U35 .B34 2021 (print) | LCC U35 (ebook) | DDC 355.020937—dc23
LC record available at https://lccn.loc.gov/2020033393
LC ebook record available at https://lccn.loc.gov/2020033394

Contents

List of Figures, Maps, and Tables		vii
Acknowledgments		ix
1	"As Is the Roman Custom": War and Mass Violence in the Roman Republic, Third and Second Centuries BCE	1
2	"Adorned with Scars": Roman Military and Political History to 146 BCE	19
3	"What the Fire Could Not Consume": Methods of Mass Violence	33
4	"The Ram Has Hammered at Their Walls": The Logic of Mass Violence	67
5	"Deterred by Fear": Defection and Deterrence in the Second Punic War	89
6	"So Much Destruction and Utter Ruin": Politics and Pragmatism in the Third Macedonian War	133
7	"He Soaked Spanish Soil with Blood": Failure and Frustration in the Lusitanian War	175
8	Conclusion	201
Appendix 1. 124 Cases of Mass Violence in Roman Warfare, c. 400–100 BCE		213
Appendix 2. 181 Cases of Mass Violence in Ancient Mediterranean Warfare (excluding Rome), c. 500–100 BCE		227

Appendix 3. The Government and Army in the Middle Republic	243
Bibliography	249
Index	271
About the Author	281

Figures, Maps, and Tables

FIGURES

Figure 2.1. A Roman legionary soldier, third and early second centuries BCE. ... 27

Figure 3.1. Roman soldiers take some women and children captive while killing others. ... 41

Figure 3.2. A Roman soldier binds the hands of a prisoner. Prisoners chained together. ... 42

Figure 3.3. Roman allies or auxiliaries behead barbarians while Roman soldiers stand guard or watch. Trojan prisoners led to Achilles for slaughter and sacrifice to the shade of Patroclus. ... 48

Figure 3.4. Roman soldiers burn enemy strongholds. A Roman soldier burns enemy villages. ... 52

Figure 3.5. Assyrian soldiers demolish Elamite Hamanu while the city burns and other soldiers bring out the plunder. Roman soldiers demolish fortifications with dolabrae. ... 54

Figure 5.1. Plan of a Hellenistic theater based on surviving examples at Morgantina and Segesta. Nineteenth-century illustration "The Theatre at Egesta." ... 105

Figure 5.2. Plan of ancient Syracuse. 107

Figure 5.3. Nineteenth-century illustration "The Death of Archimedes." 109

Figure 5.4. Plan of New Carthage. 115

Figure 5.5. Ruins of the Punic fortifications in Cartagena, Spain. 116

Figure 6.1. Reliefs from the Aemilius Paullus Monument at Delphi. 160

Figure 6.2. The ruined northeast wall, gate, and tower, at Elea, Epirus. The shattered blocks of the eastern wall at Orraon, Epirus. 165

Figure A3.1. The manipular legion in battle array. 246

MAPS

Map 2.1. Ancient Italy, with places mentioned in chapters 1 and 2. 21

Map 2.2. The ancient Mediterranean, with places mentioned in chapters 1 and 2. 30

Map 3.1. Various sites of Roman military activity, with places mentioned in chapters 3 and 4. 37

Map 5.1. The Western Mediterranean and Greece in the Second Punic War, with places mentioned in chapter 5. 94

Map 5.2. Southern Italy and Sicily in the Second Punic War, with places mentioned in chapter 5. 96

Map 6.1. Greece and the Aegean in the Third Macedonian War, with places mentioned in chapter 6. 135

Map 7.1. Spain in the mid-second century BCE, with places mentioned in chapter 7. 176

TABLE

Table 1.1. Forms of Mass Violence. 8

Acknowledgments

This book's genealogy stretches back to a quiet Iowa evening in the fall of 2008, when I sat paging through Livy and wondered aloud why the Roman legions destroyed so many cities. During the subsequent years of study and iteration, in the midst of frustrating dead-ends and revelatory breakthroughs, I have had incomparable mentors at every turn. I am particularly indebted to Constance Berman, Michael Moore, Kathleen Kamerick, Alexander Thein, Peter Green, and the late Carin Green. Special mention goes to Rosemary Moore, who has been an outstanding teacher and friend; this project simply would not exist without her wisdom and counsel.

It has been a pleasure to work with series editors Michael Barrett and Kyle Sinisi, who generously invited me to submit a book proposal, and whose comments, keen suggestions, and good humor improved my work immeasurably. I am, moreover, thankful for Susan McEachern, acquisitions editor, Barbara Stark, copyeditor, and the rest of the excellent team at Rowman & Littlefield, all of whom were patient and professional from start to finish. The anonymous reviewer also offered many astute comments, which contributed to the final manuscript. Certainly, any remaining errors and infelicities are my own doing.

My brilliant colleagues at the University of Iowa and the Nueva School created the stimulating, friendly, and generative intellectual environment that allowed this book to take root. I appreciate you all endlessly, but am especially obliged to Heather Wacha, Katherine Massoth, Yvonne Seale, Chris McFadin, Arta Khakpour, Brian Cropper, Davion Fleming, Patrick Berger, Dan Cristiani, Ali McLafferty, Barry Treseler, Sushu Xia, Christopher Miller,

Wesley Patten, Jake Fauver, and Joel Colom-Mena. I have also benefited from many inquisitive, bright, and industrious students. Special gratitude goes to former students Natalie Hope and Noah Tavares for graciously reading and commenting on draft chapters, and to Om Gokhale and Ari Nazem for finding a talented illustrator to draw figure 2.1.

My friendships have been a reliable source of encouragement and energy, particularly on those days when I could not be bothered to write another word. (And several of you continue to ask about my work, ungrudgingly and with genuine curiosity, even though it does not make for pleasant conversation—*at all*.) Caitlin Sapp, Kade Schemahorn, Tyler Phillippi, Kassia Shishkoff, Jeremy Cates, Jill Sutter, Naomi Hertsberg Rodgers, Derek Rodgers, Nate Staniforth, Katherine Staniforth, and Katie Curry made the last decade a joy, sticking around through those long stretches when I thoughtlessly dropped into a pile of books and out of existence.

Thank you also to my family, Bakers, Halls, Peacocks, Nashes, and Tates all, but especially to my parents, Joyce and Bill, and brother, Matt. The three of you have been consistently supportive, even when my work must have seemed hopelessly opaque and my explanations bordered on nonsensical.

Finally, my partner, Mandy Peacock, tirelessly accompanied me on this journey, reading drafts, listening to presentations, tramping up mountainsides and over ruins, dodging goats, and motivating me through everything. With love and appreciation, as ever, this is dedicated to you.

1

"As Is the Roman Custom"

War and Mass Violence in the Roman Republic, Third and Second Centuries BCE

Lucius Mummius halted his Roman legions outside the walls of Corinth, one of the greatest cities in Greece and the final obstacle to his imminent victory. The venerable city was nestled on the isthmus that separated the Greek mainland in the north from the Peloponnese in the south, effectively controlling north-south traffic by land and east-west traffic by sea. This superb location made it a nexus for trade and gave it immense strategic value, all reinforced by its impressive fortifications: walls of stone punctuated with eight gates; "Long Walls" that connected the city to its harbor; and, towering moodily above it all, the magnificent citadel of Acrocorinth. The city had not been captured by assault in well over a century, and the last time the Romans tried, fifty-two years prior, they had failed utterly. Kings and generals had long remarked upon Corinth's strength and splendor. Its capture would be a mighty feat, forever remembered in the annals of history.[1]

But as Mummius beheld the storied city, he felt a twinge of doubt, and he hesitated. Troublingly, the imposing gates stood open for his army of over 20,000 infantry and 3,500 cavalry. This was unusual, and he wondered if the Corinthians were inviting him into a trap. After two days of delay, satisfied that no ambush was waiting within or simply growing impatient, the Roman commander led his legions into Corinth. Although many of the inhabitants had fled during the night, the invading Romans discovered some stragglers within the walls. The legionaries had no pity for these miserable few who had refused to abandon home and hearth, and they killed any men that they encountered. They seized the women and children for subsequent sale into slavery, surely subjecting them to extraordinary violence in the process. The Roman troops eventually decided that the city was secure and peeled off to plunder the accumulated Corinthian treasure of centuries. One witness claimed that he saw soldiers dragging out precious works of Greek art and throwing them to the ground, where they served as dice boards for the victors. Other Romans put buildings to the torch, and the fires allegedly burned so hot that metals liquefied and mixed together (creating the legendary "Corinthian bronze"). As a finishing touch, Mummius had his soldiers demolish the walls and some other buildings, carrying out an order from the Roman Senate to that effect.[2]

The destruction of this brilliant city in 146 BCE reverberated through the Mediterranean world, and traces of pain and shock are plainly visible in the surviving texts. Polybius, a contemporary Greek historian, describes Corinth's fate as a "disaster," the greatest calamity that the Greeks had ever suffered. Polystratus, a Greek poet of the same period, also writes darkly of Corinth's end:

> Lucius [Mummius] has smitten sore the great Achaean Acrocorinth, star of Greece, and the twin parallel shores of the Isthmus. One heap of stones covers the bones of those slain in the rout; and the sons of Aeneas [i.e., the Romans] left unwept and unhallowed by funeral rites the Achaeans who burnt the house of Priam.* (*Greek Anthology* 7.297; trans. Paton, modified)

Corinth continued to arouse feelings of sorrow for years after its destruction. Diodorus Siculus, a Greek author writing a century later, states that

> Corinth evoked great compassion from those that saw her; even in later times, when they saw the city levelled to the ground, all who looked upon her were moved to pity. No traveller passing by but wept, though he beheld but a few scant relics of her past prosperity and glory. (Diodorus Siculus *Library of History* 32.27.1; trans. Oldfather)

Even some Romans felt deep regret about the city's fate. No less than Cicero, the famous Roman statesman of the first century BCE, plainly says, "I wish they had not destroyed Corinth."[3]†

Yet if Corinth's end was momentous and extraordinary, it was matched— and in many ways overshadowed—by the fall of another city. Mere months before Mummius marched through Corinth's open gates, Roman forces under the command of Scipio Aemilianus pushed into the city of Carthage after a three-year siege. Carthage, located in modern Tunisia in North Africa, was also a city of great fame and antiquity, and it once ruled a powerful empire stretching across the western Mediterranean. Unfortunately for the Carthag-

---

* According to tradition, the Romans were descendants of Aeneas, a legendary Trojan refugee who escaped the Greek ("Achaean") destruction of Troy ("the house of Priam"). In Polystratus's poem, Corinth's destruction is long-awaited Trojan revenge in the form of Roman wrath.

† Henceforth translations are my own unless otherwise noted in the text or footnotes.

inians, they had lost that empire in their first two "Punic Wars" with Rome.*
At the end of the Third Punic War (149–146 BCE) between these old adversaries, Roman soldiers breached Carthage's defenses and poured inside.

Once within the walls, the Romans pressed determinedly toward an inner citadel called Byrsa hill, where a Carthaginian commander named Hasdrubal made his last stand. However, the Romans had to ascend three different streets to reach this final center of resistance. Each avenue was lined with six-story edifices whose occupants intended to fight to the end, forcing the Roman invaders to struggle through buildings one by one while they simultaneously clawed their way through the lanes and alleyways below. Appian, an ancient author of the second century CE, vividly describes the sights and sounds of this brutal urban combat:

> All places were filled with groans, shrieks, shouts, and every kind of agony. Some were stabbed, others were hurled alive from the roofs to the pavement, some of them alighting on the heads of spears or other pointed weapons, or swords. (Appian *Punic Wars* 128; trans. White)

As the Roman commander, Scipio, watched the unfolding carnage, he ordered his troops to destroy the buildings along the main streets and clear away the rubble. The legionaries set to this dreadful work at once, burning and pulling down the buildings with many of their occupants still inside. Appian again narrates these "new scenes of horror":

> As the fire spread and carried everything down, the soldiers did not wait to destroy the buildings little by little, but all in a heap. So the crashing grew louder, and many corpses fell with the stones into the midst. Others were seen still living, especially old men, women, and young children who had hidden in the inmost nooks of the houses, some of them wounded, some more or less burned, and uttering piteous cries. Still others, thrust out and falling from such a height with the stones, timbers, and fire, were torn asunder in all shapes of horror, crushed and mangled. Nor was this the end of their miseries, for the [Roman] street cleaners, who were removing the rubbish with axes, mattocks, and [pitch-]forks, and making the roads passable, tossed with these instruments the dead and the living together into holes in the ground, dragging them along like sticks and stones and turning them over with their iron tools. (Appian *Punic Wars* 128; trans. White)

It went on like this for six days and nights, and slowly the Romans carved out a path to the last defenders in the citadel. But the climactic showdown never came. After ceaseless fighting, fires, and bloodshed, most of the Carthaginians on Byrsa hill surrendered in exchange for their lives. The Car-

---

* The word "Punic" is from the Latin word *punicus*, which means "Phoenician" or "Carthaginian" (Carthage itself was founded as a Phoenician colony in the ninth century BCE). The First Punic War was fought over twenty-three years, from 264 to 241 BCE. The Second Punic War (218–201 BCE) was shorter, but still a long, bloody conflict. See chapter 5.

thaginian leader Hasdrubal retreated to a lofty temple with his family and several hundred Roman deserters, but soon he abandoned his comrades and submitted to Scipio. The other holdouts cursed his name and set fire to their temple bastion, where they perished in the consuming flames. Hasdrubal's wife (unnamed in our sources) reportedly killed her children in front of her faithless husband before throwing herself into the inferno.[4]

Scipio allowed his soldiers several days to loot the city as it burned around them. While his men picked the place clean, the Roman Senate sent instructions that sealed Carthage's fate: they directed Scipio to destroy those buildings that were still standing, eliminating these last broken vestiges. The Senate further decreed that the city would remain abandoned and forbade resettlement or rebuilding. Carthage would no longer exist.[5]

Like Corinth, Carthage had been an illustrious city of great wealth and antiquity. And as with Corinth, ancient authors reacted strongly to the fall of Carthage. Polybius, who witnessed the destruction of both cities, thought this moment offered an opportunity to assess Rome's supremacy and "to see clearly whether Roman rule is acceptable or the reverse." Polybius never directly answers his own question, but he does suggest that absolute power was beginning to corrupt the Romans, and that one day they might suffer the same fate as the Carthaginians. Writing in the first century BCE, Diodorus goes further in his analysis. He argues that Rome's ruthless destruction of Carthage, Corinth, and other communities ushered in a new era of Roman rule—an era of terror. "Once they held sway over virtually the whole inhabited world," he writes, "they confirmed their power by terrorism and by the destruction of the most eminent cities." And for Sallust, a Roman author living around the same time as Diodorus, the destruction of Carthage led to the breakdown of Roman morals and institutions. In Sallust's view, the Romans' fear of Carthage had once enforced virtuous behavior and concord at Rome, especially among the ruling aristocracy. But when Carthage was destroyed, there was no more outside threat to encourage cooperation and hold the state together, and Rome's ruling class descended into unbridled ambition and avarice. This steep moral decline would ultimately destroy the Roman Republic a little over a century later.[6]

These ancient authors regarded the destructions of 146 BCE as extraordinary—either because these were shocking acts of ruthlessness, or because they marked a turning point in Roman history, or both. Some modern scholars also view 146 as a momentous year, or argue that Rome's military conduct became more brutal in this period. According to this view, the Romans became increasingly unwilling to tolerate resistance or disobedience from other Mediterranean peoples, to the point that they snuffed out two illustrious cities that dared to defy them. This argument gives special significance to the destructions of Carthage and Corinth. Yet in many ways, these twin destructions were not so extraordinary. Indeed, when placed in their historical context, Carthage and Corinth are only two notes in a larger theme.[7]

PATTERNS OF MASS VIOLENCE

Roman armies ranged across the classical world in the third century to mid-second century BCE (the so-called Middle Roman Republic*), marching from Spain to North Africa to Asia Minor and beyond. Though the legions rightly earned a reputation for battlefield prowess during these far-flung conflicts, Roman warfare was never limited to field battles against enemy armies. For almost every major war that the Romans waged in this period, the ancient sources describe the Romans sacking, burning, and demolishing cities, and enslaving, massacring, or executing populations and prisoners (see appendix 1). Additionally, both Roman and non-Roman observers characterize mass violence as a major feature of Roman warfare in precisely this period. Polybius, who wrote in the mid-second century BCE, remarks upon the Romans' "customary" use of mass killing. Narrating a Roman assault on an enemy town in 209 BCE, he writes,

> When Scipio [the Roman commander] thought that a sufficient number of troops had entered he sent most of them, *as is the Roman custom*, against the inhabitants of the city with orders to kill all they encountered and to spare no one, and not to start pillaging until the signal was given. They do this, I think, to inspire terror, so that when towns are taken by the Romans one may often see not only the corpses of human beings, but dogs cut in half, and the dismembered limbs of other animals, and on this occasion such scenes were very many owing to the numbers of those in the place. (Polybius *Histories* 10.15.4–6; trans. Paton, modified and emphasis added)

Polybius also reports an incident in 270 BCE, when several hundred prisoners of war were taken back to Rome and beheaded in the Forum (the city's main public space). Again, this execution was carried out "according to the Roman custom." Polybius's impressions of Roman warfare are not to be taken lightly. He lived in Rome for about two decades, during which he mingled with leading statesmen and witnessed the Roman army on campaign in Greece, Spain, and Africa. In short, he knew something about their military conduct, and he believed there was a Roman tendency to use massive violence against enemy populations and prisoners.[8]

It appears that Rome also earned a reputation for violence on the eastern end of the Mediterranean. The second-century BCE Jewish text First Mac-

---

* Modern scholars divide the history of the Roman Republic into three parts, largely for the sake of convenience. The "Early Republic" is generally taken as the period between the founding of the Roman Republic circa 500 BCE and the early third century BCE, when the Romans were completing the conquest of Italy. The "Middle Republic" (or "Classical Republic") spans the early third to the mid-second century BCE; it marks the period of Mediterranean expansion, when Roman armies began to wage wars outside of peninsular Italy. The "Late Republic" encompasses the mid-second century BCE down to the civil wars of the first century BCE. It ends with the Republic's collapse, the ascension of Rome's first emperor, and the beginning of the "Imperial" period. For problems with this periodization, see especially Mouritsen 2017, 105–11; Flower 2010.

cabees specifically remarks on Rome's ruthless treatment of enemies. It states that the Romans had plundered far and wide, tore down strongholds, and enslaved women and children in Greece, and that they "destroyed and enslaved" all other "kingdoms and islands" that stood against them. The third book of Sibylline Oracles, written in the eastern Mediterranean in the second or first century BCE, similarly comments on Roman violence. The Oracles characterize Rome as the latest in a series of empires, yet far more rapacious and destructive than its predecessors. In its sweeping condemnation, the text claims that the Romans plundered more wealth, waged more war, and enslaved more people than any previous empire. The Oracles also highlight Rome's capacity for violence by prophesying the city's fall: one day a champion from the east would avenge Roman barbarism, exacting three times the plunder and twenty times the slaves that Rome had stolen away from others.[9]

Later authors picked up on similar ideas, emphasizing Rome's willingness not only to conquer its foes, but to eliminate its enemies altogether. Corinth and Carthage naturally loom large here, but almost always as part of a larger pattern. As an anonymous Roman text from the early first century BCE puts it,

> By the Roman people Numantia was destroyed, Carthage razed, Corinth demolished, Fregellae overthrown. Of no aid to the Numantines was bodily strength; of no assistance to the Carthaginians was military science; of no help to the Corinthians was polished cleverness; of no avail to the Fregellans was fellowship with us in customs and in language. (*Rhetorica ad Herennium* 4.27.37; trans. Caplan)

This passage comes from a manual on rhetoric, so it is not chiefly concerned with Roman violence or warfare. (The text is actually trying to illustrate the rhetorical use of disjunction, or *diazeugma*, where a single subject governs multiple verbs.) But by tying together several famous historical examples, the author must have drawn on a theme that was already familiar to Roman audiences.

In his writings later that same century, Cicero makes a connection between Carthage, Corinth, and the Italian city Capua: each city had once threatened Rome but was eventually eliminated. Although the Romans did not physically destroy Capua, he notes, they abolished its government, leaving this great rival "weakened and disabled." Cicero also would have known that the symbolic dissolution of Capua's government was accompanied by the (very real) execution of its political leaders. Further, he and the Greek author Diodorus group together Carthage, Corinth, and Numantia (a Spanish community destroyed in 133 BCE), all communities that the Romans chose to destroy rather than spare.[10]

Ammianus Marcellinus's history, written in the fourth century CE, still paints the Roman Republic as a ruthless destroyer. In a speech he attributes to the emperor, Julian, Ammianus proudly proclaims that the Romans once destroyed all their rivals:

Our forefathers spent many ages in eradicating whatever caused them trouble. Carthage was conquered in a long and difficult war, but our distinguished leader feared that she might survive the victory. Scipio utterly destroyed Numantia, after undergoing many vicissitudes in its siege. Rome laid Fidenae low, in order that no rivals of her power might grow up, and for that same reason crushed Falerii and Veii; and even trustworthy ancient histories would have difficulty in convincing us that those cities were ever powerful. (Ammianus Marcellinus 23.5.19–20; trans. Rolfe)

All of these authors saw a theme playing right through the history of the Roman Republic, connecting the long march to Mediterranean empire with massive violence and the destruction of enemies. And while the ghosts of Carthage and Corinth often haunt their narratives, clearly, ancient writers did not see their destructions as unique. Cicero, Diodorus, Ammianus Marcellinus, and the anonymous author of *ad Herennium* place Carthage and Corinth alongside other examples of cities and peoples destroyed. Meanwhile, Polybius and the authors of First Maccabees and the Sibylline Oracles speak of mass killing, destruction, and enslavement as characteristic of Roman warfare.[11]

Given their antiquity, size, and reputations, the near-simultaneous demise of Carthage and Corinth was especially shocking to contemporaries. When these examples are contextualized, however, it becomes clear that their parallel destructions were not so extraordinary in the larger saga of Roman war and expansion: our sources suggest that mass violence was an important feature of Roman warfare by at least the third century BCE and possibly earlier. The main aim of this book is to investigate this phenomenon, and to show *why* Roman commanders employed mass violence in war. The central argument is that Roman military and political leaders frequently saw enslavement, mass killing, urban destruction, and related behaviors as effective—even necessary—ways to accomplish their objectives, and that they were willing to expend considerable resources to those ends.

Before moving ahead to the main arguments, it will be helpful to get the lay of the land. First we must define our terms. This book defines "mass violence" as widespread physical violence against groups of noncombatants (those who are unarmed or not actively defending themselves) or against communities (their inhabitant populations as well as their physical dwellings and urban structures).* Mass violence thus encompasses a spectrum of behaviors, ranging from more targeted violence, like the execution of select groups, to increasingly indiscriminate violence, like the mass enslavement

---

* Here I have borrowed and modified the definition of historian Christian Gerlach: "'Mass violence' means widespread physical violence against non-combatants, that is, outside of immediate fighting between military or paramilitary personnel. Mass violence includes killings, but also forced removal or expulsion, enforced hunger or undersupply, forced labor, collective rape, strategic bombing, and excessive imprisonment—for many strings connect these to outright murder and these should not be severed analytically" (Gerlach 2010, 1).

of entire populations (see table 1.1). And of course, this study will encounter many behaviors that lie between these two poles.

Table 1.1. Forms of Mass Violence (examples)

Forms of Violence	More Targeted	↔	More Indiscriminate
Mass Killing	Killing of a city's political leaders	Killing of a city's adult males	Killing of a city's entire population
Mass Enslavement	Enslavement of a city's political leaders and their families	Enslavement of a city's women and children	Enslavement of a city's entire population
Urban Destruction	Demolition of a city's walls, fortifications	Destruction of a city's walls and other significant buildings (e.g., temples)	Burning, demolition, destruction throughout a city's urban fabric

This definition of mass violence is deliberately broad, since "selective violence can be massive while indiscriminate violence can be limited" (to quote political scientist Stathis Kalyvas). For the sake of illustration, imagine that a Roman army captures a large city with a population numbering in the thousands, then selectively executes two hundred of the city's aristocratic leaders; meanwhile, another Roman army captures a small village of a hundred people, and indiscriminately massacres many of the inhabitants. One act of violence is selective and the other is indiscriminate, and they differ in relative scale and severity. Nevertheless, both are *mass* killings of unarmed groups, and thus I regard both as mass violence. A broad definition of mass violence is also useful because a more restrictive concept—like the contested term "genocide"—could dissociate violent behaviors that are similar and related. For example, sociologist Martin Shaw points out that mass deportation and mass killing are both means of "destroying [the] physical, social and cultural presence [of a populace] within its historic territory." Although these are not identical forms of violence, there is "analytical kinship" here. Most importantly for this study, the Romans used multiple forms of violence against noncombatants and communities, usually in combination and at different calibrations. Case in point: in 211 BCE, the Romans captured the city of Capua after its rebellion in the Second Punic War, then proceeded to execute fifty-three political leaders, sell some members of the community into slavery, and forcibly relocate others. While they killed "only" fifty-three people, this was still a mass execution, and it was just one element in a larger expression of Roman violence. To make sense of episodes like this, we must view the whole constellation of violent behaviors.[12]

With our key terms defined, a brief chapter outline is also in order. Chapter 2 sketches Rome's military and political history up through the period of Mediterranean expansion. This chapter is meant to provide some basic background for readers who are less familiar with ancient Roman history. Chapter 3 shows how Roman forces performed mass violence using swords, handtools, and fire. It concludes that urban destruction, mass enslavement, and mass killing required labor, organization, and resources on the part of Roman armies. This chapter also argues that Roman generals and political leaders normally directed these brutal military strategies. Chapter 4 explores Roman commanders' motives for using mass violence on campaign. According to ancient authors, Roman military leaders used mass violence to neutralize threats, punish perceived slights, and deter future challenges. Economic and political factors could also shape Roman decision-making since some acts of massive violence could yield profit or military glory. In most cases, the ancient texts present mass violence as an instrumental strategy, which commanders used to achieve specific aims or resolve specific challenges.

In chapters 5, 6, and 7, I examine three Roman wars where mass violence features prominently in the ancient sources: the Second Punic War (218–201 BCE), the Third Macedonian War (171–167 BCE), and the Lusitanian War (155–139 BCE). These narrative chapters do not offer radically new readings of these conflicts or exhaustive coverage of events, but they firmly situate the Romans' use of mass violence within broader historical contexts. In doing so, they highlight the decision-making of Roman leaders, and show how specific goals and challenges motivated their uses of mass violence in war. As a final note, the book is meant to be accessible for casual readers interested in Roman history, students, and experts alike. To that end, the following pages devote more attention to the ancient evidence than to scholarly debates and controversies, and modern citations are usually (but not always) limited to recent works in English that are immediately relevant to the issues at hand.

A few topics remain here. The final sections of this chapter briefly discuss ancient texts and archaeology, the main types of evidence used in this book, as well as the modern scholarship that forms the background to this study. These sections are recommended for readers who are unfamiliar with ancient history or for those interested in methods and historiography. A more knowledgeable reader—or a casual reader less interested in the minutiae of historical research—may safely skip to the next chapter.

EVIDENCE OF MASS VIOLENCE

To non-historians, it may seem like a minor miracle that we know anything at all about the ancient past. Fortunately, Roman and Greek authors produced a large body of literature and technical writing, much of which survives. For the purposes of this book, the most important surviving texts are narrative histories written by ancient authors, which preserve crucial information about antiquity that we otherwise would not know. These historical accounts must be approached with care, however. One problem is that many ancient histories do not survive or exist only in fragments; and some surviving histories were composed, not by eyewitnesses or contemporaries, but by much later authors. These late authors usually drew upon earlier historical sources, many of which are lost to us, with greater or lesser fidelity. All this means that we can only see hazy glimpses of Roman history and never quite the whole picture. Our lack of clarity is compounded by problems inherent to ancient historical writing. To put it simply, many Greek and Roman authors were more concerned with telling a good story or teaching a good lesson than writing good history (as we would understand it). Their personal agendas, politics, aristocratic outlook, and standards of writing and rhetoric could also distort their historical accounts in ways both large and small.

Ancient historical narratives present special problems for the study of mass violence and warfare. When they wrote about the capture of cities in war, many ancient authors used a set of generic scenes based on the legendary sack of Troy—scenes of wanton murder and plunder, desecration of temples, rape and abduction of women, widespread grief and lamentation, all rendered in the dramatic language of tragedy. Some ancient authors drew liberally from these *urbs capta* ("captured city") themes, repeatedly using stock imagery to depict the fall of cities. Consequently, some modern scholars argue that it is difficult to disentangle "historical" violence from generic scenes and images. Although more scrupulous Greek and Roman historians avoided *urbs capta* clichés, even they were at the mercy of faulty memories and unreliable informants. And in general, ancient warfare did not lend itself well to eyewitness testimony or clear description, with its chaotic combat, narrow perspectives, and limited communications.[13]

Still, we should avoid the temptation to throw the baby out with the bathwater. Ancient Greek and Roman authors seldom wrote wholly invented histories and, significantly, many ancient authors—or if not authors, their audiences—had experience of war, which shaped their depiction of events. At the very least, ancient descriptions of warfare needed to reflect the military practices of the time; otherwise they would not have conformed to the expectations of contemporary audiences and would not have been credible to knowledgeable readers. This same logic applies to depictions of mass violence. Even generic descriptions of violence and atrocity would not have made sense to ancient audiences unless they were consistent with actual practices in ancient warfare. In sum, the knowledge and expectations of ancient writers and readers informed narrative descriptions of violence, and constrained excesses. So, although they may be more or less "historical" in their details, ancient historical texts contain important information about mass violence in ancient warfare.[14*]

---

* A cursory glance at more recent descriptions of mass violence—from the massacres of the Vietnam War to the mass beheadings perpetrated by ISIS to the razing of towns by Boko Haram—should make it quite clear that the most horrific ancient descriptions are not as over-the-top as we would like to believe. I suspect at least some modern skepticism stems from a lack of familiarity with atrocity.

As luck would have it, several quality texts survive that shed light on Roman warfare in the Middle Republic. Foremost among these are the *Histories* of the Greek author Polybius. In 167 BCE, Polybius and about a thousand of his countrymen were taken to Rome as political hostages. During his exile, he befriended Roman senators, including Scipio Aemilianus, and accompanied the latter on military campaigns. He also wrote his *Histories* in exile, drawing upon Roman documents and informants, his own experience of Roman affairs, and Greek and Carthaginian sources. Polybius's special knowledge and circumstances make him a key source of information on mass violence in Roman warfare. On the one hand, as a Greek himself, he wanted to show his Greek readers that he was no sycophant of Rome; this made it difficult to sweep Roman atrocities under the rug, since many Greeks knew about or had suffered from Roman warfare. On the other hand, as a detainee, Polybius would not have wanted to offend his Roman friends and patrons; he was in no position to exaggerate Roman violence, cast Roman behavior in an overly negative light, or invent violent episodes for the sake of added drama. Because he had to steer a "middle course" between two audiences, his best option was to write about Roman violence in a straightforward manner and to the best of his knowledge. Furthermore, when he wrote about military violence, Polybius largely avoided the dramatic (and generic) *urbs capta* imagery. Indeed, he famously criticizes another Greek historian for "always trying to bring horrors [of war] vividly before our eyes." In Polybius's view, a "historical author should not try to thrill his readers by such exaggerated pictures . . . but simply record what really happened and what really was said, however commonplace." Generally he only wrote lengthy descriptions of atrocities when he thought the details were factual and necessary. This means that his reporting of Roman violence is relatively sober and straightforward, neither softened unduly nor trumped up for dramatic effect.[15]

The narrative of Polybius's *Histories* originally ended around 146 BCE with Rome's destruction of Carthage and Corinth, but his text only survives in fragments for the period after 216 BCE. The other major source for this period is *Ab Urbe Condita* ("From the Founding of the City"), the great history of Rome written by the Latin author Titus Livius, or "Livy," in the late first century BCE and early first century CE. Unlike Polybius, Livy was neither a contemporary nor an eyewitness to most of the history he wrote about. He drew on earlier histories to write his own narrative, but at times he would condense, embellish, rearrange, or even misconstrue descriptions of war and violence that he read in his sources. Still, Livy was not totally ignorant of warfare and he had a solid understanding of technical military matters. He also faithfully followed Polybius's military narratives, changing only "the colouring, but not the story or the terminology." Last, his descriptions of violent events are often very restrained, consisting of "a few sentences of bare narrative." When he included longer descriptions of bloodshed and devastation in his history, he frequently diverged from the usual generic images in favor of unique or interesting details. Although Livy requires caution, his descriptions of war and violence can be rich in evidence.[16]

Polybius and Livy provide the essential backbone here, but this book uses several other texts to examine mass violence in Roman warfare, including the works of Diodorus Siculus, the histories of Appian of Alexandria, and the biographies of Plutarch. These were all late authors, but they each used earlier sources, including Polybius. Some history-adjacent works, such as the *Geography* of Strabo or Valerius Maximus's *Memorable Deeds and Sayings*, also provide helpful tidbits to supplement our picture of mass violence in Roman warfare.

Aside from ancient historical narratives, archaeological materials provide an important body of evidence and will appear throughout this book. As historian and archaeologist Fernando Quesada Sanz points out, archaeology can provide evidence of violence that is "much more direct and therefore often more expressive, more horrific and more personal" than the textual evidence. One significant example is Valentia (modern Valencia, Spain), a city that the Romans reportedly destroyed around 75 BCE. Here excavators uncovered a layer of destruction containing weaponry, several bodies *in situ* bearing signs of grisly deaths, and a chronological break in habitation. Yet despite the tangible evidence we can glean from sites like Valentia, archaeology is often imprecise and is always subject to interpretation, making it difficult to connect archaeological evidence to specific historical events. Archaeology also provides a very limited picture, since most sites are only partially excavated, and the survival of ancient materials often depends on happenstance. Doubtlessly, archaeology provides information that is relevant to this study, but we must be wary when interpreting archaeological traces of violence.[17]

This summary hardly does justice to all the complications that arise when using ancient texts and archaeology. However, the main takeaway here is that the sources must be approached cautiously, and this study will often draw conclusions that are tentative. Indeed, readers unfamiliar with ancient history may be surprised at how often this book uses conditional language—"might have," "could have," "perhaps," and so forth—yet that is the nature of the beast. Certainty is rarely possible across so many centuries, using such fragmentary evidence, and sometimes it will be necessary to confront our evidence head-on in subsequent chapters.

SCHOLARSHIP ON MASS VIOLENCE

This book is the first full-length study in English in over eighty years[*] to examine mass violence in Roman warfare. Nonetheless, it does not exist

---

[*] In his 1938 PhD dissertation *Atrocities in Roman Warfare to 133 B.C.*, Mars McClelland Westington explored Roman "atrocities" (broadly defined as unnecessary injury against persons or

in a vacuum, and it is indebted to a few other studies on violence in the ancient world. Several scholars, including Catherine Gilliver, Joseph Reisdoerfer, Francisco Marco Simón, Jörg Rüpke, and Toni Ñaco del Hoyo, argue persuasively that the Romans used various forms of mass violence to achieve pragmatic ends. Their studies suggest that the Romans used mass killing, enslavement, urban destruction, and related behaviors to terrorize enemies, punish slights, or eliminate threats, among other ends. Examining "genocide" from ancient Assyria to the Roman Empire, Hans Van Wees takes a similar tack, stressing that "mass murder" was a purposeful and deliberate act: in short, ancient states used displays of massive brutality to send messages to rivals and shore up their power and status.[18]

A few scholars also reconstruct the methods of mass violence in ancient warfare, scrutinizing Greek and Latin vocabulary to understand how ancient soldiers performed violent acts. Brian Bosworth examines massacres in the Peloponnesian War (431–404 BCE) between Athens and Sparta, demonstrating that such killings would have been laborious and likely created substantial logistical problems for Greek armies. Similarly, Kathy Gaca reviews the ancient Greek practice of "andrapodizing" enemy communities. As Gaca shows, this was a violent process where an aggressive military force subdued a community's young women and non-infant children. Finally, Adam Ziolkowski investigates Roman city-sacking, arguing that most Roman sackings followed a common procedure: after breaching the walls of an enemy town, the legionaries indiscriminately killed the military-aged males at hand before they dispersed to plunder, rape, and commit individual murders (largely beyond the control of their generals).[19]

This book builds directly upon these previous studies. Like several of these prior works, it attempts to explain why Roman armies employed mass violence and emphasizes their often-pragmatic motives for doing so. However, to understand *why* Roman armies massacred, enslaved, and destroyed, it is necessary to understand *how* Roman armies carried out this violence. Therefore, taking a cue from Bosworth, Ziolkowski, and Gaca, this book also reconstructs the means and methods of large-scale killing, enslavement, and urban destruction in Roman warfare.

One final study deserves mention. Nathalie Barrandon's *Les Massacres de la République Romaine* also explores mass violence in Republican Rome, privileging "massacre" as the main unit of analysis. Barrandon's volume was released when the manuscript for this book was nearing completion, so it exercised a limited influence here. Nonetheless, it overlaps with this book in several important ways. Most notably, both Barrandon and I argue that mass violence is best understood in the context of military campaigns; these were dynamic events that cannot be boiled down to single, simplistic mo-

---

property, severe legal acts, and illegal acts that were somehow repugnant or offensive to common sentiment). Westington's work examines many important examples of mass violence in Roman warfare. However, the work is mainly descriptive, and its attempt to determine whether Roman warfare was "humane" diverges significantly from the goals of this study.

tives, and their military context is often key. Both of us also argue that mass violence stemmed primarily from the deliberate choices of Roman generals. Despite broad similarities, this book diverges from Barrandon's in essential ways. For one, the scope is different: this study focuses exclusively on the Middle Republic, giving less attention than Barrandon to the historiographical and moral-legal dimensions of mass violence and relatively more attention to Roman warfare, military strategy, and politics. Most significantly, this book parts ways with Barrandon because it examines Roman violence using a "strategic perspective"—a theoretical framework adapted from recent work in political science.[20]

A STRATEGIC PERSPECTIVE ON MASS VIOLENCE

While the Romans were doubtlessly ruthless in war, their methods would not be out of place in recent world history. The twentieth and early twenty-first centuries have been deeply scarred by widespread violence against civilians, ranging from genocide to terrorism to area bombing. Recently, political scientists have attempted to understand and interpret these behaviors. Though their studies often use different methods and arrive at different conclusions, many political scientists agree on two broad points: (1) there is a close relationship between mass violence and war and (2) military-political leaders use mass violence deliberately, to accomplish specific goals and objectives. These points of consensus are relevant to the study of mass violence in Roman warfare and deserve brief discussion here.[21]

First, many political scientists agree that the logic of war is closely related to the logic of mass violence against noncombatants. War creates a sense of impending insecurity, while also empowering military forces to destroy perceived threats. In this context, belligerents may decide that noncombatants themselves are threatening. "From this perspective," Benjamin Valentino writes, civilians "play a central, if often involuntary, role as the underwriters of war's material, financial, and human requisites. Sometimes they become the objects of war itself." In guerrilla wars, insurgencies, and conventional conflicts, military forces may attack civilians to undermine or coerce populations that support the enemy. Civilian-targeting strategies are especially common when one side in a military conflict cannot win through conventional means. In wars of territorial expansion, too, armies may attack conquered populations to remove future threats, or simply to control new territories and resources more effectively.[22]

Second, many scholars argue, military and government leaders are normally responsible for ordering and orchestrating strategies of mass violence. Such strategies are "designed to accomplish leaders' most important ideological or political objectives and counter what they see as their most dangerous threats." In war this calculus can be straightforward: leaders use mass violence to gain strategic advantages or to hasten their ultimate victory. Yet it must be stressed that purposeful violence does not mean purely "rational" violence. Leaders' ideology, beliefs, and values shape their willingness to target noncombatants and civilian infrastructure. Individual soldiers may also victimize civilians randomly or opportunistically, seeking to kill, loot, and rape for any number of reasons—irrespectively of their leaders' goals. But even when individual soldiers commit unsanctioned violence, military and political leadership usually plays an important role. For example, military commanders may tacitly allow activities like looting, since these can contribute to their larger goals (rewarding soldiers, terrorizing enemy populations, etc.). And as Valentino points out, the private motivations of individuals do not explain "why these [armed] groups were created, organized, and turned loose in the first place." The larger military and political context, which is shaped by the decisions and goals of leaders, facilitates the violent actions of groups as well as individuals.[23]

A theoretical framework can be constructed from these two points of consensus in the political science literature. This framework—which I term the "strategic perspective" (to borrow, and modify, the concept coined by Valentino)—is rooted in two basic premises: mass violence is intimately connected to the objectives and contingencies of warfare; and mass violence against noncombatants is normally a top-down strategy, directed or heavily influenced by military-political leaders. The framework is simple, but its simplicity gives it coherence and broad applicability.[24]

This book uses the strategic perspective to examine mass violence in Roman warfare during the period of the Middle Republic. It situates violence firmly within its military context and highlights the perspectives, goals, and decision-making of Roman leaders. It must be emphasized that political scientists primarily study the modern world, treating vastly different historical circumstances than this study, and they often use statistical and quantitative methods that are untenable for ancient history. Yet the main justification for adopting this theoretical framework is that Roman massacres, destructions, and enslavements took place almost exclusively amid wars and military campaigns; there is an undeniable link between Roman warfare and mass violence. Moreover, as I will show in chapters 3 and 4, Roman leaders played a fundamental role in ordering and organizing acts of mass violence, indicating that we should direct our attention to command-level perspectives and goals. Finally, Arthur Eckstein has demonstrated that approaches from political science offer a different lens with which to view the ancient evidence, complementing, rather than supplanting, the traditional sources and methods of ancient history. The same principle applies here. The theoretical framework is usually implicit in the chapters that follow, though it firmly guides the questions asked, as well as the emphasis given to command-level decisions and the military contexts of violence. In its most basic form, the strategic perspective supplies the main assumption threading through this book: as the Roman writer Seneca put it, "no one sheds blood for the sake of shedding blood."[25]

NOTES

1. Corinth's fortifications, strategic importance: Pettegrew 2016, esp. 68–73, 89–112. Previous Roman assault: Livy 32.23; Dixon 2014, 179–81. Remarks of kings and generals: e.g., Polyb. 18.11.3–7, 30.10.3; Livy 45.28.1–2.

2. Mummius's campaign: Paus. 7.16.1–7; Flor. 1.32.4–5; Livy *Per.* 52. Indiscriminate killing: Paus. 7.16.8; Oros. 5.3.6. Enslavement: Paus. 7.16.8; Just. 34.2.6; Zon. 9.31; Oros. 5.3.6; Plut. *Quaest. conv.* 9.737. Plundering: Strabo 8.6.23; Flor. 1.32.5–6; Oros. 5.3.6; Vell. Pat. 1.13.4; *CIL* 1^2.630; Pliny *NH* 35.24; Aur. Vict. *Vir. Ill.* 60.3. Burning: Paus. 7.16.7; Oros. 5.3.6. "Corinthian bronze": Pliny *NH* 34.6–7; Petron. *Sat.* 50.5–6; Flor. 1.32.6–7; Oros. 5.32.7. Demolitions: Paus. 7.16.9; Zon. 9.31. Destruction on senatorial orders: Liv. *Per.* 52. Other references to Corinth's destruction: *CIL* 1^2.636; Cic. *Man.* 5; *Off.* 1.35; *Agr.* 2.33, 87; Diod. 32.4.5, 27.1. The archaeological evidence does not suggest total obliteration; see chapter 3.

3. "A disaster": Polyb. 38.1.1–4. Poets on Corinth's end: cf. *Anth. Graec.* 7.493, 9.151. "I wish": Cic. *Off.* 1.35.

4. For the Third Punic War, see esp. App. *Pun.* 126–35.

5. References to Carthage's destruction abound: e.g., Polyb. 38.20–22; Livy *Per.* 51; Diod. 32.4.5, 32.23–26; App. *Pun.* 132–33; Zon. 9.31; Flor. 1.31.18; Cic. *Agr.* 2.51, 87; Oros. 4.23.1–6. As with Corinth, archaeological evidence does not indicate total obliteration; see chapter 3.

6. Carthage's wealth, grandeur: Polyb. 18.35.9; Hoyos 2010, 73–93; Purcell 1995, 133–34. "To see clearly": Polyb. 3.4.6–12 (Paton); cf. 36.9. "Once they held sway": Diod. 32.4.5 (Oldfather). Sallust and moral decline: Sall. *Cat.* 1–12. Historiographical significance of 146 BCE: Davies 2014.

7. 146 BCE marking a change in Roman conduct, or otherwise extraordinary: e.g., Flower 2010, 70–71; Rosenstein 2012, 211–39; Clark 2014, 141–47; Purcell 1995.

8. Forum executions: Polyb. 1.7.11–12. On Polybius, see note 15 below.

9. *Orac. Sib.* 3.175–95, 350–66, 520–44; 1 Macc. 8.1–14. Dating 1 Maccabees: Fischer 1992, 440–41; Honigman 2014, 6–7. Dating the Sibylline Oracles is contentious: Collins 1974; Gruen 1984a, 340, and 1998; Buitenwerf 2003; Jacobs 2014.

10. *ad Herennium*: Enos 2005, 331–38. Carthage, Corinth, Capua: Cic. *Agr.* 2.87. Carthage, Corinth, Numantia: Cic. *Off.* 1.35; Diod. 32.4.5. Execution of Capuan leaders: Livy 26.15.8–9; App. *Hann.* 43; Zon. 9.6; Oros. 4.17.12.

11. See also Flor. 1.32.1; Macr. *Sat.* 3.9.13.

12. "Selective violence": Kalyvas 2008, 405. For "selective" mass violence, see Kalyvas 1999; Straus 2012; Jones 2000; Bauer 1984, 213; Coster 1999, 92; Salerian et al. 2007. "Genocide" as "conceptual quagmire": Hultman 2014, 290–91; Jones 2006, 16–18. "Analytical kinship": Shaw 2013, 150; cf. 2009, 314; 2004, 149. Other broad concepts of mass violence: e.g., Straus 2015, 26–27; Downes 2008, 15. Treatment of Capua: Livy 26.15.8, 16.6; App. *Hann.* 43; Zon. 9.6; Oros. 4.17.12.

13. Sources on Republican Rome: Bispham 2010, 29–50. Literary sources and ancient warfare: Whitby 2007; Whatley 1964. Skepticism about historicity of violence: Laurence 1996, 112; Toner 2013, 137–41; Assenmaker 2013; Zimmermann 2006, 343–58. *Urbs capta* motif: Paul 1982; Barrandon 2018.

14. Ancient texts as historical sources: e.g., Lendon 2009; Bosworth 2003. Historical basis to tropes/generic elements: Bradley 2004, 304; Roth 2006, 57; cf. Levithan 2013,

10–12. For ancient descriptions of Roman violence, see recently Barrandon 2018, esp. chapters 1–2.

15. Polybius's life, methods, context: Walbank 1957, 1–16; Champion 2004, 15–18, 30–66, 204–34; Eckstein 1995b, 1–16; Marincola 2001, 104–48. Polybius's "political predicament": Champion 2004, 233; "middle course": Champion 2004, 231, discussing his larger difficulties as historian. Criticism of Phylarchus: Walbank 1957, 259–63; Paul 1982, 145–47; Barrandon 2018, ch. 1. "Trying to bring horrors": Polyb. 2.56.7–12 (Paton, modified). Polybius on violence: D'Huys 1987, 224–31.

16. Livy as historian: Chaplin 2000; Luce 1977; Briscoe 1973, 1–8. Livy's sources, aims, methods: Oakley 2010; Briscoe 2010; Tränkle 2010; Walsh 1982; Walbank 1971. Problems with Livy's description of war/violence: Erdkamp 2006; Harris 1979, 51 n.3. "Colouring": Roth 2006, 55–60; cf. Ziolkowski 1990. "Bare narrative": Paul 1982, 152; cf. Flamerie de Lachapelle 2007, 79–110. Preference for unique details: Barrandon 2018.

17. "Much more direct": Quesada Sanz 2015, 13. Valentia: Ribera i Lacomba 2006, 77–78, 80–81; Ribera i Lacomba and Calvo Galvez 1995. Problems equating archaeology with specific events: Snodgrass 1987, 41–47; but cf. Quesada Sanz et al. 2014, 259–60, and Quesada Sanz 2015, 12.

18. Gilliver 1996; Ñaco del Hoyo et al. 2009; Reisdoerfer 2007; Marco Simón 2016; Rüpke 1999 (mass enslavement); Van Wees 2010, 2016 (genocide). Cf. Purcell 1995; Gaignerot-Driessen 2013.

19. Bosworth 2012; Gaca 2010; Ziolkowski 1993.

20. Barrandon 2018.

21. Reviews of political science scholarship: Valentino 2014; Hultman 2014; Straus 2010, 2012.

22. Threat/insecurity in war: e.g., Straus 2006, 2010; Sémelin 2014; Midlarsky 2005; Shaw 2003. "From this perspective": Valentino 2014, 94. Guerrilla wars, insurgencies, civil wars: e.g., Valentino, Huth, and Balch-Lindsay 2004; Downes 2008; Sullivan 2012; Hultman 2012; Balcells 2010; Kalyvas 2006; Kim 2010. Conventional wars, wars of conquest: Valentino 2004; Downes 2008; cf. Sullivan 2012, but in civil war contexts.

23. "Brutal strategy": Valentino 2004, 3. Debates about "strategic" mass violence/ genocide: Valentino 2014, 93–98; Hultman 2014, 291–94; Straus 2012, 546–50, and 2010, 174–76. Civilian victimization to win, stave off defeat: Downes 2008; cf. Downes and Cochran 2010. Ideology and decision-making: e.g., Valentino 2004; cf. Straus 2012, 549–50; Valentino 2014, 96–97. "Non-strategic" violence: Azam and Hoeffler 2002; Mueller 2000; Kalyvas 2006; Humphreys and Weinstein 2006; E. J. Wood 2006, 2014; R. M. Wood 2014; Merger 2016. "Why these groups": Valentino 2014, 98.

24. "Strategic perspective": Valentino 2004, 3. My approach adopts Valentino's emphasis on instrumentality and leaders' decision-making, but is otherwise only broadly similar to—and much simpler—than his.

25. Eckstein 2006, 2008; cf. Burton 2011. "No one sheds blood": Seneca *Ep*.14.9.

2

"Adorned with Scars"

Roman Military and Political History to 146 BCE

Our view of Rome's origins is heavily obscured by layers upon layers of storytelling and myth making that bury the city's early history. Even in antiquity, many tall tales circulated about Rome's beginnings. The most popular origin story claimed that the city was founded in the mid-eighth century BCE* by the twins Romulus and Remus, sons of Mars and a princess named Rhea Silvia. Separated from their mother at birth and suckled by a she-wolf, the twins enjoyed a youth full of adventure, family reunions, and various hijinks, before eventually setting out to build a city of their own. The grand plan was disrupted when the twins quarreled about just where to establish their new home, a spat that ended with Romulus killing Remus in a fit of rage. The fratricidal Romulus then humbly named the new city Roma, or Rome, after himself, and became the first of Rome's seven kings. Rome grew in size and strength under its early monarchy, but the seventh king, Tarquinius Superbus (or "Tarquin the Arrogant"), became tyrannical and undermined the public's trust. His various outrages included the execution of several Roman senators, as well as forced labor programs that heavily oppressed the populace. The king's son pushed widespread dissatisfaction into open revolt when he raped Lucretia, the noble daughter and wife of leading citizens. With Lucretia's suicide, shortly before 500, a band of freedom-loving Roman aristocrats was emboldened to boot out the corrupt monarch and establish a new government: the Republic.[1]

Or so the story goes. But the Romans only began writing their history many years later, in the third century, and the first pioneering historians

---

* Henceforth, all dates in this chapter are BCE.

only dimly understood the origins of their city and institutions. Worse still, they had few sources of information to reconstruct Rome's beginnings, so they and their successors often felt free to extend, embellish, reconcile, and rationalize popular tales of Rome's early centuries. The best that can be said of such stories is that they should be taken with a very large pinch of salt. After all, many elements of the tradition—the son-of-a-god founder, for instance—are patently myth, while much of the rest is a medley of oral tradition and guesswork.

Modern scholars, paying close attention to archaeological evidence and reading the historical sources critically, offer a more cautious account of early Roman history. It seems that Rome emerged as an urban community by the mid- to late seventh century—somewhat later than the traditional date—on a cluster of hills next to the Tiber River (see map 2.1). These early Romans were one of many Latin-speaking peoples in Latium (west-central Italy), and shared a cultural identity with their Latin neighbors, but by the late sixth century, Rome had become the largest and most powerful city in the region. It is likely that the archaic city-state was ruled by some kind of monarchy, and that Roman kings were replaced by a very early version of the Republic around the year 500. Roman citizens probably had some limited role in this new government, but a few aristocratic families dominated the city and its political offices from an early date. In any event, the early Republic then embarked on a centuries-long process of experimentation, occasional bouts of civil unrest, and compromise as it slowly took form. Meanwhile, the Romans were almost constantly at war with their neighbors, and eventually they conquered all of Italy.[2]

THE CONQUEST OF ITALY

Before (and well after) the Roman conquest, Italy was not unified by language, culture, or ethnicity. Rather, the peninsula was a patchwork of peoples: in addition to the Latin-speaking communities of Latium in west-central Italy, there were Etruscan city-states just to the north of Rome; Umbrians to the north and east; various Oscan groups, like the Samnites, Campanians, Lucanians, and Bruttians, to the east and south; Iapygians in the southeastern region called Apulia; a smattering of Greek city-states along the southern Italian coast; and Gallic tribes in the far north of the peninsula.[*] All of these peoples fought among themselves and against one another, and in that respect, Rome was no different. Under their monarchy and young

---

[*] All the names here are imprecise, since (A) there was enormous variation within each of these groups of people, and (B) there were many smaller communities and tribes in between that I have neglected to mention. These catch-all terms are essentially just a convenience. The important point is not the bewildering variety of names, but simply that Italy was linguistically and ethnically diverse.

Map 2.1. Ancient Italy, with places mentioned in chapters 1 and 2. (Map by author, with base map, topography, and location data generated at Ancient World Mapping Center.)

Republic, the Romans clashed repeatedly with the Latin cities nearby and with other Italians.

Rome's early military conflicts were relatively limited in terms of scale and duration, essentially consisting of seasonal raids and counter-raids between Rome and its neighbors. By the fourth century, however, the Romans had developed a regular pattern of warfare, wherein their political leaders led out citizen armies to campaign across the Italian peninsula annually. This period of more intense warfare was encapsulated by the conquest of Veii, an Etruscan town and great rival to the north, in the year 396. The Romans' breach of Veii's walls supposedly followed a ten-year siege, creating a neat (and suspicious) parallel with the Greeks' mythic capture of Troy.

Later Romans also believed that the *stipendium* was introduced during the final war with Veii; this was regular pay for Rome's citizen soldiers, to compensate them for any income they lost on their farms while they were off fighting long military campaigns. Whatever the truth of these stories, they suggest that the scale and duration of Roman war making ramped up around this time, with campaigns getting longer and stakes getting higher.

With their more extensive war making in the mid-fourth century, the Romans came to exercise hegemony over most of central Italy. Yet Roman expansion created friction with the formidable Samnites, an Oscan people who dominated south-central Italy. The three Samnite Wars (which lasted on and off between 343 and 290) were devastating conflicts, with long sieges, multiyear campaigns, and rival armies fighting set-piece battles with thousands of troops on each side. Even after the defeat of these fierce enemies, Rome remained involved in scattered clashes with Etruscan and Umbrian communities to their north into the 260s.[3]

These increasingly large conflicts were not without setbacks, and this was not a period of unequivocal Roman victory. Between the sixth and fourth centuries, fierce Gallic tribes migrated across the Alps into the Po River valley—a region the Romans called Cisalpine Gaul*—and threatened the peoples of north and central Italy. The expansion of the Gauls led to disaster around 390, when a marauding Gallic warband tore down from the north, defeated a Roman army, and sacked Rome itself. The Romans also lost several battles to the Samnites, and they briefly faced a terrifying coalition of Samnites, Umbrians, Gauls, and Etruscans in the Third Samnite War. And perhaps most threatening of all, in 280 a foreign king named Pyrrhus invaded with a powerful army. Pyrrhus ruled a people known as the Molossians in northwestern Greece, and had been summoned to help Tarentum, a Greek city in southern Italy that feared Roman expansion but lacked the resources to fight alone. Pyrrhus was an outstanding general with a well-trained professional army, and he smashed Rome's legions in two great battles before he was finally driven from Italy. In the end, despite many defeats and failures, Rome emerged victorious from these wars and became the dominant power in Italy by the second quarter of the third century. One modern historian notes that "the speed with which [Rome] arrived at this domination in Italy . . . is quite as striking" as the speed with which Rome later conquered the Mediterranean.[4]

Roman success in Italy stemmed in large part from their willingness to absorb recent enemies and turn them into allies. Some communities that the Romans defeated in war, mostly in central Italy, received full or partial Roman citizenship. These communities lost their political independence, and they were expected to follow Rome's lead in what we might call "foreign policy." On the other hand, they were left alone to manage their own internal

---

* Cisalpine Gaul means "Gaul on this side of the Alps," i.e., the land of the Gauls on the Italian side of the Alps.

affairs and received some or all the benefits of Roman citizenship. More privileged communities even gained the right to vote in Roman elections. Other defeated peoples were turned into Roman "allies" (*socii*). Such a relationship did not signify an alliance between equal states, but rather a subordinate treaty relationship with Rome. The allied states also lost their political independence, but again, they could manage their internal affairs however they pleased. Furthermore, the Romans did not force governors or administrators upon the conquered, nor did they impose heavy tributes or other forms of economic exploitation. Instead, the primary obligation of all these conquered communities was to provide soldiers for Roman armies. As a consequence, the Romans' pool of potential manpower increased every time they defeated new enemies and absorbed them into their allied network, which in turn gave them greater strength for future conflicts. Moreover, all of Rome's Italian allies shared in the spoils of war—and not just the material plunder or slaves, but also land: at times, the Romans annexed territory from their foes in Italy, and some of the captured lands were set aside for colonies. A mix of Roman and allied colonists were settled in some of these new communities, where they would receive generous plots of land and "Latin rights."* This whole alliance system was successful both because it encouraged loyalty in the long run—the allies remained faithful for most of the Republican period—and because it required minimal management on Rome's part. This was hegemony on the cheap, and it incentivized the Italian allies to cooperate.[5]

But lest we paint too rosy a picture, Rome's domination of Italy was often brutal and frightening, at moments presaging the later destructions of Carthage and Corinth. According to tradition, the Romans destroyed Veii and enslaved many of its survivors. Ancient historians also report that the Romans killed or sold populations, burned communities, and tore down city walls in the wars of the later-fourth and early third centuries (see appendix 1). For instance, several Ausonian communities in south-central Italy were reportedly "exterminated" by the Romans. Less than a decade later, the Romans entered a final war with the Aequi, a central-Italian people who had long been a thorn in their side. Roman forces are said to have captured thirty-one Aequian towns, most of which they burned or demolished so that "the Aequian name was nearly destroyed." A similar fate befell the Gallic Senones, who inhabited northeastern Italy. When a Senonian leader killed a Roman ambassador, Rome responded by sending an army—which proceeded to kill all of the Senonian men and enslave the women and children. This merciless campaign also cleared the Senones's lands for settlement (which was perhaps the point to begin with), and the Romans soon sent colonists there.[6]

Rebellion or defection from Rome could also be disastrous. For example, our sources say that the towns Sora and Satricum defected while the Romans

---

* "Latin rights" were not equivalent to citizenship, but entailed a special status vis-à-vis the Romans, including rights of commerce, migration, and intermarriage.

were at war with the Samnites; when the Romans recaptured these communities, hundreds of leaders deemed responsible for the revolt were scourged and beheaded in the Roman Forum. And when the city Falerii rebelled, the Romans destroyed the town and forcibly removed the population to a more accessible (and less defensible) location. Ancient authors may have exaggerated some of these bloody details, particularly for early events that happened long before their lifetimes. Nonetheless, these reports of massive violence likely reflect an essential reality: Roman conquest was often ruthless, and rebellion was answered with retribution. The constant threat of Roman arms, backed by huge sources of manpower secured through the Republic's alliances, supported the conquest of Italy as much as any positive incentives.[7]

THE IMPORTANCE OF WAR IN THE ROMAN REPUBLIC

By the year 270, Rome held sway over all of Italy south of the Po River valley, using a combination of canny diplomacy and force. By this time, the government of the Roman Republic had also taken on its mature (or "classical") form. At its center was the Roman Senate, a permanent body of about three hundred men—the *patres*, or "Fathers"—who were mostly former officeholders from wealthy, influential families. The Senate was responsible for the major decisions of governing, overseeing public officials and state finances, and navigating Rome's relationships with foreign states. Aside from the Senate, the government consisted of a series of annually elected magistrates who were charged with a range of responsibilities. The most important magistrates were the two consuls, both of whom possessed *imperium*, or the power to command their fellow citizens. Though they had many public tasks, the consuls' most serious and significant duty was to lead Roman armies in war. A few other magistrates, like the praetors, could also lead armies and oversee military activities. Finally, the mass of Roman citizens played a role in republican government, voting on legislation and electing magistrates in public assemblies. Note, however, that their political power was limited, and the Roman Republic was always more an oligarchy than anything like a modern democratic republic. (For more on government in the Middle Republic, see appendix 3.)

Generally only Roman aristocrats had the means, time, and name recognition to run for any of Rome's highly competitive political offices or to serve in the Senate. And among the aristocracy, individuals from a handful of extraordinarily prestigious families won high office more regularly than their less glorious peers. Nonetheless, although noble blood certainly helped, the Roman aristocracy had no permanent or inheritable titles; even sons from the noblest families, with the most famous ancestors, could not expect automatic electoral success or immediate conferral of high status. Since success in Roman politics was ultimately decided by Roman voters, aristocratic politicians had to prove constantly that their qualities and achievements merited election.

And the Roman electorate had high standards. In a surviving funeral speech that honored a consul and leading senator, L. Caecilius Metellus, we catch sight of the qualities and achievements that Romans expected in their leaders:

> [L. Caecilius Metellus] had achieved those ten greatest things that wise men spend their life seeking. He had wished to be the best warrior, the foremost orator, the bravest general, to command the most significant military operations, to hold high offices, to possess the highest wisdom, to be considered the leading senator, to amass great wealthy by honorable means, to leave behind many children, and to be the most eminent man in the state. (Pliny *Natural History* 7.139-40, trans. Sage)

Certainly oratory, political success, and wealth turn up on this list, but war appears conspicuously in several guises; war was essential to the aspiring Roman politician. Therefore, in order to build the foundations for a successful political career, a young aristocrat would begin in his late teens by serving in the army, fighting in the cavalry and perhaps later serving as a military tribune (a junior officer). These early years of service allowed politically ambitious young men to build a reputation for bravery and self-sacrifice in battle. Weapons and armor seized from enemy champions in single combat, along with physical scars, served as proof of martial courage and were invoked with pride by Roman aristocrats. Such marks of valor could even sway political arguments. In the year 167, the senator Marcus Servilius asked the Roman people to vote in opposition to one of his political rivals, a man named Sulpicius Galba. To help persuade voters, Servilius pointed to his own battle scars and personal bravery, asserting that

> [Galba] has learned nothing except talk, and that of a slanderous and malicious sort; I have on twenty-three occasions challenged and fought an enemy; I brought back the spoils of every man with whom I duelled; I possess a body adorned with honourable scars, every one of them received in front. (Livy 45.39.16, trans. Schlesinger)

Servilius was already an old veteran when he spoke so dismissively of Galba, but younger aristocrats also hoped to acquire such manifest signs of bravery (and a reputation to match). Ideally, they would do so before they entered Rome's highly competitive political scene and ran for their first offices.

Rome's political offices were ranked in a rough sequence from lowest to highest called the *cursus honorum* ("course of offices"). After establishing himself in the army, an up-and-coming politician traditionally began his career by running for the lowest-ranked magistracies and working his way up. If successful, he would rise through subsequent elections until, if very fortunate, he reached the praetorship or even the consulship, where military opportunities were greatest. Once elected consul, a Roman politician could achieve further glory by successfully commanding armies and might even win a triumph. The triumph was a military celebration in which a victorious general paraded through Rome in a chariot, accompanied by his army, enemy prisoners, and captured plunder. This put him right in front of the citizenry, trumpeting his successes on behalf of Rome. Such victories further enhanced a consul's name and reputation, opening the door to future political success while accruing prestige for his family and his sons. Naturally, triumph-worthy victories did not come easy, and our sources make it clear that the Senate thoroughly deliberated before awarding these high honors. Apparently, the Romans understood a "good" victory as one that did considerable damage to Rome's foes. The numbers of enemies killed, cities captured, or prisoners taken could all factor into the Senate's decision to award a triumph with all its attendant preeminence.[8]

For the average Roman citizen, too, war provided opportunities for personal prestige, and martial courage (*virtus*) on the battlefield was encouraged by a whole slew of military rewards that reaped honor back home. For example, the first soldier over the wall of an enemy city would receive a gold crown in recognition of great bravery exhibited in the face of danger (assuming he did not die in the act). Upon his discharge and return to civilian life, he would wear this crown and any other military decorations in public festivals, drawing attention to his extraordinary feats. Regular Roman troops also received a cut of the booty acquired on campaign, as well as additional material rewards from their commanders. But above all, military service was a civic obligation: Roman males who met a certain minimum property requirement were expected to appear at Rome for the annual levy, when they would assemble on the Field of Mars outside the city, and some would be selected to serve in the legions (figure 2.1). They would serve for a campaign season, approximately from March through October during the conquest of Italy, though this obligation increased over time. After completion of the campaign, they returned to their farms and to civilian life. The frequency of this process meant that a large proportion of the citizen body acquired military experience through participation in one or more campaigns, and martial virtues were valued at virtually all levels of society.[9]

THE BEGINNINGS OF MEDITERRANEAN EMPIRE

With war firmly embedded in public institutions and most of Italy under its aegis, Rome began to reach out into the Mediterranean world, fighting multiyear wars using larger military forces that stayed under arms for longer periods. In 264, the Romans began a deadly struggle with Carthage over the island of Sicily. This would be the first of the three bloody "Punic Wars." Carthage was a powerful city situated in modern-day Tunisia and master of a sea empire that encompassed much of Sicily, the other islands of the western Mediterranean, and the coastal regions of Spain and North Africa. The city also commanded a fearsome navy. Yet despite their relative inexperience at naval warfare, the Romans successfully challenged Carthage for control of the seas around Sicily—building war fleets virtually from scratch—while their armies fought all over Sicily and briefly invaded North Africa. The First Punic War lasted a slogging twenty-three years, during which tens of thousands died, hundreds of ships went to the bottom, and both states nearly bankrupted one another.

Carthage blinked first, and agreed to abandon Sicily in the eventual peace treaty. They also agreed to pay Rome a large indemnity of 3,200 talents* of silver, paid annually over ten years (and possibly intended to cripple the impressive Carthaginian economy). A few years later, the Romans pushed their advantage further and took the Carthaginian-held islands Sardinia and

---

* A "talent" was a measurement of weight; the Roman talent weighed about 71 pounds while the Greek talent weighed 57 pounds. Polybius, our main source, probably sees these figures as Greek talents, so Carthage's indemnity to Rome was over 180,000 pounds of silver. See Brice 2014, 275.

Figure 2.1. A Roman legionary soldier, third and early second centuries BCE. For protection, legionaries wore a bronze helmet and body armor (mail or a bronze chest plate, depending on what they could afford). They were also equipped with a large body shield called a scutum. For armament, they carried heavy javelins (pila, sg. pilum) into battle, but the main weapon for most was the short sword or gladius. Most soldiers probably wore their swords in a scabbard on the right hip, but centurions and other officers may have carried theirs on the left like the soldier pictured here. (Illustration by Jack Friedman.)

Corsica. When Carthage protested, Rome countered by declaring war. The Carthaginians had just put down a costly rebellion in North Africa—the so-called Truceless War—and were in no position to fight. They agreed to yield the islands and, salt in their wounds, consented to pay Rome another 1,200 silver talents.

Over the next two decades, tensions between the two powers began to subside. Suspicions built up anew, however, when the Carthaginians successfully expanded their empire in Spain, recovering and then surpassing their strength from before the First Punic War. Open conflict eventually broke out again when a Carthaginian commander named Hannibal Barca invaded Italy, throwing both states into the Second Punic War (218–201). Hannibal can be fairly called a tactical genius, and he crushed Roman army after Roman army. Many Roman allies defected, and some foreign powers made treaties with Carthage. Despite these early Carthaginian successes, the Romans were aided by a seemingly inexhaustible supply of manpower and an utter refusal to surrender; these, along with a revival in their military fortunes, produced another Roman victory by 201.

The scale of the Second Punic War cannot be overstated. It undoubtedly saw the darkest days of the Roman Republic, and at many points, the Romans were pushed to the brink. Further, the war was fought in Greece, Spain, Italy, Sicily, and North Africa, by hundreds of thousands of soldiers. The war's consequences were correspondingly enormous. The Romans determined that Carthage would never again be a first-tier power. The final peace treaty stripped the Carthaginians of all their territory outside North Africa, heavily limited their military capabilities, and slapped on an even larger war indemnity. Thereafter the Carthaginians were studiously obedient to Rome, seeking to do nothing that hinted at defiance, but the Romans again became suspicious when Carthage rapidly recovered its prosperity. Perhaps the Romans were looking for a new war, or perhaps they felt that Carthage would always be a threat. The latter attitude was embodied by the Roman senator Marcus Porcius Cato "the Elder," who firmly insisted that "Carthage must be destroyed." Whatever the Romans' motive, by 149 the legions were back in North Africa. They destroyed Carthage entirely after a three-year siege.[10]

Roman interventions in northern Italy, Spain, and Greece during the Second Punic War also developed their own momentum. Between the Second and Third Punic Wars, the Romans fought to secure their hold in northern Italy, where Gallic tribes had supported Hannibal. Their wars against the Gauls were mostly won by the late 190s, but they also became embroiled in conflicts with the Ligurians, a tribal people who occupied rugged northwestern Italy. These clashes lasted until midcentury. At the same time, the Romans were struggling to maintain a permanent military presence in eastern and southern Spain, where they had fought to expel Carthaginian forces during the Second Punic War. In the decades after the defeat of Carthage, some Spanish communities rebelled against their new Roman overlords, and other communities raided Rome's Spanish domains from the north and west. The Romans soon sent armies to Spain every year, and major fighting broke out sporadically throughout the second century.[11]

Finally, the Roman legions began to appear in the Greek-speaking eastern Mediterranean, where they found an entirely different political environment. This half of the Mediterranean world had been profoundly impacted by the conquests of Alexander III, better known as Alexander the Great. In the fourth century Alexander had ruled Macedonia, a kingdom that sat in the northeast of the Greek peninsula. Despite centuries of close contact between Macedonia and Greece, the more urbanized Greeks to the south often regarded their northern neighbor as a rural backwater. They received a rude awakening when Alexander's father, Philip II, forged Macedonia into a powerful state with a strong army, with which he conquered and dominated most of Greece. When Philip was killed by an assassin's blade, his young son Alexander III took up the throne with visions of world conquest. Leading the outstanding army he had inherited from his father, he marched eastward and rapidly conquered the massive Persian Empire and even invaded northern India. By the time of his death in 323, Alexander controlled a domain stretching from Greece to Egypt to the Indus River.

Alexander's empire did not outlive him for long. His leading generals, or Successors, embarked on a series of destructive wars, vying to slice off a piece of the pie for themselves or, better yet, to take the whole thing. By the time the Romans got involved in the east, three powerful "Hellenistic"* kingdoms had emerged out of Alexander's empire, each under a different ruling dynasty: the Antigonid dynasty ruled the old kingdom of Macedonia and dominated Greece; the Seleucids controlled an empire centered on Syria and Mesopotamia; while the Ptolemies reigned over Egypt and parts of North Africa (see map 2.2). Each of these kingdoms was fabulously wealthy and powerful, and their warrior kings were animated by an ideology of conquest that sought to emulate Alexander. Caught between these major kingdoms were many smaller states and cities, which regularly squabbled with one another and with the leading powers. (Pyrrhus, the warlord who invaded Italy in 280, was a minor Hellenistic king.)

Roman attention was first drawn toward this constellation of eastern states when the kingdom of Macedonia made an alliance with Hannibal during the Second Punic War. The Romans would fight multiple conflicts with Macedonia over the course of the second century, spawning further wars with the Seleucids and other Hellenistic foes. Despite the apparent strength and resources of the Hellenistic states, the Romans prevailed in each of these confrontations: they crushed the Macedonians in 197 and completely dissolved their kingdom in 167; they humbled the Seleucids and severely

---

* "Hellenistic" is the term used by modern historians to refer to the Greek-speaking eastern Mediterranean between the death of Alexander the Great in 323 BCE and the fall of the last Hellenistic kingdom, Ptolemaic Egypt, in 30 BCE.

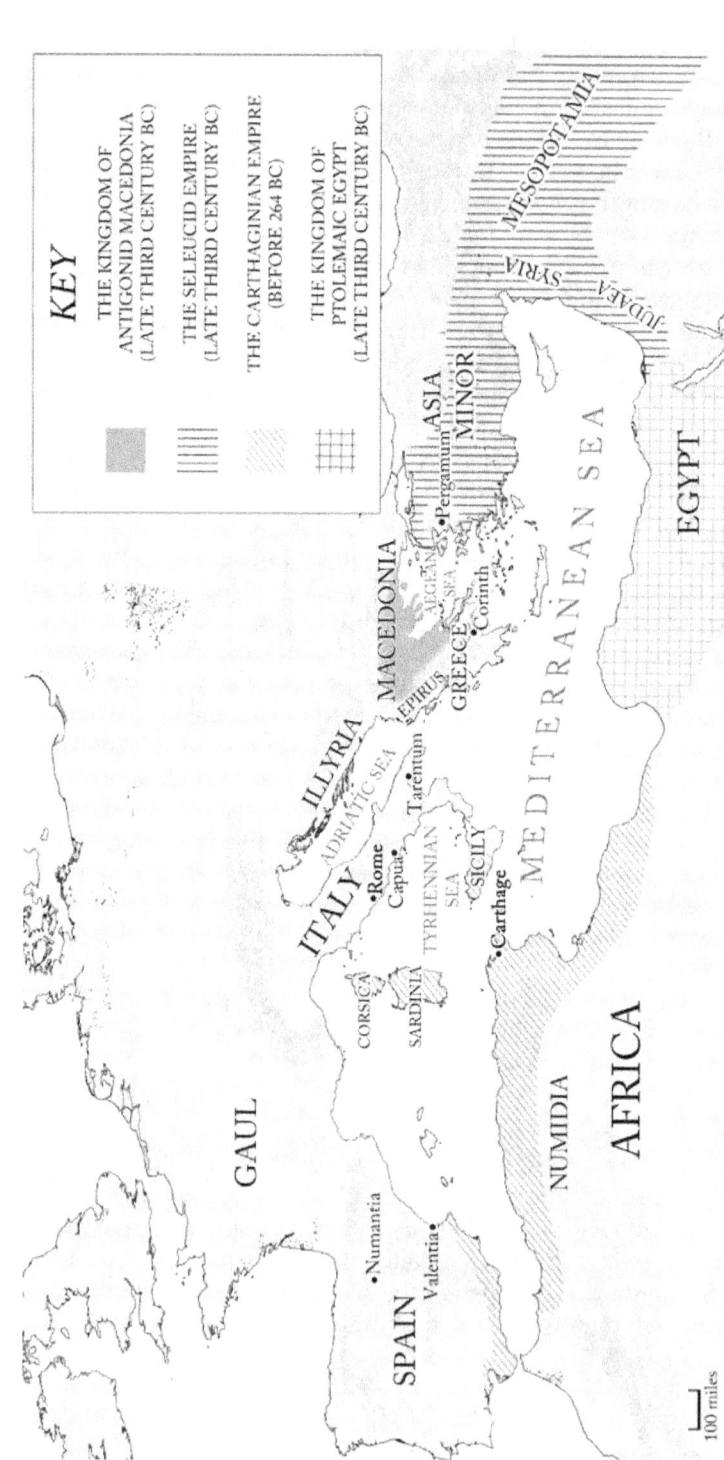

Map 2.2. The ancient Mediterranean, with places mentioned in chapters 1 and 2. (Map by author, with base map, topography, and location data generated at Ancient World Mapping Center.)

curtailed their power in 188; and in 146, when a middling Greek state called the Achaean League dared to defy Rome, they swatted the Achaeans down and destroyed their leading city, Corinth.[12]

During these wars, the Romans began to establish a permanent military presence overseas—particularly in Sicily, Sardinia and Corsica, and parts of Spain—but areas that fell under their control were slow to develop administration or taxation regimes that would be characteristic of Rome's imperial provinces in later periods. In the Greek East, moreover, they showed an astonishing disinclination to annex territory or establish a permanent presence at all. Yet while they were slow to annex or exploit foreign territories, by the middle of the second century, they had toppled or severely curtailed the strength of most of the other Mediterranean "superpowers," including the kingdom of Macedonia, Carthage, and the Seleucid Empire. Polybius, who lived through much of this era, was awestruck: "Who could be so indifferent or lazy that they would not want to know how, and with what kind of government, nearly the whole inhabited world was conquered and fell under the rule of the Romans in less than fifty-three years, something which never happened before?"[13]

It is not necessary to detail these wars and conflicts any further, and many will be discussed in subsequent chapters. It is enough here to say that during the Middle Republic, Roman armies fought across the Mediterranean and won battlefield victory after battlefield victory. However, as we saw during the conquest of Italy, the Roman way of war was not restricted to the battlefield. Ancient authors suggest that the Romans deployed massive violence, destroying enemy cities and populations, executing and enslaving prisoners, from early on. And though our sources are not always reliable for the early centuries of Roman history, our evidence becomes much richer during the period of Mediterranean expansion—and mass violence appears very prominently in our sources for this period, too. To understand this aspect of Roman warfare, we must now turn our attention to the methods of mass violence used by the Roman legions.

NOTES

1. For the traditional narrative, see esp. Livy's book 1, Dionysius of Halicarnassus's *Roman Antiquities* books 1–4.

2. Early Roman history, problems and controversies: Cornell 1995; Forsythe 2005; Raaflaub 2010, 125–46; Beard 2015, 53–130; Lomas 2018. Institutional development in the Republic: Flower 2010.

3. Early Roman warfare, expansion: Oakley 1993; Raaflaub 1996; Goldsworthy 2000, 30–45; Rawlings 2007; Armstrong 2016.

4. "The speed": Oakley 1993, 11. Threats faced by early Romans: Eckstein 2006, 118–58.

5. A good overview of Rome's *imperium* in Italy can be found at Rosenstein 2012, 73–93. See also Fronda 2010, 13–33; Cornell 1989; Bispham 2006, 2007; Scopacasa 2016.

6. Veii: Diod. 14.115.2; Livy 5.21.1–5.22.8; Flor. 1.6.10. "Exterminated": Livy 9.25.9. "The Aequian name": Livy 9.45.17. Senones: App. *Sam.* 13, *Gall.* 13; Dion. Hal. 19.13.1; Polyb. 2.19; Flor. 1.8.21.

7. Sora: Livy 9.24.12–15. Satricum: Livy 9.16.10. Falerii: Zon. 8.18.

8. Aristocratic competition: Rosenstein 1990; 2010, 365–82; 2012, 11–35. Roman triumph: Clark 2014; Pittenger 2008; Beard 2007.

9. Rome's martial values: Oakley 1985, 392–410; McDonnell 2006, 12–71; Lendon 2005, 163–211; Balmaceda 2017, 14–19. Military decorations: e.g., Polyb. 6.39; Gell. *NA* 5.6. Campaign season: Phang 2008, 89–90.

10. Carthage, Punic Wars: see e.g., Hoyos 1998, 2003, 2015; Goldsworthy 2000; Lazenby 1996, 1978; Fronda 2010.

11. Northern Italy: Roncaglia 2018, 19–38; Haeussler 2013, 91–97. Spain: López Castro 2013, 67–78; Richardson 1986, 1996; Curchin 1991.

12. Alexander, Hellenistic Age: e.g., Bosworth 1988, 1996; Shipley 2000; Errington 2008; Green 1993. Rome's eastern wars: e.g., Gruen 1984a; Morstein-Marx 1995; Eckstein 2008; Derow 1989, 2003.

13. "Indifferent or lazy": Polyb. 1.1.5. Studies of imperialism in the Republic are legion (pardon the pun), but some key contributions are Eckstein 2006; Burton 2011; Harris 1979; North 1981; Sherwin-White 1980; Rich 1993; Gruen 1984a, 1984b; Badian 1958; Erskine 2010 (discussing other periods, too).

3

"What the Fire Could Not Consume"
Methods of Mass Violence

In 200 BCE, Rome embarked on a second war with the mighty kingdom of Macedonia. Though the Macedonians' strength was on the wane since the reign of their most famous king, Alexander the Great, they still possessed considerable military resources: a fearsome army, a battery of allies, and direct or indirect control over much of the Greek peninsula. It would not be an easy fight or a quick one, and the Romans made little headway in the war's first year. Yet Roman leaders were not deterred. That November, before winter could set in, a commander named Lucius Apustius led his army resolutely toward Macedonia's western frontiers. He had orders to ravage enemy territory, and he was nothing if not effective, rapidly capturing a string of forts as his army cut deep into hostile terrain. Then he came to Antipatrea.

The fortified town of Antipatrea was perched on high ground over a narrow river gorge, from which it guarded the route further inland. Capturing this strategic center would help clear a path into Macedonia itself but, as the Roman commander surely knew, it would be difficult to conquer the place by direct assault. Apustius decided to hold off on an immediate attack, instead opening negotiations with the city's leading men and inviting them to submit to Rome's protection. It was a futile effort in the end; the locals had no interest in abandoning Macedonia for some foreign invader and were confident in the strength of their walls. They refused his invitation, and Apustius pounced. As the Roman historian Livy tersely recounts, Apustius "stormed and subdued Antipatrea by force of arms and, after killing the men of military age and granting all the plunder to his soldiers, he demolished the walls and burned the city." From there he marched to another town nearby, Codrion, which was also "strong and well-fortified." The terrified inhabitants surrendered at once rather than share in Antipatrea's ill fortune.[1]

To a modern observer, Apustius's treatment of Antipatrea may appear disproportionate, even atrocious. But modern sensibilities notwithstanding, he was well within the bounds of Roman military practice, and the Romans' willingness to mete out massive violence is not really in question (see chapter 4). Less clear is the nature of the violence itself. Livy's deceptively straightforward reports of mass killing and destruction conjure images of smoking landscapes in the reader's mind, implying the application of violence on a massive scale—perhaps an ancient version of World War II's area bombing, accomplished by the hands of Roman legionaries in lieu of high explosives. We should set aside such images for the moment. After all, Livy only tells us that the legions "burned," "demolished," and "killed," which could mean any number of things. Did Apustius's troops actually burn the whole city to the ground? Did they demolish the impressive fortifications in their entirety, using only premodern tools? How did the Romans go about finding and killing Antipatrea's adult males, and what happened to the rest of the population? Finally, was all of this violence just random mayhem committed by frenzied soldiers, or was procedure and organization involved? Livy's clipped account can answer none of these questions on its own. And this lack of transparency is not unique to Livy. Many ancient writers only supply a line or two of text to indicate that cities were "destroyed," peoples "enslaved," or populations "killed" by Roman armies. These brusque reports tell us virtually nothing about how the Romans accomplished this violence, or how these events played out historically.[2]

On the other end of the spectrum, ancient writers (including Livy!) occasionally give us more detailed descriptions of mass violence. Consider, for example, the Roman capture of Ilurgia (206 BCE) as recounted by Livy, and the Roman capture of Cauca (151 BCE) as recounted by Appian:

> **Ilurgia:** And now the hatred and resentment which had prompted the attack on the city showed itself. No one thought of making prisoners or securing plunder though everything was at the mercy of the [Romans]; the scene was one of indiscriminate butchery, non-combatants together with those in arms, women equally with men were all alike massacred; the ruthless savagery extended even to the slaughter of infants. Then they flung lighted fire-brands on the houses and what the fire could not consume was completely demolished. So bent were they upon obliterating every vestige of the city, and blotting out all record of their foes. (Livy 28.20.6–8, trans. Moore, modified)
>
> **Cauca:** [After all his demands were met, the Roman commander Lucullus] required that a garrison be installed in [the town of Cauca]. The Caucaei accepted this also, and he led in two thousand men, chosen for their bravery, whom he told to take positions on the walls after they had entered. When the two thousand had seized the walls, Lucullus led in the rest of the army, and signalled with a trumpet blast to kill the Caucaei who were of age. The Caucaei, calling

to witness the gods . . . and reviling the Romans for their lack of faith, were destroyed cruelly, with only a few of the twenty thousand men escaping through precipitous passages. Lucullus sacked the town and brought infamy upon the Romans. (Appian *The Wars in Spain* 52, trans. Richardson, modified)

At first glance, both episodes seem to provide some concrete information. In the case of Ilurgia, Livy nods to the tools and methods of urban destruction, mentioning the use of firebrands and distinguishing between arson and demolition. On the other hand, he still does not tell us how the soldiers actually demolished urban structures, how they found and killed (all?) the inhabitants, or what sort of organization, labor, and time (or lack thereof?) was involved in this merciless process. Similarly, Appian indicates that the massacre at Cauca stemmed from a central plan, implying a measure of coordination. But here, too, it is not easy to understand how the mass killing took place. Maybe we should imagine the Romans rampaging through the city, cutting down whomever they happened to meet. Or on the other hand, maybe they rounded up the males and executed them later. Or possibly they used some combination of these approaches. We just cannot say for certain. In these longer descriptions of mass violence, the additional details often heighten the drama of the scene—the rage of the troops at Ilurgia, the cries of faithlessness at Cauca—more than they tell us what happened, or might have happened.[3]

To some extent, ancient authors could afford to be vague: their Roman and Greek audiences would have known what to expect—many would have had direct experience of ancient warfare and its ravages—and they could make informed inferences when reading these texts. As one modern scholar notes, "cruelty and terror consequent on the capture of cities were experiences by no means remote from the personal knowledge of many inhabitants of the ancient world." Another historian states that violence and destruction "could be imagined by the writer's audience familiar with the destruction of parts of their own city by fire and the human violence played out in the arena or amphitheatre." Yet for us, unfamiliar with the lived experiences of the ancient world, it is necessary to go beyond any individual description of mass violence and to take a broad view of the surviving evidence. That is the aim of this chapter.[4]

This chapter reconstructs the methods of mass violence used by Roman armies. It does so by piecing together many examples of violence from our texts, identifying patterns, and scrutinizing ancient authors' language and word choices. Additionally, this chapter examines archaeological evidence as well as depictions of violence in ancient artwork. As we proceed, we will often lack useful evidence from the period of the Middle Roman Republic (circa third and second centuries BCE, the focus of this study). In such cases, this chapter uses evidence from later periods of Roman history and from other ancient societies to fill the gaps in our knowledge.* More

---

* It is true that other ancient peoples, and even Romans in later historical periods, had different military practices from the Romans of the Middle Republic. However, the fundamental

recent history also furnishes many examples of mass violence that can help us make sense of Roman behavior.[5]

Reconstructing the methods of mass violence in Roman warfare is a bit like putting together a piece of furniture without instructions—and with missing hardware, a lost table leg, and at least one cushion that definitely does not belong. But in the end, by bringing together a wide range of evidence, we can better understand how the Roman legions inflicted mass violence in war. The rest of this chapter follows a Roman army through the process of capturing, sacking, and ultimately destroying a city like Antipatrea. Along the way, we will explore city-sacking, mass enslavement, mass killing, and urban destruction in detail, examine violent episodes that took place across the Mediterranean (map 3.1), and determine what ancient sources mean when they describe these events. Note that Roman armies employed mass violence in a variety of military situations—for example, the Romans might enslave or execute prisoners taken on the battlefield—but for the sake of consistency, this chapter focuses on the treatment of captured cities and their populations.

CAPTURING AND SACKING CITIES

When the Roman legions approached a hostile town, the defenders often had a chance to avoid armed confrontation. Some towns surrendered immediately on their own initiative and were treated rather well, and Roman commanders persuaded other communities to come to terms. Yet all bets were off when a city's inhabitants closed their gates to the Romans and hunkered behind their walls for a siege. Faced with such a challenge, Roman armies had several basic options for capturing a fortified city. First, they could try to breach the city's walls by battering them down or undermining them from below; second, they could try to go over the walls using ladders, siege towers, or earthen ramps; third, they could blockade the targeted city, cutting it off from the outside world and starving it into submission; fourth, they could gain entry by stealth or trickery, sneaking inside the city at an unexpected time or from an unexpected place; fifth, they could get help from someone inside the city, who might open a gate or show them a secret entrance. All of these options carried considerable risks for Roman attackers, who often suffered heavy casualties in these operations.[6]

Despite the risks, a concerted Roman assault would overcome smaller cities rather quickly, and many towns surrendered shortly after invaders had captured or penetrated their fortifications. Yet some populations kept

---

technology and methods of mass violence seldom differed in antiquity, since there were few means available for killing large numbers of people or destroying standing structures besides fire and sword (and pickaxe). Evidence from other ancient periods and societies is therefore useful here, and can help us understand our fragmentary information about Roman warfare in the Middle Republic.

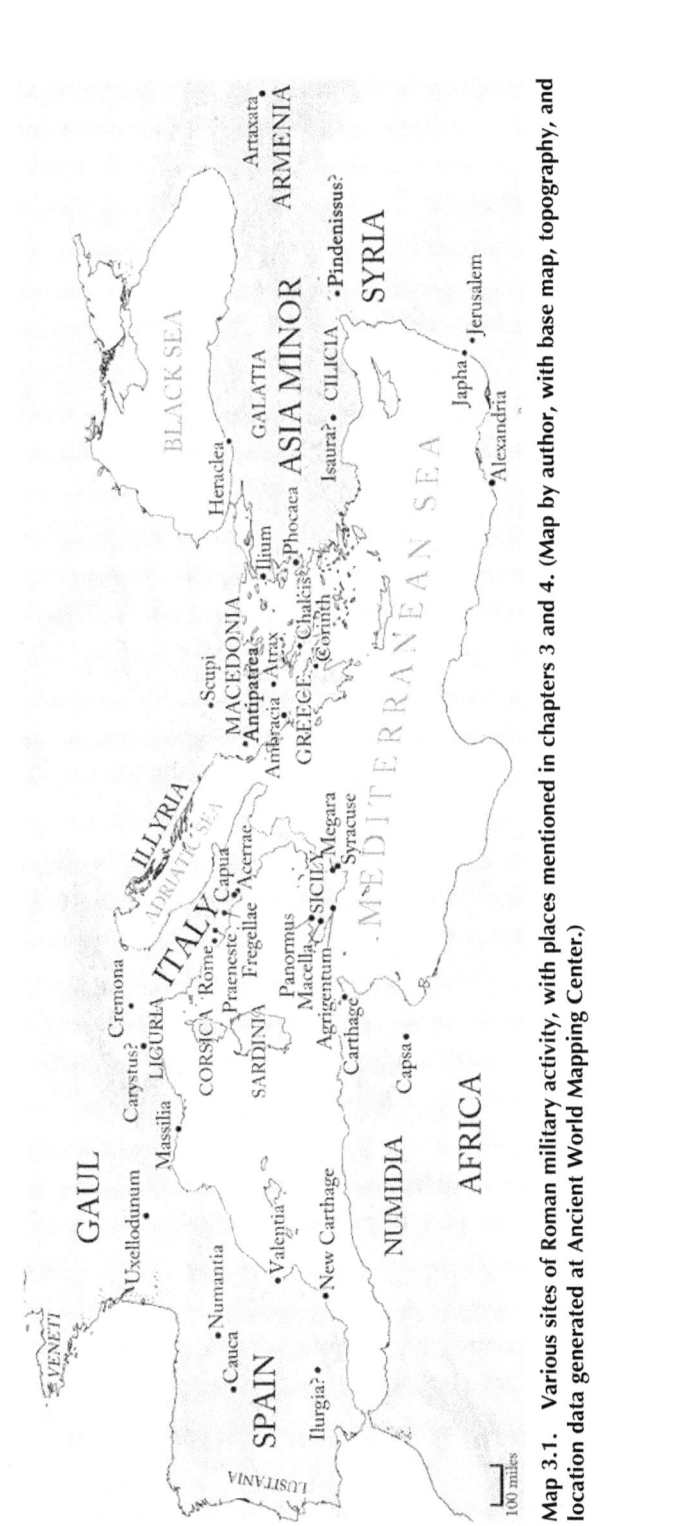

Map 3.1. Various sites of Roman military activity, with places mentioned in chapters 3 and 4. (Map by author, with base map, topography, and location data generated at Ancient World Mapping Center.)

fighting even after their walls were breached, as Roman forces learned on many occasions. During the Second Macedonian War, the Romans battered a hole in the walls of Atrax, an enemy town, and seemed to be on the verge of victory. The Roman commander believed that "if he opened a path into the city for his troops, the flight and slaughter of the enemy would follow." Instead, the defenders got into battle formation behind the opening in their walls. When the Romans tried to advance through the breach, they ran into a new wall made of men and spear points instead of stone. They were driven right back out again. In other cases, defenders occupied citadels or hills within their cities, creating a second line of defense after the Romans had broken through their outer walls. For instance, when the Roman army got inside the fortifications at New Carthage (modern Cartagena, Spain) in 209 BCE, the Carthaginian defenders still occupied a lofty hill as well as a fortified citadel. This posed a real threat to the Roman attackers, forcing them to assault both positions before they could subdue the town.[7]

Resolute defenders could also bog down the Romans in brutal urban combat, battling on from their streets, buildings, and rooftops. We have already seen how the population of Carthage forced Roman invaders to fight for every city block. The first-century historian Josephus describes a similar scenario during the First Roman-Jewish War (66–73 CE). After the Romans broke through the outer defenses of Japha, a city held by rebels, the defenders rallied in the city's narrow alleyways and continued to resist. At the same time, local women rained down improvised missiles from atop buildings, and the fighting lasted for hours before organized resistance crumbled. Defiant populations, with their men, women, and even children fighting bitterly, created extremely precarious situations for Roman invaders: unknown streets, tight alleyways, and defended rooftops threatened to turn a successful assault into a ruinous rout.[8]

In the face of these potential dangers, Roman attackers often unleashed massive violence. According to the Greek historian Polybius, it was customary for the Romans "to kill all they encountered and to spare no one" in the moments after they punched through a city's outer defenses, and Livy, Julius Caesar, and other authors provide many examples of this behavior. These massacres were probably not spontaneous. In several texts, Roman generals *order* their soldiers to attack urban populations, using trumpet blasts or other signals to convey their commands. Leaders also may have passed "kill orders" through the Roman army's extensive command structure, since officers accompanied the rank and file during attacks on cities and were expected to maintain discipline in these vital moments. Officers could conceivably direct their men against urban inhabitants, armed and unarmed alike.[9]*

---

* To give a non-Roman example, Polybius (8.30.4–9) explains how Hannibal Barca used the chain of command to massacre the Romans in Tarentum. After his army gained entry into the city, Hannibal "ordered his Carthaginian and Celtic officers to kill all the Romans they encountered"; these officers then led their troops in the effort to find and kill the Romans in the city.

These massacres could continue for a long period, especially if the targeted population was large or if Roman troops sought vengeance against obstinate defenders. Nonetheless, it is unlikely that the Romans wiped out whole populations when they burst into enemy towns and attacked the inhabitants. At Antipatrea and other towns, in fact, the Romans reportedly killed the military-aged males, thus prioritizing the most visible and immediate threats. Urban inhabitants could also hide or flee from the Romans, like the lucky few who escaped "through precipitous passages" at Cauca, and extended bouts of massacre might peter out when attacking soldiers became physically exhausted. During the siege of Jerusalem (70 CE), Josephus claims, Roman forces became worn down from the constant slaughter of the populace. Modern parallels are also instructive. Hutus who took part in the Rwandan genocide often murdered Tutsis with handheld weapons like "machetes (panga), sticks, tools, and large clubs studded with nails (masu). . . . For the aggressors, the killings were labor-intensive, exhausting work." Further, "many Hutu killers described the exhaustion that followed hours of hacking." Surely the Romans would have been susceptible to similar fatigue when they targeted urban populations, limiting their violence in Jerusalem, Antipatrea, and other cities.[10]

In any event, the Romans usually ended these urban massacres because they had served their military purpose—hastening submission, eliminating obvious threats, and securing a targeted town. Once they decided that the targeted city was secure, at a signal from their commander or on their own initiative, the legionaries would move off to sack the hapless community. (And in the chaos of combat some soldiers probably peeled away to loot earlier, and without permission from their leaders.) The historian Tacitus illustrates this transition from massacre to plundering during an outbreak of civil war violence in Rome (69 CE):

> The victors ranged through the city in arms, pursuing their defeated foes with implacable hatred: the streets were full of carnage, the fora and temples reeked with blood; they slew right and left everyone whom chance put in their way. Presently, as their licence increased, they began to hunt out and drag into the light those who had concealed themselves; did they espy anyone who was tall and young, they cut him down, regardless whether he was soldier or civilian. Their ferocity, which found satisfaction in bloodshed while their hatred was fresh, turned then afterwards to greed. They let no place remain secret or closed. . . . This led to the forcing of private houses or, if resistance was made, became an excuse for murder. . . . Everywhere were lamentations, cries of anguish, and the misfortunes that befall a captured city. (Tacitus *Histories* 4.1, trans. Moore)

Buried beneath the dramatic language, Tacitus shows how attackers initially moved through the streets, killing combatants and noncombatants that they found in the open; from there, they continued to hunt down any potential foes in order to overcome remaining threats (i.e., tall young men) or to satisfy their desire for vengeance; finally, they broke off to sack individual homes.[11]

Roman city-sacking was a disordered and terrifying process. The Latin word used to describe the sack was *diripere*, which "consisted in letting the soldiers loose, in giving them unrestricted freedom to loot, rape and slaughter" (to quote historian Adam Ziolkowsi). Moreover, both Livy and Polybius use the language of "dispersing" or "rushing" when they describe soldiers sacking cities, indicating that these events were largely unorganized, potentially chaotic, and certainly horrific. Some soldiers would break into houses and temples seeking valuables. Others would continue to kill for the sake of robbery or simply for sport. Others still would seize and sexually assault "young women or handsome youths." Roman generals had trouble controlling their men in this mayhem, though commanders could try to limit their soldiers' rapacity. Scipio Aemilianus tried to bar his troops from plundering "the gold, silver, and temple offerings" during the final assault on Carthage. When another Roman commander lost control of his soldiers at Phocaea (a town in Asia Minor), "he sent heralds throughout the city and ordered all free persons to gather in the forum, where they would not be harmed." A few commanders even punished their men for looting without orders. For the most part though, Roman city-sacking was a free-for-all that might continue for hours, days, or even longer. During this time, or after the army had its fill of plunder, the Romans began to deal with survivors.[12]

TAKING CAPTIVES

When Roman armies overran enemy cities, they frequently captured the surviving inhabitants. It is not always clear how they acquired their prisoners in these situations, but there were probably a few primary mechanisms. A city's defenders might throw down their arms and surrender to Roman invaders as a group, or a whole population might capitulate before or after the start of a Roman attack; in either case, the Romans effectively acquired a large body of prisoners all at once. Roman soldiers could also hunt down and seize the individual inhabitants of a city, on the orders of their commanders or on their own initiative, as they plundered. In such a scenario, we should imagine legionaries grabbing people in the streets and in homes, threatening them into compliance or dragging them away by force. We can glimpse something like this process on the Column of Marcus Aurelius, a Roman victory monument from the late second century CE. In several of its more chilling scenes, Roman troops roughly round up barbarian women and children. The Romans also appear to stab some of their victims, perhaps killing those who were undesirable as potential slaves or who had resisted too fiercely (see figure 3.1).[13]

Whichever way the Romans collected their prisoners, they needed to control them. This may not have been easy: many thousands of people could fall into Roman hands and, since most consuls commanded about 20,000 men (and many forces were smaller), Roman armies could become

Figure 3.1. Roman soldiers take some women and children captive while killing others. Column of Marcus Aurelius scenes 97–98, Rome. (Illustration by author.)

outnumbered by their captives. Yet even if they were heavily outnumbered, it was still possible for the legions to control large crowds of people. For the sake of comparison, ships' crews in the Atlantic and American slave trade were often "greatly outnumbered by the ranks of slaves"; ratios might be as low as four enslaved people per sailor, or as high as forty to one. These crews employed numerous methods to maintain dominance over prisoners. Ships' crews "made it their constant, and often obsessive, business to crush the spirit of resistance no matter what the cost," using intimidation, manipulation, and cruelty. Although the crews were "vastly outnumbered" by their captives, they were "well armed, generally well disciplined, and constantly on guard for the slightest signs of disaffection among the slaves." Moreover, ship captains put their captives in irons, collars, and chains, and did anything deemed necessary to prevent rebellion.[14]

The Romans lacked the firearms of these early modern and modern sailors, but they too were "well armed [and] generally well disciplined," and they took similar steps to restrain and intimidate captives. In several texts, the Romans use chains and collars to restrain their prisoners of war, or Roman commanders assign men to guard captives. Roman forces could also herd captives together into enclosures that served as ad hoc holding pens. After the capture of Jerusalem, for example, the Romans shut up their prisoners inside one of the Temple courtyards. And two different generals, on two occasions, reportedly corralled whole communities into empty army camps in Spain. Such camps were surrounded with ditches and palisades that would have fenced in the captives. Visual evidence from the imperial period also provides important clues. A scene on Trajan's Column (another monument of the second century CE) shows enemy prisoners huddled together in a wooden enclosure, with Roman troops watching over them. Another scene depicts soldiers binding the hands of prisoners. Additional monumental reliefs from the Roman Empire show captives chained, shackled, and bound together (see figure 3.2), and archaeological examples of slave shackles have been found at multiple Roman military sites. Ultimately, the Romans probably used a variety of methods to control captured people, including force, intimidation, physical enclosures, and individual restraints.[15]

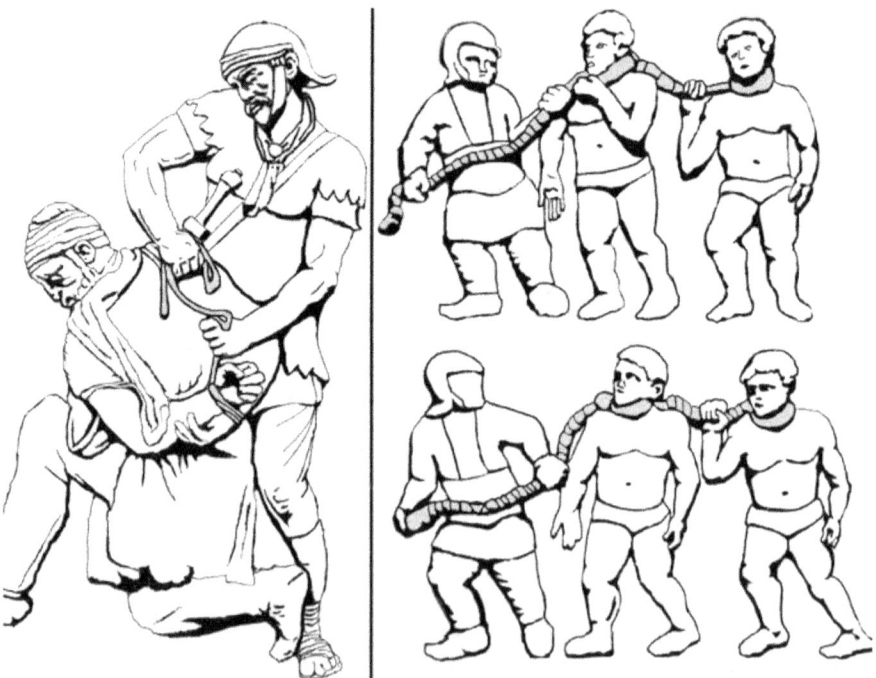

Figure 3.2. Left: A Roman soldier binds the hands of a prisoner. Trajan's Column scene 40, Rome. Right: Prisoners chained together. Marble relief from Smyrna. (Illustrations by author.)

Once the Romans had subdued a mass of prisoners, they could assert further dominance by sorting the captives into smaller groups. One sorting method involved separating the young women and non-infant children from the rest of the crowd, a process sometimes indicated by the Greek verb *andrapodizein* and its related words. Historian Kathy Gaca vividly describes this procedure:

> The precise allocation of tasks here is not uniform and many of its details remain unclear. Nonetheless, the armed fighters along with their leaders start picking through the captives somewhat like fruit scattered at their feet in order to select for forcible removal those considered ripe and ready or still green but close enough to ripening, but in any event free of daily maternal dependency—young, malleable, and female in the majority.[16]

Though Gaca is primarily concerned with the ancient Greek world, the Romans doubtlessly engaged in similar practices. Livy and Polybius both describe the Romans "dragging away" women and children, or "tearing away" maidens and youths from their parents. Roman commanders also sorted prisoners according to their tribe, city of origin, or other factors. After Mummius captured Corinth, for instance, he reportedly surrounded the survivors with his soldiers and "proclaimed the freedom of all except the Corinthians"; his troops then laid hold of the captives and "made a clear distinction between them." Moreover, according to Plutarch, Mummius interviewed the freeborn Corinthian boys to assess their level of literacy—perhaps because more educated individuals would fetch higher prices as slaves. Generals also picked out "impressive" prisoners to grace their triumphs in Rome, as when Scipio Aemilianus selected fifty such captives from the defeated population of Numantia.[17]

Some Roman generals employed more complex sorting procedures, dividing their prisoners into several different groups. When he captured New Carthage, Publius Cornelius Scipio gathered together all the surviving inhabitants. Then he separated "the citizens with their wives and children" from "the workmen." Scipio spared the former, allowing them to return to their homes. He determined that the latter group would be public slaves of Rome, promising them future freedom in return for diligent service. These individuals were also "enrolled" or "registered" with the quaestor (an officer responsible for the army's finances and supplies), indicating that there was documentation involved, and an overseer was assigned to every thirty of the new public slaves. From a third group of captives, who had probably been slaves before Scipio's arrival, he picked out young and strong looking men and incorporated them into the fleet as rowers; they too were promised future liberty if they worked hard. A similarly complicated procedure took place when Marcus Porcius Cato captured Bergium, a settlement in northeastern Spain. Cato freed members of the community who had helped him, along with their families. He then passed some of the inhabitants to his quaestor for sale into slavery. Finally, he executed some individuals who were allegedly responsible for local raiding.[18]

The whole process of taking captives and getting them under control would have required significant organization and manpower on the Romans' part, especially if they seized, restrained, sorted, and set a guard over their prisoners, who may not have surrendered willingly. Some sources suggest that all of this took time, too. For example, the population of Pindenissus in Asia Minor surrendered to the Romans on the morning of December 17, 51 BCE, and the captives were not sold until two days later. It also took several days to sort through captives after the fall of Jerusalem in 70 CE. Here the legions apprehended tens of thousands of people, many of whom died of starvation while the Romans were still sorting through them and determining their individual fates.[19]

TREATMENT OF CAPTIVES: RELEASE, ENSLAVEMENT, EXECUTION

After they had taken prisoners and gained control over them, the Romans were generally free to treat them as they wished. It must be stressed that the Romans were not systematically cruel to prisoners or to vanquished communities. In many instances, Roman generals spared defeated populations, perhaps installing temporary garrisons, but otherwise restoring communities to their previous statuses. Some commanders gave prisoners guarantees to their lives, freedom, and property. Mercy could be conditional, however: when the Romans and their allies captured the Greek city Andros in 200 BCE, the inhabitants were spared but forced to evacuate their town. In the early second century BCE, the Romans also relocated several communities in Liguria after accepting their surrender, moving them from their homelands in northwest Italy to locations further south. Such forced migrations doubtlessly involved enormous physical hardship and loss of life, to say nothing of material and psychological losses. A few Roman commanders set ransom prices for captives, too, allowing the prisoners themselves or some third party to purchase their freedom. But ransom was no guarantee to safety, since it required that someone could afford to pay. The population of Panormus (modern Palermo) learned this the hard way. When the Romans attacked this town in 254 BCE, most of the inhabitants retreated to their citadel and then negotiated their surrender. They and the Romans agreed to a ransom price, but only 13,000 people were able to pay and walk free. The remaining 12,000 could not scrape together the money, and the Roman commander sold them as slaves.[20]

Though the Romans spared and released countless people who fell into their hands, many prisoners ended up like the 12,000 unfortunates at Panormus who were sold into slavery. The enslavement of war captives was an old practice at Rome, probably stretching back to their early wars in Italy.* As with so much else, the exact mechanisms of mass enslavement are obscure to us, but we can glimpse some of the details. Most Roman commanders probably handed captives to the quaestor for record keeping and subsequent sale as slaves, a practice we have already seen at Bergium. The quaestor would then auction off captives in the field, selling them into markets that existed around the Mediterranean or to traders who followed the Roman army on campaign. Those doomed to enslavement likely passed out of Roman hands relatively quickly through these various outlets. Nonetheless, the Romans still had to invest time and labor into this process, documenting, organizing, and guarding the prisoners earmarked for sale.[21]

Commanders usually deposited the proceeds from slave auctions into the Roman treasury, but at least some generals "rewarded" their men by giving them the sale money or granting them individual slaves from the lot of captives. In the latter case, soldiers could apparently do whatever they wanted with their new slaves, including selling them or ransoming them for profit. To give one famous example, a Roman centurion gained possession of a Galatian noblewoman named Chiomara during a campaign in Asia Minor (189 BCE). The centurion raped her, taking advantage of her condition as a slave, and later ransomed her back to her family. Chiomara ordered her countrymen to cut off the centurion's head as soon as money changed hands, escaping to tell the tale to Polybius decades later.[22]

While most slaves did not get such dramatic revenge, Chiomara's story readily illustrates the potential violence and indignity that accompanied enslavement. On a personal level, passing into slavery did not simply remove one's liberty, but tore a person from everything familiar, everything that had defined their membership in a community; they effectively lost their identity as they were transmuted into property. On a larger level, selling off a whole population practically destroyed it (or a very large part of it), scattering the individuals and shattering the bonds that formed a living community. Notwithstanding the dim possibility of ransom in the future, ancient peoples "who were carried away from sacked and destroyed cities to be sold in distant places were usually lost forever." For some potential victims, the genuine dread of falling into slavery made death seem preferable. A short Greek poem, attributed to Antipater of Thessalonica (first century BCE), portrays a daughter and mother choosing suicide over slavery when the Romans captured their home city of Corinth:

---

* The Romans had a stock phrase to indicate the sale of prisoners of war: *sub corona vendere*, "to sell beneath the crown." But the phrase (and the practice of selling war captives) was so old that the Romans could not remember what it meant. According to one Roman author, the "corona" might refer to the ring of soldiers who stood around the prisoners at auction. Or maybe it meant that the prisoners wore wreaths on their heads. Or maybe those are both wild guesses, and it was neither of those things. Regardless of the meaning, the antiquity of the phrase suggests that the Romans were enslaving war captives from an early date. See Scheidel 2011, 294–97.

> I, Rhodope, and my mother Boisca neither died of sickness, nor fell by the sword of the foes, but when dreadful Ares burnt the city of Corinth our country, we chose a brave death. My mother slew me with the slaughtering knife, nor did she, unhappy woman, spare her own life, but tied the noose round her neck; for it was better than slavery to die in freedom. (*Greek Anthology* 7.493; trans. Paton, modified)

The poem presents an imagined tragedy, but several historical accounts describe populations committing suicide rather than face the horror, humiliation, abuse, and social death of slavery.[23]

It should be evident by now that the Romans often left their prisoners alive, though the price for living could be a harrowing life as human property. Another possible outcome facing prisoners was immediate execution. The Romans believed that it was their right to put captives to death according to the conventions of war. Thus they reportedly executed the leaders of several Italian cities in the fourth and third centuries BCE, particularly those leaders whom they saw as rebels and traitors. As the scope of their wars expanded, the Romans also executed the chief men in communities outside of Italy. Furthermore, several texts describe the Romans killing captured soldiers, adult males, and even entire populations, in cold blood, with many thousands killed.[24]

In order to carry out mass executions, whether of dozens or thousands of people, the Romans employed a number of killing methods. Their preferred method was beheading, as indicated by the formulaic Latin phrase *percussit securi*—literally "to strike through with an axe." Polybius and Diodorus say it was customary for the Romans to scourge prisoners with rods and then decapitate them, and Cicero states that beheading was the Romans' ancestral form of execution. There is little reason to doubt their testimony: several ancient authors describe the Romans beheading their prisoners, and there is also archaeological and artistic evidence of this practice. At the site of Valentia (modern Valencia, Spain), a city reportedly destroyed by the Romans in 75 BCE, excavators uncovered thirteen male skeletons in the city's forum within a destruction layer of ash and debris. The victims displayed "signs of torture and violent death," including "systematic decapitation and amputation of limbs." A mass grave recently excavated at the site of Scupi (in the modern Republic of North Macedonia) also illustrates this execution method, albeit in a later period. Here archaeologists uncovered 184 male skeletons dating to the third or fourth century CE, aged twenty to forty at the time of death; their bones exhibited micro-stresses from regular load carrying, which may indicate a military background. Of thirty skeletons subjected to close forensic analysis, twenty-two showed evidence of complete or partial decapitation, as well as other fatal injuries like stabbing wounds to the chest. This all "points to a possible execution of the large part of the group." And a scene on the Column of Marcus Aurelius shows Rome's barbarian allies or auxiliaries decapitating other barbarians with swords, while Roman soldiers stand guard (figure 3.3). As a final note, although ancient authors like Cicero describe beheadings with axes, the archaeological and artistic evidence suggest that Roman forces on campaign could use swords or other readily available weapons for these executions.[25]

Regardless of the specific method of execution, it is certain that exterminating a large number of captives required at least rudimentary organization and planning—particularly for a Roman army with nothing but premodern weapons at its disposal. Though the details are hazy, the evidence hints at potentially complex killing procedures, with the Romans moving captives to special killing sites, restraining them, and executing them in sequence. A Greek writer, Dionysius of Halicarnassus, describes just such a mass execution in Rome:

> Stakes were fixed in the Forum and men, being brought forward three hundred at one time, were bound naked to the stakes, with their elbows bent behind them. Then, after they had been scourged with whips in the sight of all, their necks were cut with an axe. After them another three hundred were destroyed, and then other groups of like size, a total of forty-five hundred in all. (Dionysius of Halicarnassus *Roman Antiquities* 20.16.2; trans. Cary)

Plutarch likewise describes an execution in the Roman Forum, where the executioners stripped the prisoners, bound their hands behind their backs, beat them with rods, threw them to the ground, and beheaded them with axes. The mass execution depicted on the Column of Marcus Aurelius indicates a broadly similar procedure: prisoners with bound hands are led one by one to executioners, where they are bent over to receive the sword blow. Soldiers stand guard or assist in the killing. The archaeological evidence is less clear, but investigators argue that the bodies found at Valentia and Scupi were tied up or otherwise restrained before execution.[26] There is also evidence of execution procedures in other ancient cultures. The historian Diodorus describes "the most complex slaughter" arranged by Agathocles, the tyrant of Syracuse in the late fourth and early third centuries BCE. In Diodorus's narrative, Agathocles's soldiers remove many men, women, and children from Syracuse and gather them at the seashore. Then executioners move down the line of captives and kill them in sequence:

> When a crowd, large and composed of all kinds of people, had been driven to the sea for punishment and when the executioners had taken their places beside them, weeping and prayers and wailing arose mingled together, as some of them were mercilessly slaughtered and others were stunned by the misfortunes of their neighbours and because of their own imminent fate were no better in spirit than those who were being put to death before them. (Diodorus Siculus *Library of History* 20.72.3; trans. Geer)

There is also a recurring scene in Etruscan artwork where the Greek hero Achilles slaughters and sacrifices Trojan prisoners of war, again suggesting

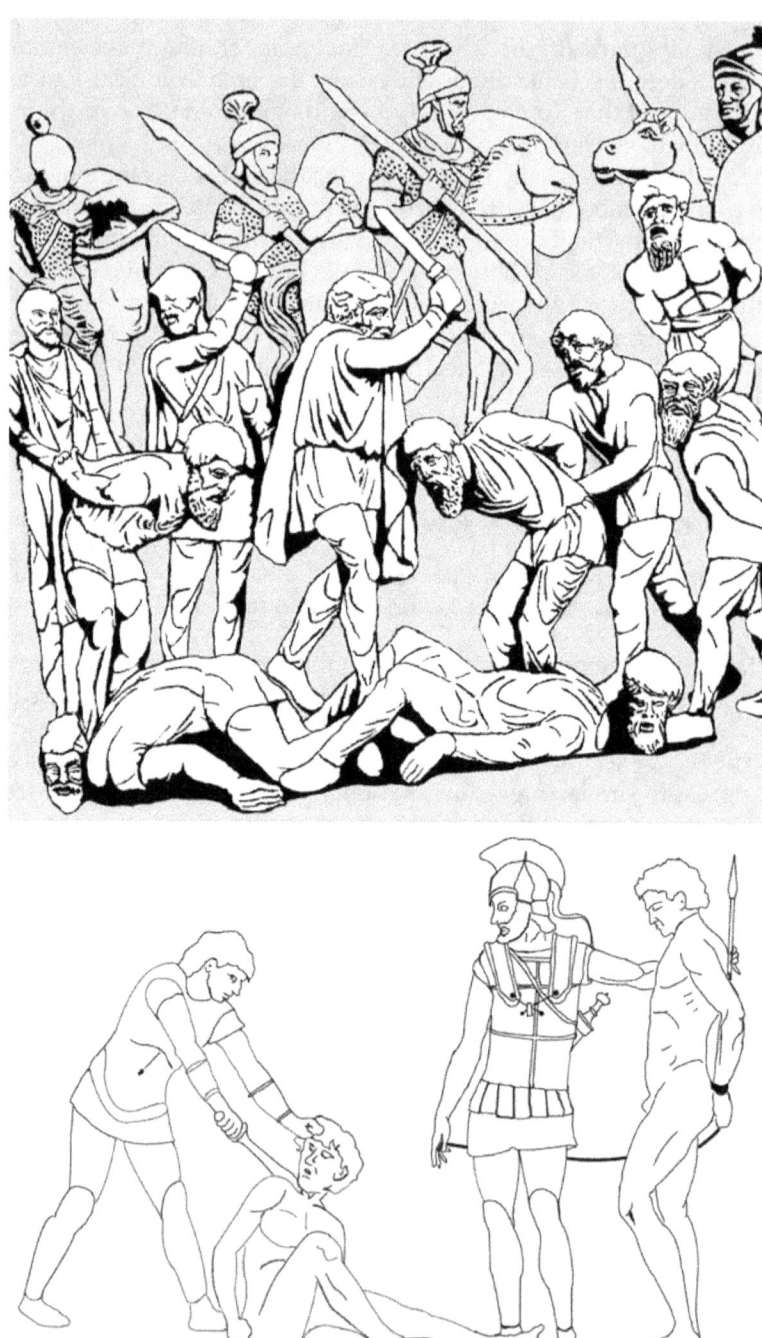

Figure 3.3. Top: Roman allies or auxiliaries behead barbarians while Roman soldiers stand guard or watch. Column of Marcus Aurelius scene 61, Rome. Bottom: Trojan prisoners led to Achilles for slaughter and sacrifice to the shade of Patroclus. François Tomb fresco, Vulci. (Illustrations by author.)

an organized killing procedure. In a fourth-century BCE Etruscan painting (from the François Tomb at Vulci), bound captives are led one by one to the "executioner," Achilles, who pulls back their heads to cut their throats. Soldiers appear to stand guard or help in the procedure (see figure 3.3).[27]

In some cases, Roman troops eliminated their prisoners in a less formalized way, simply attacking a crowd of unarmed individuals. Yet there was still planning and organization involved in these cruder mass killings. In a widely reported event that took place in 150 BCE, the Romans divided several thousand Lusitanians into three groups, each of which was sent to a different location. Roman forces approached the first group, collected their weapons, and surrounded them. They subsequently cut down the unarmed Lusitanians with swords, then repeated the whole process with the second and third groups (see chapter 7). And when the Italian city of Praeneste was captured in 82 BCE during a Roman civil war, the dictator Sulla executed the inhabitants en masse. According to Appian, Sulla marched the disarmed population onto a plain. Once the prisoners were assembled, he pulled out a few men who were "useful" to him, then separated the Romans from the other Italians. The Romans were spared, but the rest were "shot down" with javelins. In another version of the story, by the Roman writer Valerius Maximus, Sulla's subordinate Publius Cornelius Cethegus enticed 5,000 people out of Praeneste by promising their safety. Afterward, he (or his soldiers) disarmed these individuals, made them "lie on the ground," and killed them. In another scene on the Column of Marcus Aurelius, Roman troops march unarmed prisoners to a killing site, then set upon the unarmed crowd with their spears.[28]

The evidence is not clear enough or consistent enough to determine which killing procedures, if any, were standard for Roman armies on campaign, but mass executions must have always involved some organization. Indeed, the organizational methods described in the ancient sources—confinement or restraint of the condemned, division of prisoners into smaller groups, the use of special killing sites, and sequential killings—are attested to in more recent history, suggesting that such procedures are not just the imaginative license of ancient authors and artists. Moreover, on a practical level, killing procedures would have minimized prisoners' chances for resistance and escape. These were important considerations for Roman forces seeking to execute hundreds or thousands of people.[29]

As a final point, some sources imply that ancient executions were difficult and laborious, introducing special complications for would-be executioners. For instance, Tacitus implies that beheadings were physically challenging. He relates that a trembling Roman executioner needed to strike twice in order to kill an alleged conspirator, and even then he "barely" (*vix*) decapitated the man. An early Church historian, Eusebius, suggests similar difficulties. He claims to have witnessed a mass execution of condemned Christians, writing that "the executioners themselves, becoming exhausted, relieved one another in turns." He adds that the executioners' axes "were blunted, and becoming worn out they broke into pieces." And according to the late Roman author Ammianus Marcellinus, the emperor Valens ordered a series of executions that "tired the arms of the killers." In a different anecdote, Ammianus also says that Roman swords became bent and blunted as they struck repeated blows against fleeing enemy warriors.[30]

Tacitus, Eusebius, and Ammianus suggest that Roman executioners could bungle the act of killing, become exhausted, and damage their weapons. Naturally, these authors might have exaggerated for effect, with Eusebius wishing to emphasize the courage of Christian martyrs, Ammianus the cruelty of the emperor, and so forth. Nonetheless, there may be a factual basis to their reports: evidence from other historical eras confirms that mass executions were logistically difficult. Medieval and early modern executioners sometimes botched the act of beheading, much as Tacitus describes, delivering repeated blows after repeated failures. And early modern executioners report that sword blades often break or need constant maintenance after a series of killings. In more recent history, executioners reportedly became exhausted after killing their prisoners. German death squads in World War II reported physical and psychological exhaustion after executing large numbers of people. Their killing procedures included such activities as beating and shooting captives, and grave digging. Likewise, during the attack on Nanjing in 1937, Japanese soldiers found that gathering up, binding, moving, and killing large numbers of Chinese POWs was very slow work. In one instance, they bound the hands of over 10,000 captives, marched them to the Yangzte, machine-gunned the crowd, and finished off survivors at close range. The process took over two days and nights, cleanup of bodies another three, and "even then the troops may have been unable to finish the job." German and Japanese soldiers had modern firearms at their disposal, weapons immeasurably more deadly than their ancient counterparts, yet they still experienced fatigue and organizational problems. There is no reason to assume that Roman soldiers would have been immune to these same problems, whether or not they implemented careful killing procedures.[31]

While the details are murky, it is possible to make some broad claims about how the Romans executed their prisoners of war. First, the Romans' preferred method was beheading, though other forms of killing are reported as well. Second, mass executions involved planning and organization. At the very least, the Romans must have restrained or surrounded captives; there is also evidence that ancient executioners bound their prisoners' hands, divided them into groups, and killed them sequentially. Third and finally, the mass killing of captives would have been difficult, time-consuming, and labor intensive for the killers, potentially leading to physical exhaustion and damaged weaponry. The Romans could mitigate these problems by rotating out soldier-executioners—like Eusebius's killers who "relieved one another in turns"—but the important point is that mass killing required organization and the investment of military resources.

DESTROYING CITIES

When the legions conquered an enemy town and its inhabitants, they also gained control over the physical buildings of the captured community. And just as Roman armies eliminated some of their prisoners, they also damaged or destroyed some of the cities that they captured in war. Greek and Roman authors are again frustratingly vague about how the Romans destroyed physical urban spaces, or what it even means to "destroy" a city in ancient warfare, but the destruction of Antipatrea provides a good starting point. In Livy's ever-so-brief account, the Romans burn the city and demolish the walls. These two activities, burning and demolition, would have provided the primary means of urban destruction in a pre-gunpowder world, and so we will explore these methods here.

Many sources flatly state that the Romans "burned" cities, without giving additional details, but a few authors imply that there were common methods of mass arson. For example, Livy describes legionaries "throwing fire" (*ignem iniecit*) onto the roofs of buildings. Tacitus also mentions Roman troops throwing firebrands or torches inside enemy dwellings. And in numerous scenes on the Columns of Trajan and Marcus Aurelius, Roman soldiers are shown torching the roofs or interiors of houses and forts (see figure 3.4). This would have been a relatively easy way to damage or destroy structures: after the roof or interior of a building ignited, the resulting blaze would do most of the work or at least render the structure uninhabitable. Even buildings of brick or stone, with clay roof tiles, had wooden frames to feed the flames, and many ancient structures were roofed with more flammable shingles or thatch. This very basic form of arson would have allowed Roman troops to burn buildings while simultaneously sacking enemy towns, and some fires may have resulted from the vandalism of plundering soldiers.[32]

Yet despite its relative simplicity, mass arson probably required a measure of organization and cooperation on the part of Roman soldiers. In a few texts, ancient soldiers work together to burn specific enemy buildings, plying their torches in unison or piling up kindling to create more destructive blazes. Several narratives also describe soldiers burning buildings only after they had stripped them of spoils. So when Roman soldiers sacked Cremona, Tacitus says, they carried "torches in their hands, which they tossed licentiously into vacant homes and empty temples *after they had brought out the plunder*" (my emphasis). Livy writes that the Romans first "completely pillaged" the buildings at Aquilonia and Cominium, "*then they threw fire on them*" (my emphasis). Memnon of Heraclea, a Greek historian, explains that Roman soldiers burned Heraclea in Asia Minor after they had looted it. Parallel practices also appear in ancient Greek warfare. For example, Diodorus reports that hostile mercenaries "burned the houses near the marketplace" in Syracuse "after having plundered them, and now hastening against the remaining houses, they were plundering the possessions in these." Such

Figure 3.4. Left: Roman soldiers burn enemy strongholds. Trajan's Column scene 153, Rome. Right: A Roman soldier burns enemy villages. Column of Marcus Aurelius scene 102, Rome. (Illustrations by author.)

anecdotes imply coordination between fire-wielding soldiers, or even a systematic process of looting followed by burning.[33]

It makes sense that ancient soldiers would have coordinated their efforts when they burned enemy cities, seeing that overzealous arsonists might destroy valuable loot or accidentally burn buildings with their comrades inside. Cooperation would be vital if Roman attackers wanted to burn specific targets or spread devastation more widely. Analogies from more recent history also indicate that mass arson involved organization, especially if arsonists had specific objectives in mind. In 1814, for example, British commanders sought to destroy the public buildings in Washington, DC. The British officers divided their troops and tasked different men with the destruction of different structures. Their soldiers first had to scrounge for flammable materials, using anything at hand to help burn fire-resistant stone and marble buildings, then they worked together to torch their respective targets. In 1914, German forces also reportedly burned the Belgian city Leuven in a systematic fashion, passing the torch "street by street, house to house." Another account asserts that the imperial German army burned villages "by a drilled method," and did so "especially after looting." References to systematic burning (often after pillaging) pop up in other historical periods, too, lending further credence to the representations in ancient texts.[34]

Aside from mass arson, Roman armies might physically demolish buildings in enemy cities. Latin writers often use the verb *diruere*—meaning something like "to demolish" or "to raze"—to illustrate urban destruction. At times, they also distinguish between devastation caused by fire and devastation indicated by *diruere*. For example, when Livy describes the destruction of Ilurgia, he writes that the Romans "demolished [*diruunt*] buildings which could not be consumed by fire." In a similar manner, Greek writers prominently use the verb *kataskaptein* ("to raze to the ground") to describe the destruction of cities, and likewise distinguish between destruction caused by fire and the destruction indicated by *kataskaptein*. A couple of examples will suffice here. Polybius, describing a Macedonian attack on an enemy sanctuary, says that the attackers "not only destroyed the roofs by fire, but they also demolished [*kateskapsan*] the buildings to the foundations." According to Appian, a Roman commander "burned . . . the temple [of Athena], and demolished [*kateskapte*] the walls" of the city Ilium. Some ancient authors use *diruere*, *kataskaptein*, and similar vocabulary more generically to describe the destruction of cities, but these words clearly had important connotations—pointing to the non-incendiary demolition or dismantling of structures.[35]

Furthermore, both the Latin *diruere* and the Greek *katakaptein* explicitly refer to the demolition and dismantling of buildings in non-military contexts. Among many examples, Plutarch describes workmen demolishing (*kateskapsen*) a Roman home in a civilian context. And a Roman senatorial decree "against the demolition [*diruendis*] of buildings for profit" forbade people from demolishing (*diruendo*) structures in order to profit from the dismantled materials. Just as significantly, ancient writers frequently use these verbs to narrate the destruction of city walls and fortifications in Roman and Greek warfare. Since urban fortifications were often stone or brick structures, and resistant to burning, the use of these verbs must again indicate some kind of physical demolition or dismantling. Ancient writers also use special language to describe armies "pulling down" (*kathaireîn*), "throwing down" (*deicere*), "stripping off" (*kathaireîn*), or "leveling" (*adaequare solo*) walls and other buildings. The connotations are fairly obvious.[36]

Greek and Roman authors thus indicate that non-incendiary demolition was an important form of urban destruction in war. Our sources also suggest that soldiers used hand-held tools, like pickaxes, crowbars, and shovels, to demolish structures. One scene from Trajan's Column shows Roman legionaries demolishing the walls of an enemy town using their *dolabrae* (a pickaxe-like entrenching tool; see figure 3.5). A late Roman general named Belisarius additionally ordered some soldiers who were "experienced in building" to destroy the walls of an enemy cistern "with pickaxes and other implements useful for cutting stone." And a fourth-century BCE text describes how the people of Delphi in Greece gathered "with shovels and pickaxes," then went to demolish the harbor in a neighboring town, Cirrha.[37]

Ancient soldiers could use these hand-held tools to dismantle buildings from the top down. To do so, they would first climb to the top of the structure

Figure 3.5. Top: Assyrian soldiers demolish Elamite Hamanu while the city burns and other soldiers bring out the plunder. Bas relief from Nineveh. Bottom: Roman soldiers demolish fortifications with dolabrae. Trajan's Column scene 116, Rome. (Illustrations by author.)

designated for demolition. For example, Thucydides says that infantrymen and townspeople mounted the fortifications at Piraeus, the harbor of ancient Athens, in order to demolish them. Next, if the building was roofed, ancient soldiers would probably pull off shingles, tiles, or thatch, then strip off the wooden rafters; a few late Roman sources say that attackers stripped the roofs from buildings before tearing them down, and some early modern demolitions also employed this method. Once they had cleared the roof from a structure—or if there was no roof to begin with—soldiers likely got to work on the walls, prying loose the stone or brick and knocking down the successive layers of material. This method, or something like it, might be portrayed on Assyrian palace reliefs from the seventh century BCE. These reliefs show soldiers standing atop the walls of enemy cities, using picks, crowbars, and levers to pry off building materials and throw them to the ground (see figure 3.5). Medieval Arabic, early modern English, and early modern German sources also depict men demolishing fortifications from the top down, providing a neat parallel to the ancient descriptions and images.[38]

Roman armies could also use the tools and techniques of siege warfare to bring down urban structures. For instance, they might use battering rams to knock down walls or siege hooks to pull them down. Vitruvius describes an incident in which Carthaginian troops used a ram to demolish a captured fort, striking off each course of stone until they had leveled the whole structure. Julius Caesar's legionaries also reportedly used rams to knock down buildings when they were fighting in Alexandria. Ancient armies could additionally use undermining techniques from siege warfare to collapse sturdy structures. This "involved making a depression [at the foot of a] wall with the help of tools such as crowbars, pickaxes, and axes. The hollow would be gradually enlarged, and the wall strengthened with wooden props lest it should fall right on the sappers"; then the wooden supports were burned, and the structure would tumble down. Livy clearly illustrates this method: in his retelling, a team of 500 Carthaginians approached the walls of an enemy city, then hacked away at the lower layers of the masonry until a portion of the fortifications fell in. Josephus mentions the Romans using crowbars and other equipment to "pry out" (*exekúlisan*) stones from the walls of Jerusalem, seeking to weaken the lower levels and cause a wider collapse. Many armies also tunneled into the earth beneath enemy fortifications to undermine them from below.[39]

While battering rams and undermining were primarily used in sieges, Roman armies could conceivably employ these methods for all kinds of destruction. For the sake of comparison, engineers in other periods used the techniques of siege warfare to destroy structures outside of a siege context. During the sixteenth century, a military engineer named Giovanni Portinari was hired to level the Lewes Priory church in England. Portinari and his team spent weeks undermining the priory's stone walls and columns in order to bring them down. Centuries later in Atlanta, in the American Civil War, William Tecumseh Sherman's soldiers used an iron ram to destroy public buildings with strong stone walls. Similar techniques would have been familiar to a Roman army on campaign, and, like their counterparts in later periods, provided a ready-made approach for demolishing buildings.[40]

Any of these demolition methods would have been laborious, and they must have required organization, time, and manpower. Ancient authors occasionally hint at these logistical difficulties. For instance, Polybius reports that an unspecified number of Macedonian soldiers toiled day and night for three days to demolish 210 feet of a city's fortifications. And we have already seen that Livy has a team of five hundred Carthaginians working with pickaxes to undermine an unspecified length of wall. These snapshots vaguely suggest that demolition required time and labor for ancient soldiers, but that is pretty much all they tell us. For comparison, the demolition of castles in seventeenth-century England is relatively well documented and can shed some light on the ancient evidence. The 1649 demolition of Montgomery Castle, with its large tower, curtain wall, and towered gatehouse, took about four months. The strenuous work, done with "Iron Crowes Mattocks and pickaxes with other Iron tooles," involved miners and up to 150 laborers, as well as other craftsmen such as tilers and glaziers. Demolition was achieved both by undermining and dismantling the castle's structures, and the work was quite dangerous. During the demolition of Pontefract Castle—also in 1649, and also involving a large labor force and lasting for months—unplanned collapses were a constant threat. Several workers were injured in these unsafe conditions.[41]

Examples like the Montgomery and Pontefract Castles show that demolition without modern tools was difficult and dangerous work, and so it must have been for the Romans. That does not mean that ancient military demolitions were always *slow*, however. In 1992, 1,200 nationalist militants in India led thousands of others in an attack on Babri Mosque, wielding hammers, pickaxes, and other hand-held tools. The sixteenth-century mosque was primarily built of rubble masonry and occupied an area of around eighty by forty feet, yet this impressive structure was razed in about six hours. Notably, the attackers of Babri Mosque numbered in the thousands, and their intent was simply to destroy (not to reuse or recycle dismantled materials). Furthermore, hundreds of the attackers were killed or injured in the process. So even if the demolitions were relatively quick here, expediency came at the cost of much human labor and possibly increased danger.[42]

Demolitions of medieval castles in the seventeenth century and early modern mosques in the twentieth do not allow one-for-one comparisons with military demolitions in Roman warfare. Building materials and working conditions are plainly different. Nonetheless, these comparative examples suggest that demolishing large structures would be difficult, even hazardous work for ancient armies, necessitating serious investments in time, labor, or both. The resource requirements surely multiplied in the case of robust stone structures, especially if these were methodically dismantled rather than hastily destroyed. When our sources say that the Romans demolished buildings, especially fortification walls, they are describing neither spontaneous vandalism by Roman soldiers nor collateral damage from the sack. Instead, they are telling us about a process that involved organization, planning, and the direction of manpower and resources.

A NOTE ON THE EXTENT OF URBAN DESTRUCTION

When Roman armies burned and demolished cities like Antipatrea, it is doubtful that they literally wiped these places off the face of the earth with torches and pickaxes. But that does not mean our sources are exaggerating or outright lying when they describe the destruction of cities in Roman warfare. To understand what the sources mean when they claim that the Romans "destroyed" cities, and to understand the extent of this devastation, we need to examine the evidence carefully and consider the expectations of ancient writers and their readers.[43]

At times, ancient authors admit that the Romans did not completely destroy enemy towns, instead suggesting that the legions targeted specific urban structures. Our sources most frequently mention the destruction of fortifications, but they also remark upon the destruction of arsenals, granaries, and temples, as well as other public, sacred, and military buildings. Additionally, some accounts describe the *partial* destruction of cities in war. For the sake of illustration, Livy says the Greek city Chalcis was "half-destroyed" (*semirutae*) when it was burned by the Romans, and Acerrae in Italy was "partly burned" (*ex parte incensis*) when the Carthaginians set it ablaze. Even in the case of famously destroyed cities like Corinth and Carthage, some ancient authors describe the destruction of specific buildings rather than of the whole urban fabric. From the perspective of ancient writers, it seems, military destruction did not always consume entire cities.[44]

The methods of urban destruction described in Greek and Roman texts also indicate that destruction was limited. As argued above, demolition would have been laborious and time-consuming, and it is difficult to imagine an army devoting the necessary resources to raze an entire city. On the other hand, fire was a persistent, highly destructive threat in pre-modern cities across the globe. Some ancient building materials, like timber and wattle-and-daub, are very susceptible to burning. Moreover, fire can harm seemingly fire-resistant materials, given enough time and sufficient temperatures. Ancient towns containing numerous wooden structures, cramped streets, and narrow alleyways must have acted as virtual tinderboxes for an accidental blaze—or for an invading enemy force. Still, sturdy stone structures are more resistant, if not impervious, to flames, and factors like weather and the determination of arsonists could intensify or limit fire's destructiveness. Fire could be devastating, in other words, but total urban destruction was perhaps unlikely.[45]

Archaeology further indicates that the "destruction" of cities was usually less than total. For example, archaeological remains do not show that the Romans totally destroyed Carthage, though there is evidence of extensive burning and damage to buildings. Similarly, excavations at Corinth do not show that the Romans utterly obliterated the city. Archaeologists have noted the devastation of a few public structures, including a building that might have been the war or tax office, a building that could have been an arsenal, and perhaps parts of the city walls. Some sacred structures also show signs of damage, but not complete leveling. Last, at Fregellae, a city destroyed by the Romans in 125 BCE, "the archaeological record paints a grim picture, attesting to near complete and thorough devastation of the city's sacred buildings." Though the destruction here was intense, it is noteworthy that the Romans may have focused their energy on important sacred structures.[46]

Focused and limited devastation is the key to understanding urban destruction in Roman warfare. Public buildings, sacred sites, and defenses were central to the identity of ancient cities, a point iterated by many Greek and Roman authors. According to the architect Vitruvius, cities had three categories of public buildings: defensive buildings (walls, towers, gates), sacred buildings (shrines and temples), and buildings for public use (or "amenities," such as fora, baths, theaters, and walkways). The loss of any one of these would presumably challenge a city's status *as a city*. Fortification walls were especially important. They practically and symbolically represented a city's strength, safety, and freedom from outside control, and ancient authors from Aristotle to Livy emphasize fortifications as an essential element of urban identity. City walls were also symbolically important in later periods. Most notably, in the eighteenth century, French military writers equated the destruction of fortifications with the destruction of *cities*. According to one modern historian, "[t]o raze a city was equal to demolishing its defenses, since by doing so one turned it 'back' into a village." It may be that ancient Romans and Greeks had similar ideas, viewing the destruction of a city's walls as symbolically ending the city itself.[47]

In any case, the extent of Roman destruction must have varied considerably, and the literary and archaeological evidence is too ambiguous to say what was typical when the Romans "destroyed" a city like Antipatrea. For the sake of argument, however, I would wager that they normally ransacked some public buildings, burned others (along with some houses), and demolished parts of the fortifications, but they did not usually burn or raze every standing structure. Even this partial and limited destruction would have been traumatic for the victims. For survivors surveying the shattered stones and charred timbers, the broken corpses of their civic institutions, cults, safety, and culture, their community must have seemed quite thoroughly destroyed—even if many buildings remained standing or were quickly rebuilt.[48]

MASS VIOLENCE AND THE ROLE OF COMMANDERS

It is worthwhile here to return to the destruction of Antipatrea. According to Livy, the Roman commander Apustius "stormed and subdued Antipatrea by force of arms and, after killing the men of military age and granting all the plunder to the soldiers, he demolished the walls and burned the city." As we have seen, this curt narrative implies a great deal of Roman military activity. Storming the city would have been extremely risky for Apustius's troops, and they may have indiscriminately massacred some of the inhabitants after they breached Antipatrea's walls. Such massacres were mostly meant to suppress any lingering resistance, and the Romans might have killed many of the city's adult males at this point in the attack. Livy's reference to plunder implies that the Romans sacked the city, too. In which case, the soldiers would have looted, vandalized, raped, maimed, and murdered largely as individuals and on their own initiative. Livy does not mention prisoners, but since the attackers only (or mostly) killed men, it is possible that they captured the town's women and children. If so, they likely gathered the survivors together in a central location, sorted through them, put them under guard, and perhaps documented them for sale. They might have rounded up and executed the town leaders or the remaining adult men as well. Finally, with the population disposed of, the Romans destroyed the town. Livy explicitly credits *Apustius* with the burning and demolition, indicating that he ordered his troops to set fire to the city and tear down the fortifications. In doing so, his soldiers probably burned some select buildings, such as temples, public structures, and perhaps the houses, and tore down parts of the walls.

Many steps in this process would have required coordination, labor, and military resources. When assaulting the city and massacring its defenders, attackers probably worked together to locate and eliminate any obvious threats. The sack was more of a free-for-all, but the process of gathering prisoners (if such took place) was not: Roman soldiers had to cooperate in order to herd prisoners together, guard them, and ultimately pass them off to traders. If they executed captives or purposefully destroyed city structures, manpower was required for these potentially time-consuming and laborious tasks. These were not the spontaneous acts of frenzied legionaries. Given the methods of mass violence available to Roman armies, Apustius must have played an essential role, organizing manpower and resources to destroy Antipatrea's population and urban fabric.

Ancient texts often describe mass violence passively: inhabitants are just enslaved or killed, cities are just destroyed, and responsibility is not assigned to any person or group. However, when ancient authors attribute mass violence to specific parties, they usually give responsibility to generals, officers, and political leaders. In over half of the cases listed in appendix 1 ("124 Cases of Mass Violence in Roman Warfare"), responsibility is unambiguously assigned to military and political leaders. In these cases, ancient authors usually say that commanders committed the violence themselves or ordered their troops to do so: Apustius "demolished the walls and burned [Antipatrea]," Scipio sent his men "against the inhabitants of [New Carthage] with orders to kill all they encountered," and so on. In a few cases, the Roman Senate reportedly debated whether to destroy enemy communities or instructed Roman generals to destroy cities and populations.[49]

Of course, it is true that ancient authors show a "strong bias in favor of the commander's perceptions," as Josh Levithan asserts, and certainly indiscipline, rage, or lack of oversight could result in random vandalism and bloodshed on the part of Roman soldiers. Yet even the chaotic violence of city-sacking could be used by Roman commanders for their own benefit. In 49 BCE, a Roman army was on the verge of breaking through the walls of Massilia (modern Marseilles, France). The army's commander, Trebonius, essentially threatened Massilia's inhabitants with the "inevitable" sack of their town in order to compel their surrender. This underlines the point made by political scientist Benjamin Valentino: the individual motives of soldiers do not explain "why these groups were created, organized, and turned loose in the first place." Roman commanders knew what would happen when they pointed their army, like a cannon, at enemy towns, and they surely used this potent weapon for their own purposes. And city-sacking aside, it is extremely difficult to imagine soldiers spontaneously (or voluntarily) demolishing stone fortifications, rounding up thousands of prisoners and guarding them for days, or restraining and executing captives by the hundreds or thousands. Again, these are not acts of frenzy, but acts of calculation. Ancient authors are right to grant responsibility to Roman leaders in the majority of these episodes.[50]

This conclusion opens up an entirely new question: Why were Roman generals willing to devote the time, effort, and resources to employ mass violence against enemy prisoners, communities, and populations in war? The next chapter will begin to work out that puzzle.

NOTES

1. Antipatrea, Apustius's campaign: Hammond 1966, 42–43. Timing of Apustius's campaign: Warrior 1996, 61–94, 106–7. Attacks on Antipatrea, Codrion: Livy 31.27.1–5.

2. Brief descriptions of mass violence: e.g., Livy 9.25.9, 9.45.17, 10.44.2, 24.35.2, 32.15.3, 34.16.10, 42.67.9; Polyb. 3.19.12, 11.3.2, 15.4.2, 33.10.3; Diod. 23.9.4–5; Paus. 7.7.9, 7.16.7–9; Plut. *Fab.* 22.4; App. *Sam.* 13, *Pun.* 15; Dion. Hal. *Ant. Rom.* 19.13.1; Zon. 9.22.

3. Longer descriptions: e.g., Livy 24.39.1–6, 28.20.5–7, 45.34.1–6; Polyb. 10.15.4–9; Plut. *Aem.* 29; App. *Hisp.* 60; Polyaen. 8.21.1. Ancient rhetoric of mass violence ("genocide"): Van Wees 2016, 21–24.

4. "Cruelty": Paul 1982, 145. "Imagined": Laurence 1996, 112. Paul and Laurence express some skepticism about ancient descriptions of violence, yet acknowledge that they had some historical basis. Cf. Quint. *Inst.* 8.3.67–70.

5. Military technology/methods changed little in antiquity: Lendon 2005, 8–13; Levithan 2013, 13–14.
6. Ancient walls/defenses: Hansen 2006a, 73, 104; Tomlinson 2006, 79–81; Holloway 1994, 91–102. For Roman siege methods, see Levithan 2013; Davies 2006; Kern 1999, 251–351.
7. Atrax: Livy 32.17.5–32.18.2. New Carthage: Polyb. 10.15.3–7; Livy 26.46.8–9. Urban citadels risky for attackers: Onas. *Strat.* 42.18–19; cf. Aen. Tact. 1.9, 3.5, 22.2–4. See also Polyb. 7.18.7, 8.31–34; Livy 24.2.10–24.3.15, 25.24–30, 31.45.5–6; Plut. *Marc.* 18.4, *Sull.* 14.7; Hdt. 5.100.
8. Japha: Jos. *BJ* 3.302–4. See also Thuc. 2.2.4–2.4.8; Diod. 13.56.6–8, 20.58.2; Polyb. 4.57–58; Arr. *Anab.* 1.8.6–8; Livy 5.21.10; Plut. *Sull.* 9.6–7; App. *Civ.* 1.58. Dangers in ancient urban combat: Lee 2010, 143–49; Barry 1996, 55–74.
9. Killing "all they encountered": Polyb. 10.15.4–5; Livy 24.19.9, 28.20.6; Caes. *Gal.* 7.28.3; Jos. *BJ* 6.402–4; App. *Mith.* 38; Zon. 8.11. Commanders and massacre: Polyb. 10.15.4; App. *Hisp.* 52, *Mith.* 38; Tac. *Ann.* 12.17. Military hierarchy: Isaac 1998, 389–90; Hoyos 2007, 68–70, 75–76. Officers in assaults: Livy 26.46.7, *Per.* 48.20-21; Vell. Pat. 1.12.3. Officers, discipline: App. *Pun.* 15, *Civ.* 1.59. But cf. Caes. *Gal.* 7.47, 7.52; Levithan 2013, 209.
10. Angry troops: Livy 28.19.6–12; Memn. 2.35.5; Caes. *Gal.* 7.28.3. Killing males, threats: Ziolkowski 1993, 77–78, 83–86; Levithan 2013, 208–9, 214, 218–27; Livy 9.31.3, 26.46.10, 31.27.4; Sall. *Iug.* 91.6; App. *Hisp.* 52; Paus. 7.16.8. Survivors: e.g., Caes. *Gal.* 7.28.5; App. *Hisp.* 52; Memn. 2.35.6. Exhaustion in Jerusalem: Jos. *BJ* 6.414. Cf. Livy 22.48.6. "Machetes": Daly 2002, 363. "Many Hutu": Combs 2007, 257 n.286.
11. Plundering signal: Polyb. 10.15.8; Livy 5.21.13–14, 25.25.9, 26.46.10; Ziolkowski 1993, 78–80. Plundering without permission: e.g., Livy 37.32.12-13; App. *Pun.* 15. Tacitus's rhetorical language: De Souza 2008, 97.
12. "Dispersing": e.g., Livy 27.16.7; 37.32.12 (*discurre*); Polyb. 4.58.1 (*diarrein*); 7.18.9 (*ormân*). Horror of city-sacking: e.g., Polyb. 9.39.2–3; Tac. *Hist.* 3.33; Livy 29.17.15–20; Diod. 13.57.1–13.58.2; Thuc. 7.29.3–5. Diripio: Ziolkowski 1993, 70–74; cf. Yang 2017. "Young women": Tac. *Hist.* 3.33; cf. Livy 29.17.15. Commanders lose control: Livy 37.32.12; Plut. *Luc.* 19.3–5; Caes. *Gal.* 7.47–52. Cf. Ps.-Sall. *Epist. ad Caes.* 3.4. "Gold, silver": App. *Pun.* 133. "Sent heralds": Livy 37.32.12; cf. Diod. 14.53.2. Punishing looters: App. *Civ.* 1.59, 1.109, 4.73, *Pun.* 15, 133; Plut. *Pomp.* 10.7. Looting for days: e.g., App. *Pun.* 133.
13. Whole communities surrender: Livy 7.27.7, 21.51.2, 31.45.6, 31.46.16, 32.17.1–2, 42.8.1; Diod. 23.18.4, App. *Hisp.* 59, 96–97, *Civ.* 1.94; Oros. 4.21.10. Soldiers surrender (in cities and battlefield): Livy 5.32.3, 10.43.8, 26.46.8–9; Polyb. 18.26.10–11; Memn. 2.35.7. Orders to take captives: Jos. *BJ*. 6.414; Diod. 13.62.4. Soldiers take captives independently: Zon. 8.11; Tac. *Hist.* 3.33; Diod. 16.19.4. Column of Marcus Aurelius: Beckmann 2011; Ferris 2009; Dillon 2006. Forcible capture: Gaca 2010.
14. Many captives: e.g., Polyb. 1.29.7, 10.17.6; Diod. 23.18.5; Livy 10.17.8, 27.16.7, 41.11.8, 41.28.4, 45.34.5; Plut. *Fab.* 22.4, *Aem.* 29.4–5; Strabo 7.7.3. Cf. Ziolkowski 1990, 15–36. "Greatly outnumbered": Walvin 2013, 94. Ratios in Atlantic/American slave trade: O'Malley 2014, 56. "Constant business," "vastly outnumbered," "well armed": Taylor 2006, 68–69. Chains, etc.: Taylor 2006, 83.
15. Chains, restraints in texts: Polyb. 3.82.8, 20.10.7–8, 21.5.3; Livy 26.14.8; Jos. *BJ* 4.629; Dio 49.39.6; Arr. *Anab.* 1.16; Diod. 20.13.1–2, 20.69.5, 34.2.36; Hdt. 5.77. Guards/overseers: Livy 38.23.3; Polyb. 10.17.10 (and 3.85.3 for a Carthaginian example); Jos. *BJ* 3.537–539, 6.419; possibly App. *Pun.* 130. Jerusalem: Jos. *BJ* 6.415. Camps in Spain:

App. *Hisp.* 60, 100; Livy *Per.* 49. Trajan's Column enclosure: Joshel 2010, 82, 84. Imperial reliefs: Bradley 2004; Joshel 2010, 88–91. Archaeological/shackles: Thompson 1993. See also Thompson 2003, 217–44.

16. Gaca 2010, 137.

17. Dragging away: Livy 38.43.4–5; 29.17.15; Polyb. 9.39.2–3. Captives divided by origin: Caes. *Gal.* 7.89.5; App. *Civ.* 1.94. Mummius/Corinth: Zon. 9.31; Plut. *Quaest. conv.* 9.737. Captives for triumph: App. *Hisp.* 98; Beard 2007, 118–19.

18. New Carthage: Polyb. 10.17.6–15. Bergium: Livy 34.21.5–6. See also Jos. *BJ* 3.537–542, 6.416–419.

19. Pindenissus: Cic. *Att.* 5.20. Jerusalem: Jos. *BJ* 6.416–419.

20. Sparing the defeated: e.g., Livy 28.3.16, 32.17.1–2, 32.24.7; Polyb. 10.17.7–8. Garrisons: Polyb. 1.55.5–10; Livy 29.8.1–5, 36.13.3–4. Andros: Livy 31.45.6; cf. Zon. 8.7. Ligurians deported: Livy 40.38.1–7, 40.41.3, 42.22.5–6. Suffering in deportation: e.g., Patrick 2005; Douglas 2012. Ransom: Polyb. 9.42.5–8 (cf. 22.8.9–10); Livy 10.31.3–4, 10.46.11, 32.17.2. Panormus: Diod. 23.18.5.

21. Sale of captives old practice: Scheidel 2011, 294–97. Quaestor: Polyb. 10.17.10; Livy 27.19.2, 34.21.5–6; Dion. Hal. *Ant. Rom.* 10.21.6; Joshel 2010, 84. Auctioned on the spot: Joshel 2010, 84; Scheidel 2011, 296; Beard 2007, 118. Markets: e.g., Strabo 14.5.2; Livy 41.28.8; Aur. Vict. *Vir. Ill.* 57.1–3. See also App. *Hisp.* 85; Sall. *Iug.* 44.5; Strabo 4.6.7; Cic. *Quinct.* 24; Dio 42.14.3–4; Pliny *NH* 35.200.

22. Profits to treasury: Cic. *Att.* 5.20.5; Livy 26.40.13; Plut. *Fab.* 22.4; Polyb. 11.3.1–3; Joshel 2010, 84–87. Profits to soldiers: Sall. *Iug.* 91.6; Livy 45.34.5–6; Plut. *Aem.* 29.3. Soldiers get captives: Caes. *Gal.* 7.89.5; Zon. 8.11. Chiomara: Polyb. 21.39.1–7 (=Plut. *Mor.* 258 E–F); Livy 38.24; Val. Max. 6.1e.2; Flor. 1.27.6. For a recent and thorough analysis of mass enslavement practices in Roman warfare, see Wickham 2014.

23. Indignities: Diod. 13.58.1–2. Future ransom: Livy 32.22.10. "Lost forever": Chaniotis 2005, 113. Social death: Joshel 2010, 89–95. Communities dispersed: Van Wees 2010, 245. Suicide: Livy 28.23.1–2; Polyb. 16.32–34; Diod. 17.28.3–5. See also Pritchett 1991, 219–23.

24. Right to execute captives: Caes. *Gal.* 7.41. Leaders executed in Italy: Livy 7.19.2–3, 9.16.10, 9.24.12–15, 26.15.8-9, 29.8.2; Diod. 19.101.2; Oros. 5.18.26. Leaders executed outside Italy: Livy 26.40.13, 39.32.3–4, 41.11.8, 43.4.10; Flor. 1.21.4; App. *Hisp.* 68; Diod. 33.17.3; Obs. 51; Gran. Lic. 35.28; Aur. Vict. *Vir. Ill.* 70.4; Caes. *Gal.* 3.16.4; Dio 39.43.5; Oros. 6.8.17. Soldiers executed: Polyb. 1.7.9–12; Dion. Hal. 20.4.6–8; App. *Sam.* 21, *Civ.* 1.93; Livy 34.21.5–6, *Per.* 88; Strabo 5.4.11; Oros. 5.21.1; Flor. 2.9.24; Val. Max. 9.2.1; Dio fr.109. Adult males killed: App. *Civ.* 1.94; Plut. *Sull.* 32.1; Val. Max. 9.2.1. Whole populations killed: Livy 7.19.2; App. *Hisp.* 60, 100; Zon. 8.11.

25. Customary beheading: Polyb. 1.7.12; Diod. 19.101.3; Cic. *Pis.* 34.84; cf. Forsythe 2005, 150. Prisoners explicitly beheaded: Livy 7.19.2, 9.16.10, 9.24.12–15, 26.40.13, 26.15.8, 39.32.3–4, 41.11.8, 43.4.8–13, 45.31.14; Diod. 19.101.3; Polyb. 1.7.12; Dion. Hal. 20.4.6–8; App. *Sam.* 21, *Hisp.* 68; Flor. 1.21.4. Cf. Livy 24.20.5–6, 25.7.14; Diod. 34.2.21. "Torture and violent death": Ribera i Lacomba 2006, 80. "Systematic decapitation": Alapont Martín et al. 2009, 13 (translated from Spanish). Scupi: *Paleopatologia* Editorial Staff 2013; Jovanova et al. 2017, 6–40. Column of Marcus Aurelius: Beckmann 2011; Ferris 2009.

26. Plutarch on beheadings: Plut. *Publ.* 6.2–3. Execution procedures: Val. Max. 9.2.1; App. *Civ.* 1.94, *Hisp.* 60, 100; Plut. *Sull.* 32.1. Valentia and Scupi: see previous note.

27. "Complex slaughter": Diod. 20.72.2. See also Thuc. 3.85, 4.47; Achilles scene: Spivey and Squire 2004, 54; Bonfante and Swaddling 2006, 54.

28. Lusitanians: App. *Hisp.* 59–60; Livy *Per.* 49; Val. Max. 8.1.abs.2, 9.6.2; Cic. *Brut.* 89–90; *Orat.* 227; Suet. *Galba* 3.2; Oros. 4.21.10. Praeneste: App. *Civ.* 1.94; Val. Max. 9.2.1. Column of Marcus Aurelius: Beckmann 2011, 147–48. Other episodes: e.g., Livy 24.39.1–6; App. *Hisp.* 100.

29. Organized mass killing in other periods: Saunders 1971, 65 (Mongols); Browning 1992, 55–70; Wette 2006, 116 (German killing squads in WWII); Kenji 2007, 79–82; Honda 1999, 128–29 (Japanese in Nanjing).

30. Logistical difficulties: Bosworth 2012, 19–20; Zimmermann 2013, 178–88; Barrandon 2018, ch. 4. Trembling executioner: Tac. *Ann.* 15.67. Christians executed: Eus. *Ecc. Hist.* 8.9.4. Valens's executioners: Amm. 29.1.40. Swords bent, blunted: Amm. 16.12.52–54.

31. Botched beheadings: Dechambre et al. 1884, 457–58; Harrington 2013, 85–88. Blades damaged: Croker 1857, 534–35; Janes 2005, 75; Honda 1999, 130–31; Hatzfeld 2005, 39. German death squads: Sakowicz 2005, 49; Klee, Dressen, and Riess 1999, 59–69, 97–98; Langerbein 2004, 48. Nanjing executions: Kenji 2007, 79–82.

32. Antiquity of mass arson: e.g., "The Lament of Ur," ETCSL t.2.2.2. Burning roofs/interiors: Livy 6.33.4, 10.44.2, 28.20.7, 31.23.6; Tac. *Hist.* 3.33, 3.71; Polyb. 5.9.3; Sen. *Ira* 1.2.1. Columns of Trajan, Marcus: Thill 2011; Beckmann 2011; Ferris 2009, 119. For historical elements in the images, cf. Bradley 2004. Ancient roofing: Gerding 2013; Ulrich 2007, 123–78.

33. Specific buildings: App. *Pun.* 128–129; Livy 5.21.10; Polyb. 4.62.2, 4.67.3, 5.9.2; Arr. 3.18.11; Plut. *Alex.* 38; Diod. 17.72.5–6. Gathering kindling: Thuc. 2.77.3–5; Curt. 5.7.7. "Brought out": Tac. *Hist.* 3.33. "Threw fire": Livy 10.44.2. Heraclea: Memn. 2.35.8. Cf. Jos. *BJ* 6.252–66; Plut. *Luc.* 19.4. "Near the agora": Diod. 16.20.3.

34. Burning Washington: Pitch 1998, 106–21; Gleig 1821, 126–32. "Street by street": Davis 1914. "Drilled method": Mirman 1917, 18. "Systematic" burning in other periods: e.g., Dobkin 1972, 167; Olásolo 2008, 70, 117; Lu 2004, 141, 147; Buchanan 2009, 9; Hewitt 1999, 296–98.

35. Distinguishing *diruere* from burning: Livy 28.20.7, 7.27.8, 42.54.6; Cic. *Cael.* 78; Oros. 7.9.6; Just. 6.5.8. Distinguishing *kataskaptein* from burning: Aesch. 3.123; Diod. 13.57.6; Jos. *BJ* 6.625, 7.328. "Burned the temple": App. *Mith.* 53. "Not only the roofs": Polyb. 5.9.3; cf. Polyb. 4.65.4. Cities *diruere*: Livy 24.35.2, 42.8.3, 42.63.11, 42.67.9, *Per.* 52; Just. 34.2.5–6; Vell. Pat. 1.12.5; Trog. 34 pr.; Pliny *NH* 34.12. Cities *kataskaptein*: Zon. 7.7, 7.13, 8.7, 8.18, 9.3, 9.22, 9.29, 9.30; Dion. Hal. *Rom. Ant.* 5.49.5; Diod. 23.9.4, 32.4.4–5, 32.18.1, 33.17.3; Polyb. 3.19.12; App. *Illyr.* 2.8, *Hisp.* 32, *Pun.* 1, 133–36; Strabo 6.4.2, 8.6.23, 9.2.30, 14.5.2; Plut. *Aem.* 22.4. See also Hansen and Nielsen 2004, 120; Steinbock 2013, 312–20; Gaignerot-Driessen 2013, 294–95; Conner 1985; Edlund-Berry 1994, 18; Roller 2010; *Thesaurus Linguae Latinae*, s.v. "*Diruo*"; Liddell, Scott, et al. 1996, s.v. "*kataskaptō.*"

36. Workmen demolish a house: Plut. *Pub.* 10.3. Senatorial decree: *CIL* X 1401 = *FIRA* I2 45. Non-military uses or contexts for *diruere, kataskaptein*: Nep. *Con.* 4.4; Sall. *Cat.* 20.12; Cic. *Cael.* 78; Suet. *Gai.* 60; Val. Max. 4.1.1; *Syll*[3] 344. *Diruere* walls, fortifications: Livy 8.20.7, 9.41.6, 31.27.4, 34.17.11, 45.34.6; Front. *Strat.* 1.1.1, Aur. Vict. *Vir. Ill.* 47.2; Oros. 4.23.6, 5.3.6. *Kataskaptein* walls, fortifications: Thuc. 4.109.1, 8.92.10; Xen. *Hell.* 2.2.23; Dem. 19.325; Diod. 12.28.4, 12.81.1; App. *Mith.* 53; Zon. 9.31. "Pulling down" (*kathairein*) or "stripping off" (*periairein*) walls, fortifications: Thuc. 1.56.2, 1.101.3, 1.117.3, 3.50.1; Xen. *Hell.* 3.2.30, 5.2.7; Diod. 13.59.4, 16.60.1; Polyb. 22.12.3; App. *Hisp.* 41, *Mith.* 30, *Civ.* 1.96; Strab. 9.1.15; Plut. *Cat. Mai.* 10.3, *Aem.* 6.3. Walls

"thrown down" (*deiecti*) or "leveled" (*adaequat solo*): Livy 1.29.6; 8.14.5; Curt. 7.5.33; Just. 5.8.5; Oros. 6.6.3–4.

37. "Experienced in building": Proc. *Bell.* 6.27.5; cf. 6.27.18–21. "Shovels and pickaxes": Aesch. 3.122–23. For *Dolabrae*/etc. in demolitions: Caprino et al. 1955, 84; Tac. *Hist.* 3.20; Plut. *Tim.* 22.1; Ath. 13.48; Val. Max. 1.3.4. Trajan's Column: Campbell 2002, 68.

38. Climbing to the top: Thuc. 8.92.10. Stripping roofs: Libanius *Or.* 30; Vict. Vit. 1.8. Early modern stripping roofs: Blackie 1885, 69. Assyrian reliefs: Russell 1992, fig. 38; Frankfort 1996, fig. 206; Barnett 1976, plate 66. Cf. Baker 2015, 53–54. Other historical examples: Gabrieli 2009, 203; Morris 2017, 233; Mintzker 2012, 150–51.

39. Rams: Livy 33.17.8; App. *Mith.* 36. Hooks: Polyb. 21.27.4; Veg. *Epit. Rei Mil.* 4.14; Caes. *Gal.* 7.86.5. Carthaginian ram: Vitr. 10.13.1. Caesar's demolitions: Campbell 2002, 285, with sources. "Making a depression": Nossov 2005, 123. Undermining: Livy 21.11.7–8; Jos. *BJ.* 6.222. See also Curt. 9.5.19; Polyb. 5.100.2–6, 10.31.8–11, 21.28.3–16.

40. Portinari: Knowles 1976, 267; Goring 2003, 27; Cooper 2006, 136. Sherman: Hoffman 2007, 241–42.

41. Difficulty of mass demolition in antiquity: Hansen and Nielsen 2004, 137; Purcell 1995, 137. Macedonian demolition: Polyb. 5.100.4. Montgomery Castle: Thompson 1987, 186, 193; Rátkai 2015, 238. Pontefract Castle: Rakoczy 2008, 270–71.

42. Babri Mosque: Friedland and Hecht 1998, 102; Awasthi and Mahurkar 1992; Ramakrishnan 1993.

43. Doubts about extent of destruction: Hansen and Nielsen 2004, 122; Snodgrass 1987, 41–42.

44. Fortifications destroyed: Livy 8.14.5, 8.20.7, 9.38.1, 9.41.6, 27.16.9, 31.27.4, 34.17.11, 45.34.6; Front. *Strat.* 1.1.1; Plut. *Cat. Mai.* 10.3, *Aem.* 6.3; App. *Hisp.* 41, *Mith.* 30, 53, *Civ.* 1.96; Polyaen. 8.17.1; Zon. 9.17, 9.31; Oros. 4.23.6, 5.3.6, 6.6.3–4; Paus. 7.16.9; Strabo 9.1.15, 16.2.40. Other buildings: Livy 31.23.6–7; App. *Mith.* 41; Plut. *Sull.* 14.7; Livy 9.28.5; Vitr. 5.5.8; App. *Mith.* 53; Obs. 56b. Sacred sites: Rutledge 2007. Chalcis: Livy 31.24.3. Acerrae: Livy 27.3.6. See also Livy 31.10.3, 42.54.6; Jos. *AJ* 12.251; Jos. *BJ* 7.1–4; Plut. *Cleom.* 25.1–3, Memn. 2.35.8. Carthage/Corinth partly destroyed: Paus. 7.16.7–9; Oros. 4.23.6; 5.3.1; 5.3.6; Dio fr. 72.1.

45. Fire in pre-modern cities: Goudsblom 1992, 114; Kuretsky 2012, 27–28; Canter 1932. Building materials: Thompson 1980; Adam 2003; Ulrich 2007, esp. ch. 6–9. Susceptibility of timber: Curt. 5.7.6. Wattle-and-daub: Vitr. *De Arch.* 2.8.20. Stone: Vitr. *De Arch.* 2.7.2; Jackson and Marra 2006, 426; Hanson 1998, 75. Weather: *Dig.* 9.2.30; Thuc. 2.77.4; App. *Civ.* 1.94; Livy 42.63.9. Determined arsonists: Thuc. 2.77.3–4.

46. Carthage: Lancel and Morel 1992; Docter et al. 2003, 2006; Davies 2014, 182 n.9. Corinth: Gebhard and Dickie 2003, 262–65; Romano 2003; Wiseman 1979, 494–96; James 2014, 25; Scranton 1951, 176–77; Carpenter et al. 1936, 126. "Grim picture": Rutledge 2007, 187. Fregellae: Crawford and Keppie 1984, 32. See also Ribera i Lacomba 2006, 77–78, 80–81; Ribera i Lacomba and Calvo Galvez 1995, 19–40; Alapont Martín et al. 2009, 16; Barrandon 2018, ch. 4.

47. Ancient urbanism: Vitr. *De Arch.* 1.3.1; Paus. 10.4.1; Billows 2003, 197; Osborne and Wallace-Hadrill 2013, 49–65. Importance of walls: Livy 1.7.2–3, 9.4.12; Arist. *Pol.* 7.1331a; Hansen and Nielsen 2004, 135–37; Camp 2000; Becker 2007, 198, 204–5; Gabba 1970, 463. "To raze": Mintzker 2011, 34–35. Slighting castles: Thompson 1987, 147–50. Partial demolition of walls: Strabo 12.3.31; Jos. *BJ.* 4.117. See also Conwell 2008, 104–5; Lawrence 1979, 115.

48. Cities not obliterated: Hansen and Nielsen 2004, 122; Gaignerot-Driessen 2013, 295.

49. Senate debating destruction: Livy 9.26.3, 26.16.7; Plut. *Cat. Mai.* 27.1–4. Senate ordering destruction: e.g., Livy 45.34.1, *Per.* 52; Plut. *Aem.* 29.1; App. *Pun.* 135.

50. "Strong bias": Levithan 2013, 210. Massilia: Caes. *B. Civ.* 2.12–13. Cf. Livy 28.19.6–9; Sall. *Iug.* 68.3–4; Caes. *Gal.* 7.17. "Why these groups": Valentino 2014, 98. Barrandon 2018, ch. 6, likewise argues that responsibility for mass violence lay especially with Roman commanders.

4

"The Ram Has Hammered at Their Walls"

The Logic of Mass Violence

Marcus Tullius Cicero had limited experience of warfare and little interest. He had built his political career in Rome as a successful lawyer and powerful orator, preferring the wheeling and dealing of the capital to the drudgery of business out in the provinces. Still, when he was assigned to govern the province Cilicia in 51 BCE, he was as eager as any Roman to rack up a few military honors. He found his opportunity not in the form of a great foe—like the expansionist Parthian Empire to the east—but in the rugged communities of the Amanus Mountains. Located at the edge of his province, about where Asia Minor meets Syria, these mountain tribes were probably not a serious threat to Rome's interests. However, they acted as hotbeds for banditry and raiding, and Roman arms had not yet pacified them. Cicero led a surprise march into their territory sometime in October, capturing several strongholds and devastating the countryside. To cap off his campaign, he directed his army against a fortified town called Pindenissus. Surrounding it with earthworks, fortifications, and artillery, he compelled the inhabitants to surrender after a fifty-seven-day siege. In the end, the Romans had demolished or burned most of the settlement, and Cicero sold the captives as slaves.[1]

This was a relatively minor military action on the edge of the Roman world, and normally such small-scale campaigns probably went unrecorded. However, as chance would have it, a large collection of Cicero's letters survives, including several he wrote during his governorship of Cilicia. Addressed to friends and senatorial colleagues, these documents give us unparalleled first-person insight into a Roman military campaign as it happened—and unparalleled insight as to why a Roman commander decided to attack, and ultimately destroy, a specific community.

Cicero first mentions Pindenissus in a letter dated November 26, 51 BCE, addressed to a younger senatorial colleague named Marcus Caelius Rufus. After breathlessly describing his campaigns' success so far—rattling off enemies killed or imprisoned, strongholds captured and burned—Cicero writes that he has been involved in a siege for over three weeks:

> I drew off my army to the most disturbed part of Cilicia. There for the past twenty-five days I have been assailing a very strongly fortified town called Pindenissus with earthworks, mantlets, towers, and with such great resources and energy, that the only thing now lacking for the attainment of the most glorious renown is the credit of taking the town; and if, as I hope, I do take it, I will then at once send an official dispatch [to Rome]. (Cicero *Letters to Friends* 2.10; trans. Shuckburgh, modified)

Two points are worth noting here. First, Cicero emphasizes the strength of the town and the vigor of his siege operations. Second, he openly hopes to reap glory from the undertaking and plans to petition the Senate for recognition of his efforts.

The town fell a few weeks later on December 17. Two days after its capture, Cicero wrote to Titus Pomponius Atticus, a prominent Roman businessman and close friend, to tell him about the whole affair. In the letter, he sourly admits that no one has heard of Pindenissus, complaining that he could not "turn Cilicia into an Aetolia or a Macedonia"; in other words, his province was not an ideal outlet for impressive military victories against impressive enemies. Nonetheless, Cicero gives another quick rundown of his campaign before coming to the siege itself:

> I was at Pindenissus, the most strongly fortified town of the Free Cilicians, never peaceful within living memory. The people were fierce and brave, and furnished with everything necessary for standing a siege. We surrounded it with stockade and ditch, with a huge earthwork, mantlets, an exceedingly lofty tower, a great supply of artillery, a large body of archers. After great labour and preparation I finished the business without loss to my army, though with a large number of wounded. I am spending a merry Saturnalia,[*] and so are my soldiers, to whom I have given up all the spoils except captives: the captives were sold on the third day of the Saturnalia (19th December), the day on which I write this. The sum realized at the tribunal [from the sale] is 120,000 sesterces.[†] (Cicero *Letters to Atticus* 5.20; trans. Shuckburgh, modified)

Once again, Cicero stresses the strength of the foe and the difficulty of the siege, though here he also remarks on the enemy's lack of fame—which allows him to bring up the glory (or lack thereof) that he might gain from this enterprise. He makes it an additional point to mention the plunder his army

---

[*] The Saturnalia was an important Roman festival dedicated to the god Saturn.
[†] Sesterces, or *sestertii*, were silver coins worth one-quarter of a *denarius*; the *denarius* was the standard Roman silver coin after 211 BCE.

took, indicating that they sacked the city, as well as profits reaped from the auction of prisoners.

With Pindenissus in the bag, Cicero whisked off messages to other senators and magistrates of Rome. In these letters Cicero was "officially" reporting his victory and trying to wrangle accolades from the Senate, such as a public thanksgiving (*supplicatio*) or even that highest of military honors, a triumph. In one of the surviving letters, addressed to Marcus Porcius Cato ("the Younger"), Cicero again runs through the particulars of his campaign before coming to the attack on Pindenissus. And unlike his letters to Rufus and Atticus, he carefully lays out his logic for attacking this particular community:

> I led the army away to Pindenissus, a town of the Free Cilicians. And since this town was situated on a very lofty and strongly fortified spot, and was inhabited by men who have never submitted even to the kings, and since they were offering harbourage to deserters, and were eagerly expecting the arrival of the Parthians, I thought it of importance to the prestige of the [Roman] empire to suppress their audacity, in order that there might be less difficulty in breaking the spirits of all such as were anywhere disaffected to our rule. (Cicero *Letters to Friends* 15.5; trans. Shuckburgh, modified)

Then, as in the other letters, he describes his siege operations and the difficulties weathered by his army before the city surrendered. He further adds that "every region of their town had been battered down or fired," and that a neighboring community called Tebara—"no less predatory and audacious"—submitted and handed over hostages "after the capture of Pindenissus."

Fascinatingly, Cicero's letter to Cato does not just explain why he *attacked* this particular place: he also explains his decision to "suppress" (*comprimere*) this enemy, to "break their spirit" and the spirits of others (*animi frangerentur*), and by implication to remove this challenge to Roman authority. He is in effect explaining why he chose to assault and to destroy this obscure town. We must remember that Cicero was not a dispassionate reporter. His interest was to make this little victory seem as impressive and important (and justified) as possible. Yet what matters here is not that he may have exaggerated his achievements. What matters is how he chose to explain his attack on Pindenissus, since his explanation needed to make sense to an audience of leading Romans. And in a nutshell, Cicero reasoned that the town was strategically located and well fortified, and acted as a potential resource for Rome's enemies; it had shown insufficient deference to Rome, for which it needed to be punished; and its suppression would provide a frightful example to others who might resist Roman authority. And lurking behind all of these factors are other implied motives, including profit: the 120,000 sesterces that he deposited in the treasury after selling the population, and the plunder that he gave to his army. More importantly, perhaps, there was glory to be gained, too, with Cicero's not-so-subtle hints that he wanted to leverage this military accomplishment into a triumph.

As illustrated in chapter 3, mass violence frequently required time, effort, and organization on the part of ancient armies, indicating that it was the product of command-level decision-making rather than random frenzy. Cicero's letters reinforce that impression by laying out explicit and implicit reasons that he *decided* to attack an enemy community, destroy its physical structures, and remove its population. And while his letters are unique for the individual insight they provide, many other ancient writers tried to explain why the Romans used mass violence—giving us rich and tantalizing hints about the motives for this behavior.

Following along with that theme, this chapter explores Roman leaders' motives for using mass violence in war, as those motives appear in the ancient texts. It is worth stressing that our Roman and Greek sources rarely provide the gripping first-person details found in Cicero's letters, especially for the wars of the Middle Republic, which are the main focus of this study. The sources are also marred by several deficiencies. For instance, many ancient writers probably did not know the motives for Roman behavior, so they simply guessed. Other authors might have been trying to exonerate Roman heroes or condemn Roman villains, and shaped their depiction of violence to conform to their agendas. Yet even with these and other drawbacks, our sources have much to offer. For one, they were far closer to Roman warfare than we can ever hope to be. They were witnesses to Roman war making, historians of Rome living in the ancient Mediterranean world, and even military commanders themselves. So whether or not they accurately explain Roman motives in every instance, they are the best starting point for understanding why Roman commanders used mass killing, enslavement, and destruction as tools of war. We should pay attention to what they say.

In addition to narrative sources, we possess another kind of evidence with a different perspective altogether. When Roman commanders returned from military campaigns, they often memorialized their achievements on celebratory monuments, many of which included boastful inscriptions on stone. Amazingly, some of these inscriptions survive. While they do not recount *motives* for military action, they do show what the Romans *valued* in their military victories and what they hoped to gain from war. Therefore, they may point to other factors that encouraged the use of mass violence. With the necessary caveats in mind, we may now turn to the evidence.[2]

MOTIVES FOR MASS VIOLENCE: MILITARY NECESSITY AND PRAGMATISM

According to Cicero's letters, Pindenissus could threaten Roman control in Cilicia with its strong fortifications, strategic position, and ability to act as a safe harbor for Rome's enemies. He claimed that he preemptively eliminated Pindenissus to quash this threat. In other ancient texts, a similar strategic logic motivates Roman commanders to destroy cities, fortifications, and populations in war. In one vivid episode, we are told that the consul Gaius Marius attacked and captured Capsa, a town in North Africa, in 105 BCE; there he killed the adult males, sold the rest, and burned the city to the ground. According to the Roman historian Sallust, Marius committed these violent acts

> not because of avarice or cruelty, but because the place was of advantage to [the enemy leader] Jugurtha and difficult of access for us, while the people were fickle and untrustworthy and had previously shown themselves amenable neither to kindness nor to fear. (Sallust *The Jugurthine War* 91.7; trans. Rolfe, modified)

Said another way, Marius believed that this fortified city was a boon to Rome's foes, the Numidians, and that he could not win over its inhabitants. Destroying Capsa deprived the enemy of this military asset and removed a threat to the Roman war effort.

The Romans might also destroy cities and fortifications to prevent the emergence of future threats. The Roman author Cassius Dio says that the Romans destroyed the walls and buildings at Corinth "out of fear that some people might again join forces with it as their strongest city." Cicero similarly speculates that Corinth was destroyed because of "the convenience of its position in particular," which "might encourage someone to make war in the future." That same reasoning may explain an incident in 195 BCE, when Marcus Porcius Cato "the Elder" destroyed the fortification walls of several towns in Spain. According to the ancient authors Livy, Frontinus, and Aurelius Victor, Cato was trying to prevent future rebellion in the region. Appian adds that these communities became "easier to attack" (or "more accessible," *euéphodoi*) after Cato destroyed their walls. With both Corinth and the Spanish towns, whole or partial urban destruction removed threatening strongholds and prevented future trouble from these fortified centers.[3]

In some reports, the Romans use preventative mass violence against enemy populations, not just against physical cities. For example, Cato enslaved the populations of several communities in northeastern Spain after they had twice revolted; he did this "so that they would not disturb the peace more frequently," Livy says. Another incident took place in 150 BCE, when a commander named Servius Sulpicius Galba massacred and enslaved several Lusitanian populations in western Spain. In a speech he gave in Rome, Galba argued that his actions were preventative: "he had discovered that [the Lusitanians] had sacrificed a horse and a man according to their custom and planned to attack his army under cover of the truce" (see chapter 7). Livy attributes a similar motive to Lucius Pinarius, who massacred the inhabitants of Enna in Sicily. Pinarius commanded a Roman garrison at Enna during the dark days of the Second Punic War, and he suspected that Enna's leaders were planning to betray the Romans and join the Carthaginians. To prevent the city falling into enemy hands, he ordered his troops to massacre the population (see chapter 5).[4]

The historian Tacitus, who wrote during the imperial period, sometimes attributes mass violence to an even more hard-nosed pragmatism. When a Roman army attacked a Black Sea town in 49 CE, Tacitus claims, their commander did not want to bother with prisoners—so the assault troops were "given the signal to kill" the inhabitants. Tacitus also says a Roman general razed Artaxata, a city in Armenia, in 58 or 59 CE because he did not have enough men to garrison the place. Thus it was more advantageous (or glorious) to destroy it and deprive the enemy of a stronghold.[5]

At times, our texts are not explicit about Roman motives, but they still imply that mass violence was motivated by a sort of military pragmatism. For example, Livy stresses the strength of Antipatrea's fortifications and the strategic positioning of Chalcis, two Greek cities that the Romans attacked and devastated in the Second Macedonian War (200–197 BCE). Strong urban centers like these allowed an enemy to control the surrounding countryside or seas and acted as bases to repel invaders. More generally, ancient cities served as "vital sources of manpower, nodes of administration and communications, centres of economic activity, and the like," and so they were important to whoever controlled them at a given moment. When the Roman commander Apustius destroyed Antipatrea's walls, burned the city, and killed the military-aged males, he effectively cleared an obstacle out of the Romans' path as he pushed further into enemy territory; he also took all of these advantages away from his adversaries. Similarly, at Chalcis, the Romans killed or drove out the fighting men, and burned the arsenal and granaries, buildings of obvious military significance. Clearly, the Romans could think about mass violence along strategic and pragmatic lines, and such motives probably lurked silently beneath many other episodes of violence and destruction.[6]

MOTIVES FOR MASS VIOLENCE: PUNISHMENT AND DETERRENCE

Cicero's assertion that Pindenissus deserved punishment because of its "audacity" reflects a wider Roman tendency: ancient sources often claim that Roman violence was punitive, avenging perceived offenses with displays of massive force. This punitive logic was even baked into the conventions of Roman warfare. Custom dictated that enemy cities must submit to the Romans "before the ram has touched the wall"—or rather, before the Romans launched a full-scale attack. If the Romans were forced to attack, the subsequent siege and assault operations were often extremely costly, both in terms of lives and materiel. From the Roman perspective, there was an incentive to repay those costs, or at least to unleash their soldiers in order to punish stubborn resistance.[7]

In numerous accounts, Roman commanders also employ mass killing, mass enslavement, or both to punish communities that had harmed Rome's ambassadors. In 283 BCE, a chieftain of the Senones (a Gallic tribe in northeastern Italy) reportedly murdered some Roman envoys. In response, Appian tells us, a Roman consul marched against them, killed their military-aged males, and enslaved their women and children. Roman envoys were also sent to the Illyrian queen Teuta in 230 BCE to protest Illyrian piracy. The Illyrians supposedly killed one or both of these ambassadors, provoking a Roman declaration of war that led to a quick Roman victory. According to Florus, a Roman consul beheaded captured Illyrian chieftains to "atone" for the murder of the Roman ambassadors. And Livy, Strabo, and Cicero all agree that Corinth was destroyed, at least in part, because the Corinthians had mistreated Roman ambassadors.[8]

Rome could also answer perceived betrayals with mass violence. As discussed in the previous chapter, the Romans often executed the leaders of cities that rebelled or defected to an enemy. For example, the Sicilian city Agrigentum revolted and joined the Carthaginians during the Second Punic War. When the consul Valerius Laevinus recaptured it, he had the leading men scourged and beheaded, and sold the rest of the population (see chapter 5). Several communities in Epirus (northwest Greece) similarly switched sides in the Third Macedonian War (171–167 BCE), going over to Rome's enemies in the middle of the conflict. The Roman Senate decided on an especially severe punishment in response. Acting on senatorial orders, the consul L. Aemilius Paullus sacked seventy Epirote settlements "that had defected," tearing down their walls and selling their populations into slavery (see chapter 6). Evidently the Romans were not willing to tolerate perceived breaches of faith.[9]

In some cases, outbursts of vengeful violence simply resulted from Roman anger. Near the end of the Second Punic War, Polybius reports, the Carthaginians violated an armistice by attacking first some Roman supply ships and then some Roman envoys. In response, the Roman commander P. Cornelius Scipio attacked several Carthaginian cities, refusing their offers of surrender and enslaving their populations "to make manifest the anger that he had for the enemy because of the treachery of the Carthaginians" (see chapter 5). In Polybius's retelling, Scipio was so enraged at Carthaginian behavior that he insisted on swift, brutal reprisals.[10]

Brutal reprisals could also have an exemplary function for the Romans, terrorizing other real or potential foes. As Cicero puts it, violently suppressing Pindenissus would make it easier to "break the spirit" of other peoples who might challenge Rome. Other authors echo Cicero's sentiments. According to Polybius, the Romans massacred the inhabitants of enemy cities "to inspire terror" and encourage surrender. Livy also suggests that Roman commanders used terror to hasten surrender. M. Claudius Marcellus destroyed Megara in Sicily, he writes, in order "to terrify other cities" that were resisting the Romans (see chapter 5). Likewise, Julius Caesar surely sought to spread terror when he enslaved the Veneti in northwestern Gaul and executed their leaders. Venetian leaders had supposedly imprisoned Roman envoys without just cause, so the whole tribe needed "to be punished more severely so that the rights of ambassadors would be preserved more diligently by the barbarians in the future." Caesar similarly justifies his harsh treatment of the town Uxellodunum, one of the last pockets of resistance during his conquest of Gaul. He "judged that their stubbornness deserved a great punishment," because their defiance might encourage additional communities to revolt. Once he captured the town, he resolved that "others would be deterred by an exemplary punishment" and cut off the hands of all the fighting men, leaving them alive so that the penalty would be "more conspicuous" to others. Like Caesar, the Roman commander (and future emperor) Titus used displays of violence to send a stark message. When he ordered his troops to demolish the walls of Jerusalem in 70 CE, he deliberately left several impressive towers intact "to demonstrate to posterity what kind of city it was, and how well fortified, which the Roman valor had subdued."[11]

Terroristic violence could also serve as a form of control, frightening subordinate peoples so that they would *remain* subordinate. For the Roman historian Trogus, the Romans destroyed Corinth "so that by this example other cities would be afraid to revolt." Diodorus goes further, asserting that the Romans widely used terrorism to sustain their control over the Mediterranean in the mid-second century BCE:

> Once [the Romans] held sway over virtually the whole inhabited world, they confirmed their power by terrorism and by the destruction of the most eminent cities. Corinth they razed to the ground, the Macedonians (Perseus for example) they rooted out, they razed Carthage and the Celtiberian city of Numantia, and there were many whom they cowed by terror. (Diodorus Siculus *Library of History* 32.4.5; trans. Oldfather)

In Diodorus's formulation, then, terror eventually became the substance of Rome's dominion.[12]

According to another view, the Romans used mass violence to eliminate the threat of competitors. Indeed, Polybius writes, some Greeks believed that the Romans destroyed Carthage to eliminate a present or future challenge to Rome's supremacy:

> [Some Greeks argued that] to destroy this source of perpetual menace, this city which had constantly disputed the supremacy with [the Romans] and was still able to dispute it if it had the opportunity and thus to secure the dominion of their own country, was the act of intelligent and far-seeing men. (Polybius *Histories* 36.9.4; trans. Paton, modified)

Likewise, Cato the Elder famously insisted that "Carthage must be destroyed" because of the city's quick recovery after its second war with Rome, its proximity, its wealth, and the threat it had posed, still posed, or would pose in the future. About a hundred years after Cato's death, Cicero repeated many of the same points. As he writes, Carthage posed an unmitigated danger to Roman dominion due to its resources, wealth, harbors, fortifications, location, and even its very existence. Cicero also contends that the Romans crushed cities like Carthage, Corinth, and Capua because these places could sustain imperial ambitions. Suppressing these cities was thus a form of deterrence, preventing future challenges to Roman dominance through the destruction of real or potential rivals.[13]

MOTIVES FOR MASS VIOLENCE: GREED AND GLORY

In the reports of ancient authors, Roman commanders used mass violence to neutralize threats, punish slights, and deter challenges. Aside from these motives, which are relatively clear in our sources, there were more implicit factors that could encourage the use of mass violence in war. First, the Romans could glean material rewards from some forms of violence, especially from the sale of prisoners or city-sacking, and these potential profits may have shaped commanders' choices. For instance, Cicero mentions that he gave his soldiers all the plunder from Pindenissus except for the captives, and that the sale of the latter fetched 120,000 sesterces for the Roman treasury. These rewards were not dazzling by any means—Rome's more impressive victories could net more than a thousand times as much—and there was probably little personal benefit to Cicero. Indeed, recent research indicates that only a few commanders were significantly enriched by war, and many, if not most, war profits were given to the soldiers, deposited in the treasury, or used to recover expenditures. On the other hand, generals like Cicero could use a portion of their war booty for various public uses, allowing them to enhance their prestige and profit *politically* (if not personally and materially) from warfare. As historian Sandra Joshel puts it, commanders might use "the money gained from the auction of captives for material manifestations of their glory: what was not given to the state was spent on games, distributions, and buildings." This politicization of war plunder is quite clear from inscriptions on Roman victory monuments. One monument, which celebrated a victory over the Carthaginians in 260 BCE, proclaims that the Roman commander captured 3,600 gold and 100,000 silver coins and gave the Roman people "naval plunder." Similar inscriptions record loot that commanders took from enemy towns (several of which they had sacked, destroyed, or enslaved) and gifted to Roman communities. In still other inscriptions, successful generals announce that their military victories financed new buildings for the gods and citizens. Such self-advertisement allowed commanders to put their achievements, and the benefits these provided to the Roman people, directly in public view, providing a key advantage in Rome's enormously competitive politics.[14]

However, even if commanders like Cicero wanted to benefit from war profits, the impetus for seeking loot often came from rank-and-file soldiers rather than their generals. In fact, Roman troops clearly expected opportunities for plunder, especially after a difficult siege or assault operation. Commanders must have felt immense pressure to oblige them, since soldiers denied such opportunities might disobey orders or even mutiny. Some legionaries were so avid for plunder that they sacked cities (or tried to do so) against orders, putting morale and discipline at stake. What is more, Roman soldiers were also Roman voters, and they could shape the outcome of political decisions back home. The soldiers' latent political power was amply illustrated in 167 BCE, when the consul Aemilius Paullus returned victorious from the Third Macedonian War and requested a triumph as his due reward. Although the Senate voted in Paullus's favor, the final decision lay with the citizen voters. Unfortunately for Paullus, many of his men thought he had been stingy with loot, and their grousing nearly derailed his triumphal bid. The vote went Paullus's way in the end, but just barely, and only after a well-respected senator berated the assembled voters.[15]

Pressure from the soldiers was very real, and it could encourage Roman leaders not only to court the favor of their troops, but also to sell captives or engineer city-sacking opportunities to reward their armies. As noted above, Aemilius Paullus sacked and enslaved seventy communities in Epirus at least in part because these cities had betrayed Rome. But in addition, this immense display of violence was designed to reward the legions at the end of a difficult campaign. As Plutarch puts it, Paullus had "an order from the senate to enrich the soldiers who had fought with him in the battle against Perseus [the Macedonian king] with booty from the cities in Epirus"; and Paullus duly gave his men the profits from the sackings and enslavements. (Though as we have seen, it still was not enough to stop their complaining!) A similar desire to appease the soldiery may lurk behind other choices to sack towns or sell captives. For example, after selling the women and children of Capsa, Marius gave the profits to his troops. As a consequence, his soldiers were "enriched" and "praised him to the skies." Perhaps Marius had shrewdly considered the military and political advantages of a grateful army.[16]

For some Roman commanders, the desire to gain profits for their soldiers or for their own uses probably incentivized the use of mass violence. Yet only a few forms of mass violence, like city-sacking and mass enslavement, would yield any material rewards at all. As shown in chapter 3, urban destruction and mass killing could require significant labor and resources, and would have been less than materially profitable. Even mass enslavements may not have been as lucrative as one might imagine. When thousands of captives hit the market at the same time, the suddenly abundant supply probably forced prices down. And many Roman commanders would need to sell off their prisoners as quickly as possible, since captives would consume provisions and slow down the army's march. Historian Nathan Rosenstein rightly notes that "[t]he need for a quick sale would have tended to weaken . . . the bargaining position of a general seeking to profit from the sale of prisoners and to reduce the price that buyers were willing to pay." So if Roman leaders sought material benefits from war—a question that continues to occupy modern historians—mass violence may not have been the best way to achieve those benefits.[17]

But there was something else that mass violence could accrue for Roman commanders, something else that they could leverage for political gain back in Rome. Significantly, Cicero's letters place less emphasis on the acquisition of wealth than they do on the glory he might attain from Pindenissus's capture. Throughout the Republican period, military success was one of the most important avenues for political success in Rome. If an aristocrat-politician like Cicero won a significant victory, there was always a chance that the Senate and people of Rome would reward him with public festivals and honors celebrating his achievement; these honors would subsequently increase his prestige and his chances for further political gains. In this context, it makes perfect sense that Cicero would try to squeeze as much public recognition as possible out of his campaign in Cilicia, and the desire for glory may have influenced his decision to attack and destroy Pindenissus in the first place.

Although the textual sources seldom mention the pursuit of glory as an explicit motive for mass violence, our evidence *does* show that the Romans equated the killing, enslavement, and destruction of foes with military achievement and victory. Valerius Maximus states that Roman commanders needed to kill at least 5,000 enemies to qualify for a triumph. Similarly, Diodorus says that Roman generals were honorably proclaimed *imperator* ("commander") after their armies had killed 6,000 foes. And the Latin writer Gellius claims that "bloodless" victories were only worthy of a lesser reward, called an ovation, rather than the more prestigious triumph. Roman commanders also publicly boasted about the number of enemy cities they had captured or destroyed, even exaggerating for added effect. Cato the Elder bragged about capturing "more cities than he spent days in Spain" during his consulship (four hundred towns, reports his biographer Plutarch), and Polybius asserts that Tiberius Gracchus destroyed three hundred cities in Spain. In a telling response, Posidonius says that Polybius was exaggerating—like the Romans do in their triumphs.[18]

Although these bloody metrics probably were not hard-and-fast rules that the Romans used to judge military achievement, these authors were still getting at something very real. Roman commanders certainly reported the number of enemies they had killed and captured, and the number of cities they had taken or destroyed, in their bid to win triumphs. And the Senate took those figures into consideration when debating whether or not to issue honors. With that in mind, authors like Valerius Maximus and Gellius were trying to give shape to the Roman concept of victory—a concept that closely connected military success, enemy defeat, and the safety of Rome with casualties inflicted, cities captured (or destroyed), and prisoners taken.[19]

Surviving fragments of victory monuments and dedications to the gods give us remarkable insight into this Roman idea of victory. Several inscriptions proudly announce the massive violence that the Romans meted out to their adversaries: one proclaims that L. Mummius destroyed Corinth; another that Tiberius Gracchus killed and captured 80,000 Sardinians; one that P. Servilius sold the captives he took from Isaura; one that C. Duillius took the city of Macella by force, sank and seized dozens of enemy ships, and led Carthaginian captives in his triumph; and so on. In their own writings, Cicero and Julius Caesar emphasize many of the same things. We have already seen how Cicero bragged about destroying Pindenissus and selling the inhabitants, and Julius Caesar often carefully enumerated the foes he killed, captured, and enslaved during his conquest of Gaul. Significantly, Caesar did not make clear distinctions between combatants and noncombatants when tallying these totals, and there is little reason to suspect that other victorious generals did, either.[20]

In the triumph itself, which was one of the greatest marks of recognition that a Roman commander could receive for his military achievements, bald claims of enemies killed and cities captured were transformed into graphic imagery. During the triumphal celebration, the victorious general rode through the streets in a chariot, the centerpiece of a parade that included his marching soldiers, enemy captives, and dazzling plunder seized from the vanquished. As Rosenstein puts it, triumphs "offered the citizens a flattering picture of themselves as a powerful, conquering people while simultaneously displaying to them in vivid terms those they had vanquished." A key part of the display was the violence wrought against Rome's rivals. After his victory in the First Roman-Jewish War, Titus's triumph included paintings of brutal conquest:

> Here was to be seen a prosperous country devastated, there whole battalions of the enemy slaughtered; here a party in flight, there others led into captivity; walls of surpassing compass demolished by engines, strong fortresses overpowered, cities with well-manned defences completely mastered and an army pouring within the ramparts, an area all deluged with blood, the hands of those incapable of resistance raised in supplication, temples set on fire, houses pulled down over their owners' heads, and, after general desolation and woe, rivers flowing, not over a cultivated land, nor supplying drink to man and beast, but across a country still on every side in flames. (Josephus *The Jewish War* 7.143–45; trans. Thackeray)

Images of mass destruction and death were literally paraded before the citizenry in this spectacular celebration of military glory. Though the author, Josephus, was describing a scene in the early imperial period (71 CE), Republican triumphs also included visual representations of campaigns and conquered cities, and sometimes ended with the execution of prisoners. The themes and messaging would have been similar.[21]

All Roman commanders were part of a military and political culture that made an explicit, highly visible connection between devastation inflicted on enemies and success in war. This cultural framework likely influenced the decision-making calculus of Roman generals, including their understanding of how victory was achieved and how enemies were best defeated. There are even a few tantalizing hints of this logic in the sources. For the sake of example, one Roman commander sent his soldiers to pursue a routed enemy because "he reasoned that the war would be completed if the greatest possible number of foes was killed or captured." Achieving victory required that he destroy as many of the enemy as he could.[22]

MOTIVES FOR RESTRAINT: PRAGMATISM AND NORMS

The fact that commanders could employ mass violence for any number of reasons indicates that the Romans did not have qualms about this conduct, nor did they regard massacres, mass enslavement, or wholesale urban destruction as innately criminal acts. In some contexts, such brutal treatment of the enemy was worthy of reward. For the Romans, like other ancient peoples, believed that victors in war enjoyed "rights of conquest"—effectively the right to do whatever they wanted to conquered foes.[23]

Nonetheless, the Romans did not visit massive violence upon their enemies as a matter of course, nor did all their defeated foes end up like the ill-fated inhabitants of Pindenissus, carted off into slavery while their homes burned to ashes. There were in fact several brakes on Rome's military conduct, motives for restraint that could stay the hand of Roman generals. At one level, it could be quite useful to treat defeated enemies with mercy. As the ancient military theorist Onasander argues, enemies are less willing to fight to the bitter end (and more willing to surrender quickly) when they expect fair treatment. Some Roman commanders evidently agreed with this line of thinking. For example, Julius Caesar famously cultivated a reputation for clemency, and some of his foes surrendered in expectation of mercy during both the conquest of Gaul and the civil wars. During the Second Punic War, Scipio routinely released Spanish prisoners unharmed. Polybius and Livy attribute this mercy to his humane character while admitting that his goodwill was also political: at that point in the war, most Spanish tribes were allies or subjects of Carthage, and Scipio wanted to encourage defection and rebellion (see chapter 5). Similarly, when he was marching through neutral and potentially unfriendly territory in Greece in 198 BCE, the consul Titus Flamininus ordered his troops to treat the territory like it was their own. He too wanted to win support from the locals.[24]

Aside from naked pragmatism, the Romans developed powerful cultural norms that regulated their war making. Roman authors repeatedly stress that enemies deserved mild treatment when they surrendered voluntarily, just as those who surrendered under duress or fought too hard deserved punishment. Although it was customary to punish cities that did not submit "before the ram has touched the wall," it was also customary to spare those communities that yielded before the Romans attacked. As one Roman commander put it, "towns are sacked after they are captured [by force], not after they surrender [willingly]." Cicero goes further, elevating this custom to moral abstraction. In his treatise *On Duties*, Cicero lays out a series of ethical guidelines for public officials. When it comes to warfare, he writes, Roman leaders should spare those they conquer and offer protection to those who throw down their arms, "even after the ram has hammered their walls." He further claims that "great men" should act in an "upright and honorable" manner, avoiding cruelty and recklessness even when sacking cities, and they should only punish guilty individuals while sparing the majority.[25]

It is doubtful that many Roman generals observed Cicero's strict ethical standards (Cicero included . . .). Yet the Romans did practice a ritual of formal surrender that shaped their interactions with the defeated, and shepherded both parties toward a nonviolent outcome. In the ritual known as *deditio*, or *deditio in fidem*, a foreign community would "surrender to the good faith" of the Romans. In the ritual itself, the Roman commander asked a series of formulaic questions to representatives of the surrendering community, then formally received their submission. Livy describes the ritual formula when he narrates the legendary surrender of Collatia, a central Italian town (late seventh/early sixth centuries BCE):

> The [Roman] king [Tarquinius Priscus] asked, "Are you the representatives and spokesmen sent by the People of Collatia to surrender yourselves and the People of Collatia?"
> "We are."
> "Is the People of Collatia its own master?"
> "It is."
> "Do you surrender yourselves and the People of Collatia, city, lands, water, boundary marks, shrines, utensils, all appurtenances, divine and human, into my power and that of the Roman People?"
> "We do."
> "I receive the surrender." (Livy 1.38.2–3, trans. Foster, modified)

Polybius confirms Livy's description of the *deditio* ritual, noting that "the Romans gain possession of everything [that had belonged to the surrendering state,] and those who surrender remain in possession of absolutely nothing." On paper, this exchange is extremely harsh, and one can immediately sympathize with the many peoples who put themselves at the mercy of Rome and an uncertain future. And in practice, *deditio* could lead to anything, up to and including the annihilation of the surrendering party. Still, we should not let the language of the *deditio* ritual distract us from the usual outcome. After accepting a community's formal surrender in the *deditio*, the Romans might impose an indemnity (i.e., a payment of war reparations) or some other penalty, but they usually handed everything back to the surrendering people. A historical *deditio* from Spain survives on a bronze inscription and illustrates the usual nonviolent outcome:

In the consulship of C. Marius and C. Flavius [104 BCE]. The people of Seano . . . surrendered themselves to L. Caesius, son of Gaius, *imperator*.* L. Caesius, son of Gaius, *imperator*, after he accepted their surrender, referred to his advisory council what demands they considered ought to be imposed on them. On the advice of the council, he ordered that they hand over all captives, horses, mares which they had captured. All these they surrendered. Then L. Caesius, son of Gaius, *imperator*, ordered that they be free. He handed back to them such lands and buildings, laws and other things which were theirs on the day before they surrendered, which were in existence at that date, for so long as it pleased the people and senate of Rome. (*Tabula Alcantarensis*, trans. Richardson, modified)

The language indicates that Caesius could do whatever he wanted to this surrendering Spanish community. In the end, he simply forced them to return "captives, horses, [and] mares," which presumably they had taken in raids, before giving back their lives, laws, and property. The transaction was apparently bloodless.[26]

The Romans often asserted that they won their empire through moderation and fair dealing, choosing, in Virgil's immortal words, to "spare the defeated and crush the proud." There is some truth to this, and Roman norms around surrender and *deditio* probably prevented violence at the end of many military encounters. But norms are not equivalent to "laws of war," ancient military conventions were not always followed, and in the end there were no firm limitations on the actions of Roman generals. Even when Roman commanders were publicly criticized for their harsh treatment of a foreign people, Rome's military norms could not restrain violent behavior in any meaningful way.[27]

Consider, for example, a scandal that burst forth into Roman public life in 173 BCE. In that year, the consul Popilius Laenas was campaigning against a people in northwestern Italy called the Ligurians. At one point in the campaign, Laenas approached a town called Carystus, where a large force of warriors had congregated. This town belonged to a Ligurian tribe called the Statellates, who were not at war with Rome; they were actually one of the only tribes in the region that was not openly hostile. When they saw the Roman army approaching, however, the Statellates responded by drawing up their warriors for battle. Laenas eagerly engaged them and won the ensuing fight. After their defeat in the field, the Statellates surrendered unconditionally to the consul, performing a *deditio*. Laenas then disarmed them, sacked their city, and sold the population as slaves.

From Laenas's perspective, he had done a good deed, and the Senate "should have decreed honors to the immortal gods for the successes he had enjoyed in war." After all, he had utterly defeated a foreign state in battle, killing thousands and capturing thousands more for sale into slavery. Such things were frequent markers of military success for the Romans, and so they should be for him. But most of his colleagues in the Senate felt otherwise:

---

* *Imperator* simply meant "commander" in this period.

> It seemed outrageous to the Senate that the Statellates—who alone of the Ligurians had not made war on the Romans, who even on this occasion had been attacked although they had not begun a war, who had entrusted themselves to the good faith of the Roman people—were harassed and destroyed with every form of extreme cruelty; and that so many thousands of innocent persons, calling upon the Roman people for protection, had been sold—a fate which established the worst possible precedent and issued a warning that no one should ever dare in the future to surrender. (Livy 42.8.4–6; trans. Sage and Schlesinger, modified)

In other words, the problem with Laenas's behavior was not just a moral one, though this too was surely a concern, the problem was also political and strategic. He had attacked the *only* Ligurian tribe still at peace with Rome and then enslaved them after their formal surrender, potentially discouraging others from surrendering in the future. The outraged senators decreed that Laenas should restore the enslaved Ligurians to freedom, hand back their weapons, and return as much property as possible.[28]

Laenas was furious at the Senate's response, failed to cooperate with their directives, and even attacked the Statellates *again*. The Senate, incensed at his insubordination, eventually supported the creation of a special court to prosecute the wayward consul for his misdeeds and exact some justice at last. But the whole episode barely left a mark on Laenas's career: his trial fizzled out with no resolution (the presiding magistrate quietly let the matter drop), and Laenas was elected to the prestigious office of censor in 159 BCE. The aggrieved party also did not receive complete compensation. Although some of the enslaved Ligurians were eventually freed, the Romans resettled them further south, rather than return them to their own lands.[29]

Another controversy erupted when the consul Fulvius Nobilior sacked the city of Ambracia. During Nobilior's consulship in 189 BCE, Rome was at war with a federation of communities in central Greece called the Aetolian League. Ambracia, a member of the League, was a well-fortified city of strategic significance, and Nobilior put it under siege. His army launched fierce operations against the city, but Ambracia's population and some Aetolian defenders bravely fought off their attacks, never succumbing to a Roman assault. Instead, ambassadors from third-party Greek states brokered negotiations, convincing the Aetolian and Ambracian defenders to surrender. The Aetolians made their own peace with Rome, and the population of Ambracia placed itself at the mercy of the Roman consul. The Ambracians must have been sorely disappointed when Nobilior let his soldiers sack their city.[30]

Back in Rome, a senator named Aemilius Lepidus smelled blood in the water, sensing in Nobilior's lack of scruples an opportunity to damage a political rival. So Lepidus "introduced ambassadors from Ambracia into the Senate, having suborned them to make accusations against Fulvius." In Livy's account, the Ambracian envoys paint a pitiful picture for the assembled senators:

While they were at peace and had performed the orders of the previous consuls and were ready to render the same obedience to Marcus Fulvius [Nobilior], war had been declared on them, and first their fields had been laid waste and fear of plunder and slaughter held before the city, so that they were compelled by that fear to close their gates; that then they were beleaguered and besieged and that every form of war had been waged against them—slaughter, fires, destruction, plunder of the city; that their wives and children had been carried off into slavery, their property taken from them, and, what disturbed them most of all, the temples throughout the city had been stripped of their ornaments; the images of the gods, or rather the gods themselves, had been torn from their seats and carried away; bare walls and door-posts, they said, had been left to the Ambracians to adore, to pray to, and to supplicate. (Livy 38.43.3–5; trans. Sage, modified)

This tragic story turned up the hackles of some Roman senators. However, Nobilior's friend and political ally, C. Flaminius, stood up and offered a firm defense. He argued that the consul had done nothing out of the ordinary to Ambracia. Rather, he had just done the things that "are usually done when cities are captured," carting away loot from a defeated enemy like so many commanders before him. Furthermore, he asserted, Nobilior deserved a triumph for these achievements so he could proudly and publicly display all the plunder he had captured from Rome's adversaries. Flaminius and Lepidus bickered for two days without resolution, but Lepidus eventually pushed through a decree ordering the return of Ambracian property. Shortly thereafter, he passed another measure, which stated that Ambracia had not been captured by force (insinuating that it had been captured by less-than-glorious means). All of this suggests that Lepidus was as (or more) concerned about the glory gained by his rival, Nobilior, as he was about the rough treatment of Ambracia.[31]

Upon his return to Rome, Nobilior made essentially the same argument as his ally Flaminius, adding that Lepidus was pursuing a private vendetta at the expense of public welfare. He underlined his military achievements to drive his point home, describing how his army endured fifteen days of hard fighting at Ambracia and inflicted 3,000 casualties upon the enemy. Most senators were persuaded by his logic because, in the end, he was awarded a triumph for successfully ending the war. In the triumph, naturally, he paraded massive piles of Ambracian loot before adoring crowds.[32]

In later chapters, we will examine a few other cases in which commanders' conduct abroad came under scrutiny at home. For now, it is sufficient to draw three major takeaways from the cases of Popilius Laenas and Fulvius Nobilior. First, it appears that neither commander saw anything particularly wrong about his actions. In fact, they both asserted that they deserved *reward* for plundering and enslaving their foes, since the Romans often measured military success relative to the damage inflicted upon their enemies. Just as Cicero shamelessly described his destruction and enslavement of Pindenissus—after he had compelled the inhabitants to surrender—Laenas and Nobilior were quite ready to report their deeds to the Senate and people of

Rome. They probably did not anticipate any backlash, and they and their defenders did not regard their treatment of the Ligurians or Ambracians as terribly out of bounds.

Second, neither Laenas nor Nobilior suffered serious repercussions for their actions, and the same can be said for other generals who received similar criticisms or none at all (see especially chapters 6 and 7). Our sources describe many other commanders destroying, enslaving, sacking, or killing whole communities after their surrender, without mentioning any political consequences back in Rome. This indicates that such conduct was not normally seen as problematic, or at least it was not problematic enough to warrant a public response. Most of Rome's military leaders enjoyed a free hand to wage war as they saw fit.[33]

What made Laenas's and Nobilior's cases different, and the third main takeaway, is that political concerns loom as large in these episodes as moral-ethical considerations. Certainly Laenas was censured in part because he enslaved the Ligurians after accepting their *deditio*, thereby violating the moral principle of Rome's good faith. But the Senate also feared that his harsh treatment of a hitherto neutral people would harm Rome's relationship with other tribes, or discourage other communities from surrendering in the future. And Nobilior was targeted by an avowed political rival, who saw an opportunity to avenge earlier slights and obstruct Nobilior's bid for a triumph. In other words, political concerns were as important here as moral or ethical ones, and politics pushed acceptable (if non-normative) actions to the point of public censure.[34]

To conclude this section, Roman military leaders exercised virtually unlimited power over the defeated and were free to determine their fate, as the *deditio* ritual starkly demonstrates. However, pragmatic considerations as well as military norms could encourage Roman commanders to avoid or limit their use of violence. Treating enemies with moderation, in the fashion of Scipio, Flamininus, and Caesar, could return both immediate and long-term benefits. Expectations of mercy might persuade enemies to come to terms more rapidly, while a reputation for clemency could encourage surrender, loyalty, or even alliance. These would have been real concerns for Roman commanders on campaign, shaping their choices to use or restrict violence, weighing its benefits against its potential drawbacks.

Likewise, conventions around the treatment of surrendered enemies encouraged some commanders to restrict their use of violence. And at least some senators believed that there were limits to acceptable military conduct. Otherwise, there would have been no grounds for the accusations against Nobilior and Laenas, and no audience willing to hear those accusations. But Laenas's and Nobilior's cases also show that military norms were weak forces of restraint. Norms seldom led to censure and even more rarely led to any sort of punishment in the breach. Given the disagreements between Laenas, Nobilior, and their political rivals, customary ideas about acceptable and unacceptable behavior must have been flexible, exercising only a limited influence on command-level decision-making.

CONCLUSION: MASS VIOLENCE AND THE GENERAL'S JUDGMENT

Cicero's letters from Pindenissus give us special insight into a Roman general's thinking as he attacked and ultimately destroyed an enemy community. While his first-person account is extraordinary, other ancient sources indicate several factors that encouraged (or discouraged) mass violence in Roman warfare. On one level, mass violence was a pragmatic response to military circumstances. According to this logic, Roman military and political leaders might destroy towns, fortifications, and populations in order to neutralize a present or potential threat. Roman leaders also used mass violence to punish and "make an example" of communities that had resisted too long or somehow offended Rome. While these punitive displays could stem from Roman anger, they also served a sort of strategic function—namely, terrorizing others into surrender or deterring future challenges to Rome. Economic gain could also incentivize Roman violence. Some commanders used city-sacking or the sale of captives to reward their troops or fund "material manifestations of their glory" back home. The military and political culture of the Roman aristocracy also created incentives for mass violence. Successful generals were able to translate the devastation of cities, or the killing and capturing of large numbers of people, into political rewards. Indeed, the Romans apparently regarded victories as more glorious when they did more damage to the enemy. This connection between violence and achievement arguably shaped the decision-making of Roman generals, including their understandings of such "rational" matters as strategy, tactics, and objectives. Finally, restraint could garner its own benefits. In practical terms, the moderate treatment of recently hostile enemies created positive incentives for surrender and gave commanders another tool for undermining resistance. Roman military norms around the treatment of defeated enemies also discouraged purposeless violence. Although breaching these norms rarely led to punishment, most enemies who performed a *deditio* (ritual unconditional surrender to Rome) were treated well. Evidently, some commanders balanced the potential benefits of mass violence against conventional standards of military conduct and Rome's reputation, among other things.

This great variety of potential motives indicates, above all, that mass violence was *instrumental*. From the perspective of ancient Roman and Greek authors, Roman leaders used mass violence as an adaptable tool that might resolve strategic dilemmas, neutralize threats, encourage surrender, punish insults, avenge betrayals, vent Roman anger, deter other enemies, produce material gain, reward soldiers, score political points, and reap military glory. And as Cicero's letters from Pindenissus suggest, a commander's choices were pushed and pulled by multiple motives working simultaneously, from the unique challenges encountered on campaign to personal political goals.

The best way to understand such dynamic choices is to move from the general to the specific. The next three chapters examine three wars during the Middle Republic in which the Romans repeatedly used massacre, executions, large-scale enslavement, city-sacking, and urban destruction on campaign. These chapters analyze violence using the "strategic perspective": thus they firmly situate the Romans' use of mass violence in the context of warfare, and highlight the perspectives, goals, and decision-making of Rome's military and political leadership. For this period, unfortunately, we have nothing like Cicero's letters giving us first-person accounts of command-level decisions. However, by viewing mass violence within the larger framework of Roman military campaigns, it is possible to give context and depth to the decisions of Roman generals—and to show how mass violence served as a specific means to specific ends.

NOTES

1. Cicero in Cilicia: Cic. *Fam.* 2.10, 15.4, *Att.* 5.20; Goldsworthy 1996, 95–97; Beard 2007, 187–99.

2. Interpreting or guessing at motives for violence: e.g., Polyb. 10.15.5; Cic. *Off.* 1.35. Condemning violence: e.g., Val. Max. 9.2.1; App. *Hisp.* 60. Excusing or rejecting accusations of violence: Livy 24.30.3–7, 38.9.13.

3. Corinth's strategic value: Polyb. 18.11.4–6, 18.45.5–6, 30.10.3. "Out of fear": Zon. 9.31. "Convenience": Cic. *Off.* 1.35. Demolishing Iberian walls: Front. *Strat.* 1.1.1; Livy 34.17.8–11; Aur. Vict. *Vir. Ill.* 47.2. "Easier to attack": App. *Hisp.* 41. Cf. App. *Hisp.* 99; Livy 9.41.6.

4. "Disturb the peace": Livy 34.16.10. "A horse and a man": Livy *Per.* 49; cf. Cato fr. 105. Enna: Livy 24.38–39.

5. "Given the signal": Tac. *Ann.* 12.17. Artaxata: Tac. *Ann.* 13.41.

6. Antipatrea: Livy 31.27.2–4. "Sources of manpower": Billows 2007, 305.

7. "Before the ram": Caes. *Gal.* 2.32, *B. Civ.* 2.12; Cic. *Off.* 1.35.

8. Senones: App. *Sam.* 13, *Gall.* 13. Illyrians: Flor. 1.21.4; see also Polyb. 2.8.5–13; App. *Ill.* 7; Dio fr. 49.2–5; Zon. 8.19. Corinth: Cic. *Man.* 5; Livy *Per.* 52; Strabo 8.6.23; cf. Polyb. 33.10.3.

9. Beheading Italian leaders: Livy. 9.16.9–10, 9.24.13–15, 26.15.8; Diod. 19.101.2. Agrigentum: Livy 26.40.13. "That had defected": Livy 45.34.1. Epirotes punished: Livy 45.34.1–6; Plut. *Aem.* 29.1; Eutr. 4.8.

10. "To make manifest": Polyb. 15.4.2.

11. "Inspire terror": Polyb. 10.15.4. "Terrify the others": Livy 24.35.2. Cf. Livy 24.39.7, 26.40.14, 28.20.8–12, 31.27.5, 32.15.2–3. "Punished more severely": Caes. *Gal.* 3.16. "Deserved a great punishment": Caes. *Gal.* 8.39. "Exemplary punishment": Caes. *Gal.* 8.44. "Demonstrate to posterity": Jos. *BJ* 7.2 (Whiston).

12. "By this example": Just. 34.2.6.

13. Cato and Carthage: Flor. 1.31.4–5; Pliny *NH* 15.74–76; Livy *Per.* 49; Diod. 34.33.3; *Rhet. Her.* 4.20; Zon. 9.30; Cic. *Off.* 1.79. See also Plut. *Cat. Mai.* 26.1–27.3; App.

Pun. 69. Imperial rivals: Cic. *Agr.* 2.87. Cf. *Mur.* 28.58, *Man.* 20.60; App. *Pun.* 88; Vell. Pat. 1.12.5–7.

14. Commanders' booty: Rosenstein 2011a, 133–58; Kay 2014, 29–35; Churchill 1999, 85–116. "Material manifestations": Joshel 2010, 84. "Naval plunder": Riggsby 2006, 118 (#3). Plunder given to Roman communities/financing new construction: Riggsby 2006, 217–21 (#s 2–8, 11, 12, 14, 15, 18, 19, 20, 22, 26).

15. Soldiers' expectations of plunder: Rosenstein 2012, 110; Harris 1979, 43, 47, 102–3. Plundering against orders: Caes. *Gal.* 7.47, 7.52; App. *Pun.* 15; Livy 37.32.11–13. Paullus opposed by soldiers: Livy 45.34.7, 45.35.5–45.39.19. Plut. *Aem.* 30.2–32.1; Pittenger 2008, 246–74.

16. Commanders seek soldiers' favor: e.g., Plut. *C. Gracch.* 2.2–3; Sall. *Iug.* 96.2; Livy 43.1.3. "Order from the Senate": Plut. *Aem.* 29.1; cf. Livy 45.34.1. "Enriched," "praised him": Sall. *Iug.* 92.2.

17. "Quick sale": Rosenstein 2011a, 147–48. Cf. Kay 2014, 30. Large sales lower prices: Aur. Vict. *Vir. Ill.* 57.1–3; Plut. *Aem.* 29.5, *Luc.* 14.1–2; Jos. *BJ* 6.384. Economic motives for war: Badian 1968, 17–20; Harris 1979, 54–104; Sherwin-White 1980, 177–81; North 1981, 1–9; Gruen 1984a, 59–82; Rich 1993, 38–68; Raaflaub 1996, 273–314.

18. 5,000 enemy dead/triumph: Val. Max. 2.8 init.1. Cf. Oros. 5.4.7. 6,000 enemy dead/*imperator*: Diod. 36.14.1. "Bloodless": Gell. *NA* 5.6.21. Cato in Spain: Plut. *Cat. Mai.*10.3. Gracchus in Spain: Strabo 3.4.14.

19. Casualty metrics: Pittenger 2008, 104–14; Beard 2007, 209–10. Military glory and mass violence: Marco Simón 2016, 223–25, 237–38; Rüpke 1995, 232.

20. Inscriptions: Riggsby 2006, 217–21 (#s 3, 10, 15, 21, 25, 27). Caesar enumerates casualties: e.g., Caes. *Gal.* 1.29. Cf. Cic. *Prov.* 32–35. Caesar's non-distinction between combatants, noncombatants: Bellamore 2012, 41. See also Gilliver 2005, 67–71; Marvin 2012, 51; Konstan 2004, 75–77, 88–91.

21. "Flattering picture": Rosenstein 2012, 29. Imperial triumphs: Itgenshorst 2005, 219–26. Images etc. in Republican triumphs: Livy 26.21.7, 37.59.3; Strabo 3.4.13; Pliny *NH* 35.21–22, 135; App. *Pun.* 66; Quint. *Inst.* 6.3.61; Vell. Pat. 2.56.2; Cic. *Off.* 2.28; *Pis.* 25.60. Triumphal painting: Holliday 1997. Killing captives in triumphs: Beard 2007, 130–32.

22. "Greatest possible number": Livy 38.23.3; cf. App. *Hisp.* 98.

23. Permissive attitudes to violence: e.g., Barrandon 2016 and 2018, ch. 7; Bellamore 2012, 38–49; Isaac 2004, 215–24. Rights of conquest: Livy 9.1.5, 26.31.2, 38.43.8–11, 39.4.11–13; Cic. *Verr.* 2.1.57, 2.2.50. See also Kern 1999, 323–31; Gilliver 1996, 222–30.

24. Benefits of clemency: Gilliver 1996, 231–34, noting Onasander; see Onas. *Strat.* 35.4, 38.1–6, 42.24–26. Caesar: Gilliver 1996, 220–22; Dowling 2006, 20–24. Scipio: Livy 26.49.7–10, 27.19.3–7; Polyb. 10.18.3–15, 10.34.1–10, 10.35.1–3. Flamininus: Livy 32.14.5–7; Plut. *Flam.* 5.2–3; cf. Livy 36.21.3–4, 44.7.5, 44.31.1.

25. "Towns are sacked": Livy 37.32.12. Cicero on morality in war: Cic. *Off.* 1.35, 1.82; cf. *Off.* 2.26.

26. "Gain possession": Polyb. 36.4.2–3; cf. 20.9–10. *Deditio* generally positive: Burton 2011, 114–58. Cf. Burton 2009, 248–50; Eckstein 2009, 276; 1995, 267; 1994, 86; Dmitriev 2011, 268–73; Ferrary 1988, 74–78; Eilers 2002, 34–35.

27. Roman moderation in war: e.g., Livy 5.27.6–8, 8.13.16, 26.49.7–8, 45.8.5; Gell. *NA* 6.3.33, 47, 52; Diod. 32.4.4–5; Polyb. 18.37.7–8, 27.8.7–8, 36.9.5; Cic. *Off.* 1.35, 82. "Spare the defeated": Virg. *Aen.* 6.853. Weakness of ancient conventions: Pritchett 1991, 203–42; Eckstein 2006, 37–117, 204–5; Roth 2007, 397.

28. Statellates: Livy 42.7–9. "Should have decreed": Livy 42.9.3.

29. Laenas's career: Broughton 1951, 445. Special court: Livy 42.22.7–8. Fate of Statellates: Livy 42.22.5. See also Burton 2011, 326–28; Dmitriev 2011, 258–59.

30. Ambracia: Polyb. 21.26.1–21.30.14; Livy 38.3.9–38.9.13; Zon. 9.21; Polyaen. 6.17.1.

31. "Introduced ambassadors": Livy 38.43.2. Flaminius vs. Lepidus: Livy 38.43.7–38.44.6.

32. Nobilior's defense, triumph: Livy 39.4.5–39.5.17. Cf. Pittenger 2008, 196–212; Briscoe 2008, 155; Linderski 1996, 390–96.

33. Mass violence post-surrender: Livy 43.1.1–3, 44.45.7, 45.26.3–11, 45.34.1–6, *Per.* 60; App. *Hisp.* 52, 59–60, *Pun.* 74–80, 129–32, 134; Diod. 32.18.1; Polyb. 36.9.1–17; Sall. *Iug.* 91.5–7. Cf. Lintott 1972, 635–37. Commanders' freedom of action: Eckstein 1987. Commanders' authority over defeated: Barrandon 2018, ch. 7. Cf. Sen. *Epist.* 95.30.

34. Politics and ethical censure: Yakobsen 2009, 71; Dmitriev 2011, 259.

5

"Deterred by Fear"

Defection and Deterrence in the Second Punic War

Over 50,000 Romans and Italians lay dead as the sun set on August 2, 216 BCE.[*] Next to an abandoned town called Cannae in southeastern Italy, their blood drenched a plain that was only five or six square miles. This was a shocking turn of events for the Romans, who had been confident of victory mere hours before. Earlier that day their army of 80,000 infantry and 6,000 cavalry, the greatest Roman force ever assembled, had marched against the Carthaginian general Hannibal Barca. The Romans had greatly outnumbered Hannibal's invading army of about 40,000 infantry and 10,000 cavalry, which itself was a mosaic of Libyan, Numidian, Gallic, and Iberian troops. To take advantage of their numbers, the consuls Gaius Terentius Varro and Lucius Aemilius Paullus had packed their hordes of Roman and Italian footmen into a dense, deep formation, counting on the strength of the legions to punch through the center of the enemy battle line and win the day. Hannibal, however, was an innately gifted general with hardened veterans at his command, and anticipated the typically aggressive Roman plan. When the two sides entered combat, the heavy Roman formation predictably pushed the center of Hannibal's line further and further back—but at the same time, Hannibal slowly withdrew his center in the face of Roman pressure. The stratagem worked. As the Romans pushed back the Carthaginian center, it bowed inward. Eventually, Hannibal's left and right wings marched forward and encircled the Roman flanks. The legionaries suddenly found themselves surrounded on the left and right, and their forward momentum ground to full stop. While the infantry battle was unfolding, the

---

[*] Henceforth, all dates in this chapter are BCE.

Carthaginian cavalry drove off their Roman and Italian counterparts, freeing them up to attack the Roman footmen in the rear. With that, Hannibal had totally enveloped the oversized Roman force. The legions soon devolved from an organized body of soldiers into a panicked, demoralized crowd, and the Carthaginians slaughtered them. By nightfall most of the Romans and Italians who fought at Cannae were dead or captured.[1]

A handful of Romans managed to escape, with 4,000 infantry and 600 cavalry making it to the nearby town of Canusium. Most of their leadership was dead, including the consul Aemilius Paullus, two quaestors, twenty-nine military tribunes, and several other senators; only four military tribunes and some young men from Rome's aristocratic families arrived safely with the other survivors. As a few of these junior leaders deliberated their next move, another youth came upon them in despair. They were "clinging to a lost cause for no purpose," he told them. "The Republic is left for dead: some of the young nobles, who are led by Marcus Caecilius Metellus, are looking to the sea and to ships so that they can desert Italy and flee to a foreign ruler." All of the young men were thunderstruck at this news, and some proposed that they summon a full council to discuss the situation. Then one of the surviving tribunes sprang into action. His name was Publius Cornelius Scipio.

In his late teens or early twenties, the ambitious Scipio had made a name for himself two years earlier, during the first military engagement of the war—a cavalry skirmish at the Ticinus River in northern Italy. In this battle, Hannibal's horsemen routed a Roman force under the command of Scipio's father, the elder Publius Cornelius Scipio, and in the middle of the fighting, the elder Scipio was wounded, isolated, and surrounded by foes. The younger Scipio was commanding a cavalry squadron at the time, and he urged his comrades to help him rescue his father. When the men around him faltered, he daringly charged into the thick of enemy forces alone. His fearlessness shamed the rest of his squadron into action, and together they drove off many foes and saved his father's life. The younger Scipio thus gained an "acknowledged reputation for bravery."[2]

Hours after Rome's greatest defeat at Cannae, Scipio again acted with impulsive audacity. This was the time for action, not for idle talk: "Those who want to save the Republic should take up arms immediately and go with me," he thundered at his comrades. With a few followers, Scipio stalked to the place where Metellus was lodging and barged inside, finding the alleged traitor along with some other young men assembled there. Drawing his sword, he thrust it over the head of the conspirators and uttered an oath and a threat:

> I solemnly swear that I will not abandon the Republic of Rome, nor will I allow any other Roman citizen to do so; if I knowingly break my oath, then you, O Jupiter Optimus Maximus, visit me, my home, my family, and my estate with utter destruction. I require you, Caecilius [Metellus], and all who are here pres-

ent, to take this oath. Whoever will not swear, let him know that this sword is drawn against him. (Livy 22.53.10, trans. Roberts, modified)

Metellus and his compatriots were sufficiently frightened and immediately surrendered.

The story is exciting but dubious, and it may well be an invented tale of patriotism to offset the horror of Rome's greatest defeat. Nonetheless, the account of wavering loyalties, planned betrayals, and threats of retaliation neatly illustrates, in microcosm, the crisis that Rome faced in the Second Punic War. Between 218 and 216, Hannibal won three major battlefield victories, killing or capturing tens of thousands of Roman troops. Like Metellus and his co-conspirators, many of Rome's Italian allies believed that the Republic's days were numbered and looked to their own interests. Communities across the peninsula deserted the Romans and sided with the Carthaginian invader, and Rome's grip on Italy began to loosen.[3]

But the Romans were notoriously stubborn, and they reacted to the hand that their foe had dealt them with ferocious determination. Their chief strategy after Cannae was to avoid direct confrontation with Hannibal while limiting his gains. They knew they could not beat him in battle, and they did not try again until later in the war. Instead, they turned against their turncoat allies with a singular intensity: like Scipio in the story of Metellus's near desertion, Roman generals leveled a sharp sword against those who dared to betray the Republic, beheading many rebel leaders and massacring and enslaving disloyal populations. While Roman commanders in the Second Punic War employed mass violence to accomplish a range of objectives, its overwhelming purpose was to beat back the tide of defections and regain control of wayward allies, using a combination of terrifying punishment and stark deterrence. And given the unparalleled scale of the war, the extent and severity of Roman reprisals was extraordinary.

ROME, CARTHAGE, AND HANNIBAL IN THE LATE THIRD CENTURY

Hannibal's invasion of Italy was a shock, but the Second Punic War did not come out of thin air. After their loss to Rome in the First Punic War (264–241), the Carthaginians moved forward with plans to recover their strength, wealth, and standing as a leading power. In 237, the Carthaginian government sent a general named Hamilcar Barca to Spain with a large army and a mandate to conquer. Though Carthage had long used Spain as a recruiting ground for mercenaries, and the southern coast hosted several Punic colonies, the peninsula was then dominated by several indigenous groups—including the Iberians in the south, the Lusitanians in the west, and the Celtiberians in the center and north. All were tough and practiced in warfare (see chapter 7). Yet Hamilcar was a talented commander, having fought well against the Romans in the First Punic War and having crushed a major rebellion in North Africa. He was up to the challenge and successfully conquered much of Spain until he was killed in 229/228. The project was then picked up by his son-in-law, Hasdrubal "the Fair," and, when Hasdrubal was assassinated in 221, by Hamilcar's son Hannibal Barca. Hannibal was only in his mid-twenties when he took command in Spain, but he had been at Hamilcar's side since he was a boy, effectively growing up in the army and learning the ropes of command from his father and brother-in-law. He probably also learned about Hamilcar's experiences fighting the Romans in Sicily, and, if we believe our ancient sources, he inherited his father's lingering anti-Roman bitterness. Under this succession of capable commanders, most of southern Spain had fallen under Carthaginian sway by 219, giving Carthage direct access to plentiful Spanish manpower and mineral wealth.

Carthage's rapid resurgence put it on a collision course with Rome. Messy controversies about the causes of the Second Punic War can be set aside; the important point is that Roman leaders became alarmed at the recovery of Punic strength. On more than one occasion, Roman ambassadors traveled to Spain and tried to set limits to Carthage's expansion, and in 220, they warned Hannibal to keep his hands off a Spanish town called Saguntum (to which Rome had extended promises of protection). Hannibal would have none of it. As far as he was concerned, the Romans were in no position to make demands in Spain, which had become Carthage's back yard. In spring 219, he defiantly led his army against Saguntum, capturing and sacking it in direct contravention of Roman demands. At that, the Roman Senate sent representatives to Carthage to insist that they give up Hannibal and his advisors or accept a new declaration of war. Carthaginian leaders refused to be kicked around any longer, and stood on their rights to do what they pleased as an equal power. The Romans refused to see them as such, and the two states began another terrible contest of arms.[4]

Hannibal was ready. In June of 218, he and his veteran army set out on an epic five-month march from Spain, through southern Gaul, and over the Alps. His forces were severely reduced by the arduous journey, but by winter they arrived in Cisalpine Gaul and almost immediately embarked on a string of brilliant successes. In late 218, Hannibal repelled the consul Publius Cornelius Scipio in a cavalry skirmish at the Ticinus River. Shortly thereafter, Hannibal met the consul Tiberius Longus at the river Trebia, where he lured Roman forces across the river's freezing waters and made short work of them in battle. Afterward, many Gallic tribes in northern Italy, only recently subdued by Rome, flocked to the Carthaginian invader and brought fresh warriors to his banner. The following year, Hannibal's army swept south into Etruria and led the consul Gaius Flaminius into an ambush alongside the shores of Lake Trasimene, annihilating his legions. Next, the Romans named a senior senator, Quintus Fabius Maximus, as dictator.* Fa-

---

* In the Early and Middle Roman Republic, "dictator" did not carry the negative connotations that it does today. The Romans would appoint a dictator only in dire emergencies, grant-

bius's strategy was essentially to avoid engaging Hannibal unless presented with a decisive opportunity, in the meantime monitoring and occasionally harassing the enemy's army. The "Fabian strategy" was deeply unpopular at Rome (Fabius was derisively called "the Delayer," *Cunctator*), but his approach provided breathing space to rebuild Roman forces and prepare for a crushing victory. What followed instead was Cannae, where Hannibal wiped out the greatest Roman army ever put in the field (see map 5.1).[5]

As devastating as Cannae was, Hannibal "was not waging a war of extermination with the Romans, but fighting for honor and power." Indeed, his objective was to revise the balance of power between the two states with a strategy that centered on Rome's allies. He apparently recognized that alliances were the key to the Republic's strength, and hoped that the Italian allies would begin to desert Rome if he defeated every army that he fought. Hannibal also tried to expedite these desertions by showing kindness to non-Roman Italians. After the battles of Trebia and Trasimene, he gathered his Italian captives and explained that he had not come to fight them, but to *free* them from Rome. Then he released them to carry this goodwill back to their home cities and precipitate rebellion. Once stripped of friends and much of its manpower, Rome would be forced to accept terms that would significantly reduce its strength while proportionately elevating Carthage. Following Cannae, the third battlefield defeat that the Republic suffered in as many years, Italian communities finally began to switch sides, and Hannibal's plan took shape.[6]

HANNIBAL'S NEW ALLIES

As we have seen, the Roman Republic came to dominate the Italian peninsula in the fifth, fourth, and early third centuries. By the time of the Second Punic War, most Italian communities were expected to follow Rome's lead in foreign affairs and to provide troops for the Republic's armies. Yet despite their subordinate status, the Italian allies (*socii*) had several incentives for cooperating. Conquered communities largely retained their internal autonomy, and some also received Roman citizenship or special privileges. The Romans shared the profits of their military campaigns, too, and supported friendly Italian regimes against internal political challenges. Further, Rome's aristocracy intermarried with preeminent families in various Italian communities, creating tight bonds between leaders. All of these factors encouraged shared interests.

However, as the historian Michael Fronda has shown, at least some of the *socii* continued to see themselves as independent powers, each pursuing its own agenda in less-than-ideal circumstances. Many openly chafed under

---

ing this individual supreme authority for a short term (usually six months) in order to rescue the Republic from disaster.

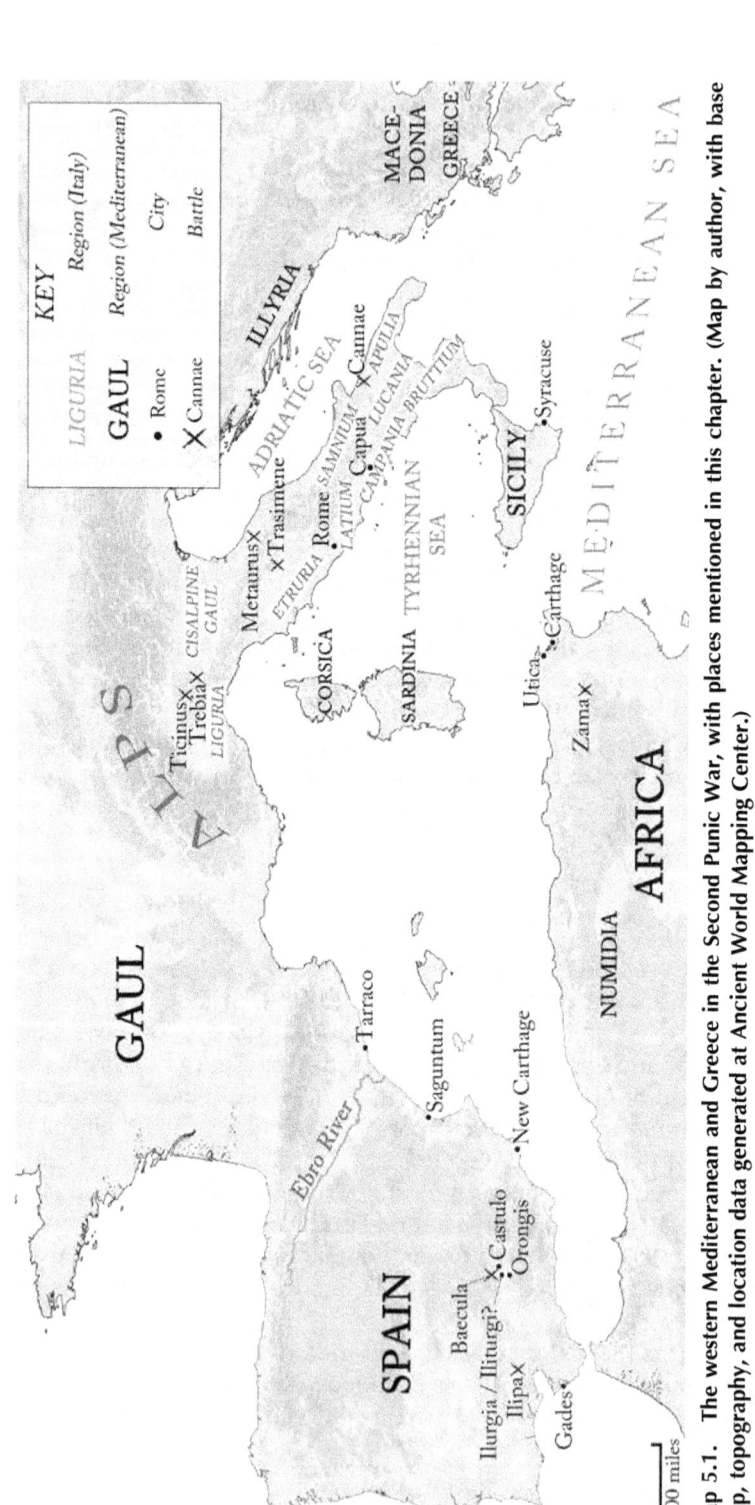

Map 5.1. The western Mediterranean and Greece in the Second Punic War, with places mentioned in this chapter. (Map by author, with base map, topography, and location data generated at Ancient World Mapping Center.)

Roman domination as late as the mid-third century, when a couple of cities in Etruria attempted to revolt. Normally, of course, Rome's overwhelming manpower advantage discouraged such open resistance. Roman and Latin colonies, established across the peninsula, also ensured that the Republic had loyal strongholds situated at control points, disrupting coordination among potential rebels and weakening any appetite for an uprising. As a consequence, successful rebellion was virtually impossible unless many different communities joined forces or "an outside force altered the equation." Hannibal provided the latter.[7]

The years after Cannae saw a steady trickle of Italians pass over to Hannibal's side. In the immediate aftermath of the battle, several nearby communities in Apulia (southeastern Italy) defected. From there, Hannibal marched into Samnium, a district where there had been frequent and fierce resistance to Rome, and many Samnite tribes joined him as well. But the greatest post-Cannae prize by far was Capua. Wealthy Capua, sited about one hundred miles south of Rome as the crow flies, was the largest city in the Campania region of Italy (see map 5.2). More significantly, the Capuans and other Campanians had been Roman citizens (without the vote) since the fourth century, and Capuan aristocrats had marriage ties to leading Romans. These were close allies and Roman citizens, and they deserted Rome to join Hannibal. Other towns in Campania followed suit. In southern Italy, the Bruttians and many Lucanians were in Hannibal's column by 215, as were some of the Greek cities in the southwest, like Locri and Croton. A few years later, a coterie of local noblemen helped Hannibal take control of Tarentum, the largest Greek community in the southeast. Other Greek cities in southern Italy soon revolted too.[8]

Hannibal also found friends outside of Italy. Across the Adriatic Sea, King Philip V of Macedonia caught word of Hannibal's crushing victories. Though the Macedonians had no previous dealings with Rome, there was reason enough to kick the Romans while they were down. Roman forces had crossed the Adriatic Sea in 229 and 219 to campaign in Illyria, a region to the west of Philip's kingdom. Rome's involvement there, however minimal, was none too welcome to Philip, who had his own designs on the region. By 215 he and Hannibal worked out an alliance. Their agreement probably did not envision substantial military cooperation, though it gave Philip the go-ahead to pursue his interests in Illyria. On Hannibal's end, "it provided a convenient second front to embarrass the Romans, without any real commitments, and its propaganda value among the Greeks of south Italy was considerable."[9]

The Greek cities of Sicily also wavered. The Romans annexed much of Sicily in the First Punic War, and in the 220s they began sending an annual praetor to watch over the western half of the island—the first step in the creation of overseas "provinces," as we would understand the word. However, they allowed Hiero, the ruler of Syracuse, to control the eastern part of the island, and he remained a loyal Roman ally in the intervening years.

Map 5.2. Southern Italy and Sicily in the Second Punic War, with places mentioned in this chapter. (Map by author, with base map, topography, and location data generated at Ancient World Mapping Center.)

When Hiero died in early 215, his grandson Hieronymus took the throne. Hieronymus was a mere teenager who reigned under the guidance of advisers, several of whom were eager to use Rome's dire straits in Italy to revive Syracuse's standing in Sicily. The young ruler opened negotiations first with Hannibal and then with Carthage, but was murdered in 214 before he could make much of an impact. Then a pro-Carthage faction in Syracuse, led by two brothers named Hippocrates and Epicydes, won over the nearby Sicilian city of Leontini. With a few thousand loyal soldiers, they proceeded to raid western Sicily and killed some Roman troops there. Rome's control of the island became increasingly tenuous.[10]

It is difficult to overstate the grimness of Rome's position in the year of Cannae and for several years thereafter. Livy was not far off in saying that "any other people would have been overwhelmed by such a great disaster." The Republic had been bruised and bloodied as never before; many of its subjects and allies turned their backs, and foreign powers took advantage of its weakness. Still, even with these huge gains, the flaws in Hannibal's strategy were soon apparent: everywhere that Hannibal nabbed new allies, some communities stayed faithful to Rome.* This meant that the Romans had friendly bases throughout the peninsula, giving them enough traction to recover and counterattack.[11]

PUNISHING "THEIR FAITHLESS ALLIES": ROME RETALIATES IN ITALY

The Roman Senate met the fires breaking out seemingly everywhere by recruiting unprecedentedly huge fighting forces, deploying anywhere from 14 to 25 legions annually for the remainder of the war. Although these legions were probably understrength, there were something like 60,000 to 80,000 Roman soldiers in service down to 209 (when the heavier levies began to wind down), along with naval forces and roughly equivalent numbers of their still-loyal allied troops. Even on a conservative estimate, historian Dexter Hoyos argues, "close to one in every three military-age Romans and a similar proportion of loyal Italians" were fighting for the Republic in the years of highest recruitment; "[n]o other ancient state was remotely capable of such an effort." Moreover, the Romans partly set aside their normal rules of political competition, which emphasized power sharing and prevented individual aristocrats from holding multiple consulships in succession. Instead, the voters kept electing a handful of experienced men as consuls, while the Senate repeatedly extended the tenure of commanders in the field as "proconsuls" and "propraetors."† Both efforts meant that experience and continuity accrued to Rome's military leadership.[12]

With these forces at their disposal and experienced leaders at the helm, the Senate and its commanders employed a fresh strategy in Italy. They would largely steer clear of Hannibal, given his proven tactical superiority. Instead, they dispatched their massive manpower resources all over the peninsula to shore up those allies who remained loyal, recapture lost towns, and brutally punish defectors or suspected waverers. This strategy often focused on Campania between 216 and 211, where the rebels were geographically closest, and betrayal had cut deepest. In 215, shortly after Capua defected,

---

* Italian communities that stuck with Rome did not necessarily do so out of loyalty. Rather, they preferred the more distant hegemony of Rome to the emergence of a rival closer to home and newly empowered by Hannibal. Additionally, some Italian leaders owed their political ascendancy to the Romans and had good reason to stick with them. For this dynamic, see Fronda 2010.

† When the Senate extended the command of a consul or a praetor for a year, these officials would then become "proconsuls" or "propraetors," literally someone commanding "in place of" (*pro*) a regularly elected magistrate.

a Roman commander named M. Claudius Marcellus suspected that "more than seventy" people at the city of Nola were planning to rebel. He had the suspects arrested and beheaded. That same year, a contingent of Roman troops worried that a Campanian town called Casilinum was about to join Hannibal. Casilinum was an important stronghold and commanded a major river crossing, so the soldiers decided to kill many of the locals and seize part of the town. Hannibal soon captured the place anyway, but the Romans retook it in 214—again massacring many of the inhabitants in the process.[13]

Roman armies operating further south and east employed similar tactics. When one general captured three rebellious Samnite towns, he beheaded the leaders allegedly responsible for defection, sold 5,000 captives into slavery, and gave the rest of the plunder to his men. The aforementioned Marcellus ravaged an area of Samnite countryside so thoroughly "that he restored their memory of past Samnite disasters." In 214 the former dictator Fabius Maximus killed or captured 25,000 people when he retook rebel towns in Samnium, Apulia, and Lucania, and had three hundred deserters executed back in Rome. The following year Fabius's son captured Arpi in Apulia and had his soldiers kill all the Carthaginian troops that they found in the city.[14]

Notably, the Romans' use of mass violence could backfire in these years. In 212 the Romans were holding hostages from Tarentum and Thurii—Greek cities in southern Italy—to ensure the good behavior of their home communities. When they discovered that the hostages were trying to escape, they executed them all. According to Livy, "the harshness of this punishment outraged two of the most famous Greek cities in Italy." Tarentum soon joined Hannibal, and Thurii followed not long after because of "their anger against the Romans" for killing the hostages.[15]

Still, the overall pattern was one of Roman momentum, and by 212 the Romans had struck back against rebel communities in Campania, Samnium, Lucania, and Apulia, recapturing or garrisoning several strongholds. That year, the consuls Appius Claudius Pulcher and Quintus Fulvius Flaccus, along with one of the praetors, turned their full attention to Capua—the city whose defection was the most serious and shocking. They surrounded the place with earthworks, fortifications, and an army over 40,000 strong. Just before they trapped the city, they gave the Capuans an ultimatum: those inhabitants who left Capua before the Ides (15th) of March would retain their liberty and property, but anyone who remained or left after that date would be regarded as an enemy. The Capuans were still confident—perhaps because they expected Hannibal to come to the rescue—and they hurled back insults and threats. The Roman commanders replied by closing the siege ring and choking the city off from the outside world.

The Senate made the capture of Capua their chief priority in 211 and extended the commands of Claudius Pulcher and Fulvius Flaccus to continue the siege. The beleaguered Capuans made several attempts to strike back at their besiegers and even scored minor successes with their skilled cavalry, but the Romans soon gained the upper hand in these skirmishes. Before long, the chastened Capuans were unwilling to venture outside their fortifications. Meanwhile, Hannibal received many desperate messages from his Capuan allies and made many attempts to relieve the city. He tried attacking the Roman siege works from the outside while the Capuans sallied out and attacked from the inside, but the Romans stood fast and repelled them both. Hannibal also tried to provoke Fulvius and Claudius into battle, but the Roman commanders would not come out from behind their defenses. Finally he made a feint on Rome itself, but the city was too well protected, and he did not peel the besiegers away from Capua as he had hoped. When all his efforts had failed, he went back south to continue operations there. Capua was alone.[16]

The Capuans, along with the Carthaginian garrison stationed there, fell into enraged despair and tried to send one last letter to Hannibal begging for help. Selecting about seventy Numidians to deliver their message, they had these soldiers pose as deserters and sent them to the Roman camp. But the deception was detected. The Roman commander Fulvius ordered his men to arrest the false deserters, beat them with rods, cut off their hands, and send them right back to Capua. The city was already starving after months of siege, and "the sight of so severe a punishment broke the spirit of the Capuans." After a furious discussion, Capuan leaders decided to give up the fight, although several leading aristocrats chose to commit suicide rather than fall into Roman hands. They may have anticipated the coming retribution.[17]

A day after the mutilation of the false deserters, the Capuans opened a gate to signal their surrender. Roman forces entered the city unopposed and seized Capuan soldiers and senators, picking out fifty-three of the latter whom they considered most responsible for rebellion. They sent twenty-five of these doomed individuals to a nearby town, Cales, and the other twenty-eight to Teanum. According to Livy, the two Roman commanders disagreed about how to treat the rebel leaders. Claudius was supposedly inclined toward mercy, or at least wanted to refer their fate to the Roman Senate. Fulvius, however, was concerned that the captives would make false accusations against other Italian communities, thus inflaming the already-fragile relations between Rome and the loyal allies. Rather than run this risk, he acted on his own. Racing by night with a body of cavalry, he entered the gates of Teanum at daybreak. He had the Capuan leaders dragged out from wherever they were wasting away, and collected them in the forum. The locals must have been gathering in the marketplace as the city came to life that morning, so we should imagine all of this taking place in front of a crowd. Fulvius clearly wanted an audience. (According to the ancient Greek orator Aeschines, such public humiliation was worse than death itself: "It is not death that men fear," he wrote, "but rather a shameful end. How pitiable is it to look into the face of a laughing enemy and to hear the insults" of the crowd?) There in the open marketplace, Fulvius had the condemned men tied to stakes, beaten with rods, and beheaded. No sooner had their heads rolled than Fulvius sped on to Cales and again gathered the Capuan captives in public. At the last moment a rider approached bearing a letter from the Senate. Fulvius worried that the letter would order him to stop the executions—so he pocketed it and had the other Capuans scourged and beheaded too.[18*]

Two other rebel towns in Campania, Atella and Calatia, surrendered in the aftermath of the executions, presumably acting out of a combination of hopelessness and fear. The Romans also beat and beheaded seventy of their leading men. Of the other Campanians, many were imprisoned, expelled from their territory, or sold as slaves, depending on their culpability for the revolt. The Roman Senate then debated what to do with Capua itself. Some argued that a city "powerful, nearby, and unfriendly" should be destroyed, extinguishing that threat forever. After discussion, however, the senators ultimately decided not to burn the buildings or demolish the walls. Instead they would end the city's civic existence. Its territory and structures would become Roman property; although it would remain a center of population, and its resident foreigners, artisans, and freed slaves could continue to dwell there, it would no longer possess political institutions and would remain under the direct administration of Roman prefects. Livy praises the Senate for its clemency, but Cicero equates Capua's fate directly with Carthage and Corinth, two cities that were physically destroyed in 146. And we can understand why he makes the comparison: much of Capua's population was forcibly removed, its leaders executed, and the city's government resolutely abolished.[19]

The fall of Capua well illustrates Rome's strategy in Italy after Cannae, in which brutal, retaliatory violence was a key element. On the one hand, the Romans used mass violence to remove threats. Executions and mass enslavements eliminated seditious leaders and untrustworthy populations, in turn depriving Hannibal of bases, supplies, and other support from rebel communities. On the other hand, Roman actions were punitive. From the Roman perspective, the defectors were traitors and needed to be taught an abject lesson; it is no coincidence that Roman generals often singled out and executed the ringleaders of rebellion. And of course, the "punitive" and "pragmatic" purposes of mass violence were mutually reinforcing. By punishing defector communities and their leaders in highly visible ways, Roman commanders sought to make an example of them that would deter further defections and deny Hannibal additional Italian support.[20]

At times, mass violence was counterproductive, most notably when the Romans executed hostages from Thurii and Tarentum and precipitated the rebellion of both cities. Overall, however, Rome's terroristic tactics and threats of retribution became more credible, immediate, and effective as the conflict wore on. Thus, when Capua was under siege and despairing of help from Hannibal, they opened their gates after seeing the horrifically

---

* Livy reports a couple of versions of this story, so one could reasonably question the dramatic details here. In every version, however, Fulvius beheads the prisoners, and on this our other sources agree.

mutilated Numidians. Likewise, Atella and Calatia surrendered after the fall and punishment of Capua. In large part, the threat of Roman violence became more credible and more effective because Hannibal was unable to offer consistent protection. As the legions put pressure virtually everywhere they chose, Hannibal could not afford to split up his army into small garrisons to guard his new allies. Nor could he move to multiple places at once to challenge the many armies mobilized against his coalition. He was forced to dash around Italy shielding the allies he already had, preventing cities from flipping back to Rome (and occasionally using his own brutal punishments to keep them in line), and trying to win over new collaborators. He simply could not apply pressure as widely or as hard as his foe.* His new allies also proved utterly incapable of challenging the Romans in battle, and armies composed of rebel Italians were swatted aside even when they were commanded by veteran Carthaginian officers. Livy accurately reflects Hannibal's larger dilemma after the fall of Capua: "the enemy was forced to acknowledge how much power the Romans had to punish their faithless allies and how helpless Hannibal was to defend those he had received under his protection."[21]

"WHAT THEY DESERVE TO HAVE SUFFERED": ROME RETAKES SICILY

While the Romans were busy bringing the disaffected Italians back into line, their commanders in Sicily were struggling to maintain control of the island. When we left off with Sicily in 214, the pro-Carthage faction at Syracuse, led by the brothers Epicydes and Hippocrates, had taken control of a town called Leontini. From there they raided the western half of the island, killing some legionary troops who were guarding local farmland.[22]

By this time, the consul M. Claudius Marcellus had arrived in Sicily and taken command of Roman forces there. Marcellus comes off in our sources as a hardened and hard-hearted veteran. During his first consulship in 222, he had rushed alone out of a battle line to fight a Gallic chieftain in single combat. Galloping forward on horseback, he speared the Gallic leader through his breastplate and knocked him to the ground with the force of the blow, following up with two quick strikes as his opponent writhed on the earth. He then leapt off his horse, stripped the armor from his enemy's corpse, and committed the spoils to Jupiter. In that same war, he led Roman forces to victory against the Gallic Insubres. And earlier in the war with Hannibal,

---

* Although the Romans often decided to punish defectors, some commanders showed mercy to rebels who surrendered quickly. Some communities also went back to Rome of their own accord (e.g., Livy 25.1.2, 27.15.2–3). Nonetheless, "there was a general pattern to the settlements imposed by Rome on formerly rebellious communities, which typically entailed harsh terms and severe punishments" (Fronda 2010, 308).

Marcellus had ordered the execution of leaders at Nola, on mere suspicion that they were planning to rebel. He did not take betrayal lightly, it seems.[23]

Shortly after his arrival in Sicily, Marcellus stormed Leontini and captured it. Epicydes and Hippocrates escaped, but the consul beheaded 2,000 Roman deserters he found there. Afterward, a rumor spread "that an indiscriminate massacre of soldiers and townspeople had occurred, and no adults . . . were left alive; that the city had been sacked, the property of the rich given away." Livy forcefully denies that any of this took place, claiming that the report confused Marcellus's execution of deserters with a general massacre of the population, but it is likely enough that Marcellus's men massacred some of the Leontinians. Not only were such massacres common when the legions stormed cities (see chapter 2), but also Livy admits that Marcellus's men assaulted the city with "ardor" and "anger" because of the recent deaths of their comrades in western Sicily. Marcellus may have even ordered his army to massacre the inhabitants. As we saw at Nola and as we will see in Sicily, he unflinchingly employed mass violence to deter betrayals and terrorize cities into surrender.[24]

In any case, if Marcellus hoped his swift and severe strike at Leontini would discourage new defections, it had the opposite result. Syracusan soldiers became disaffected when they heard reports of a Roman massacre, allowing Hippocrates and Epicydes to suborn them and seize power in Syracuse. This was a significant loss for the Romans. Sited at an important juncture for trade, Syracuse was a large city equipped with good harbors and extensive fortifications, including land and sea walls, towers, and gates. In addition, some of Syracuse's neighborhoods were protected by their own defenses, and the city sported not one but two urban strongholds: a fortress called Euryalus on the western, land-facing side of the city, and a citadel on a small peninsula called Ortygia on the southeast, water-facing side. The city had never been captured, plus the Syracusans had an ace up their sleeve: one of the most brilliant scholars of the ancient world, Archimedes, helped design and organize machines and artillery for the city's defense. In Polybius's words, the Romans "did not reckon with the ability of Archimedes, or foresee that in some cases the genius of one man accomplishes much more than any number of hands."[25]

Marcellus had his command extended as a proconsul in 213, and early that spring, he commenced operations against Syracuse. With two or possibly three legions, he led an assault on the city's seaward walls with a fleet while another commander assailed the landward side with an army. Both attacks were complete disasters. The Roman land assault was answered by artillery, much of it designed by Archimedes, which fired various projectiles at high speed and at different ranges. Huge stones smashed through Roman forces, mowing down men and throwing the ranks into confusion, while a storm of smaller missiles created their own havoc. On the sea side of the city, Archimedes's machines dropped heavy beams onto the attacking Roman ships, plunging several down into the depths. Contraptions that looked like giant claws or cranes wrenched vessels out of the water, dropping them back down into the sea or smashing them against the cliffs, throwing men and debris this way and that. The Syracusans frustrated all attempts to break through, and the Romans seemed to be "fighting against the gods." Thwarted, Marcellus called off direct assaults and settled down for a siege, hoping that Syracuse's large population would rapidly consume its food and starve.[26]

While these events were unfolding, more Sicilian communities declared for Carthage. Marcellus determined that he needed to stem the flow of defections, so he left behind two-thirds of his army to continue the blockade of Syracuse while he took the remaining soldiers to attack three other towns. Two of these towns surrendered when Marcellus approached; a third, Megara, resisted. After capturing the city by assault, "he sacked and demolished it." Livy's language (*diruit*, "demolished") may indicate that Marcellus's troops tore down the walls and other buildings, alongside the usual violence of city sacking. Marcellus's purpose was clear: he hoped the treatment of Megara would "terrify the others, especially Syracuse."[27]

Again he miscalculated, and the Roman position only worsened. Around the time that Megara fell, a Carthaginian relief army arrived in the south of the island under the command of one Himilco. It consisted of 25,000 infantry, 3,000 cavalry, and 12 war elephants, posing a serious danger to Roman operations. Marcellus recognized that events threatened to spin entirely out of his control and raced south to prevent more towns from going over to the enemy. But he was too late, and the arrival of Carthaginian forces precipitated the defection of two additional cities. One of these cities, Agrigentum, was well fortified and commanded a major east-west route along the southern coast, as well as several roads into the island's interior. Its loss was a heavy blow, and Marcellus lacked the manpower to engage the newly arrived Carthaginian army. A Punic fleet of fifty-five warships also arrived in Syracuse's harbor around this time, potentially providing naval power and supply lines to the besieged, and further souring Marcellus's outlook.

The Romans did enjoy some good fortune in the face of these setbacks. During their march back to Syracuse, Marcellus's troops unexpectedly fell upon a Syracusan army and mauled it badly. Additionally, another legion arrived as reinforcements, and the Carthaginian fleet proved itself completely inconsequential by sailing away. But these Roman gains notwithstanding, Himilco's Carthaginian forces moved inland and encouraged several communities to defect in the interior. The town Morgantina, about 52 miles (85 km) northwest of Syracuse, betrayed its Roman garrison and went over to the Carthaginians. Other unnamed towns followed suit, and Livy claims that several Roman garrisons were driven out or butchered. The loss of Morgantina, like that of Agrigentum, was a significant strike against the Roman effort in Sicily and raised further concerns about their hold over the island. Morgantina sat on an important juncture of two major routes, one connecting the north and south coasts and the other going from the interior toward the eastern coast; it was also centered in local grain fields.[28]

About twelve miles to the northwest of Morgantina, there was another strategic town that threatened to join the invaders: Enna. Towering about 3,000 feet above sea level, Livy describes Enna as "situated in a place that is lofty and steep on every side" and "impregnable." Cicero emphasizes its central position, calling it the "navel of Sicily," and notes that it was sacred to Demeter. To their observations, one should add that Enna was near both major roads connecting the Sicilian interior to the north coast, and it commanded one of Sicily's principle agricultural regions. Enna's strategic significance is underlined by its important role in later periods: it became the main base of a slave revolt about eighty years later, in the 130s, and in the Middle Ages it served as an administrative hub for first the Byzantines and then the Arabs—and was fiercely contested by both parties—allowing them to project control over the center of the island.[29]

When Morgantina fell, the Romans still held Enna with a garrison. But the garrison commander, Lucius Pinarius, became increasingly concerned as nearby towns kept defecting to the Carthaginians. His fears may have been well founded. According to Livy, Enna's leaders demanded that Pinarius hand over the keys to the city gates and surrender control of the citadel, where the Roman garrison was stationed. Pinarius, who had maybe 2,000 soldiers under his command, protested that he could not leave his post since he had been stationed there on Marcellus's orders; however, he urged them to send envoys to the proconsul and make their request directly. When Enna's leaders refused, Pinarius asked that they summon the citizen assembly "so that it might be known whether this was the demand of a few men or of the entire city." They conceded, agreeing to call a public meeting for the next day. In Livy's retelling, Pinarius expected the Ennaeans to denounce the Romans in their assembly, force him to hand over the gate keys, and give the city to the Carthaginians. With his suspicions growing darker, he returned to the citadel and mustered his troops. Livy gives Pinarius a rousing speech—certainly fictitious, but perhaps reflecting his sentiments at the time—in which the commander warns his men that they must act quickly and decisively, since it would be their blood or the Ennaeans'. Then he laid out his plan: he would try to stall for time in the Ennaean assembly while his men got into position around the meeting place; eventually he would give his legionaries a signal, upon which they would "raise a shout from all sides and, falling upon the crowd, cut down all with swords."[30]

The following day, the Ennaeans duly gathered in their theater for the public confrontation with Pinarius. The theater, like similar structures in other Greek cities, would have provided ample seating and good acoustics to accommodate a large gathering. Enna's theater does not survive, but we can guess at its appearance from the remains of contemporary theaters in Sicily (figure 5.1). It likely featured tiered seating built into the slopes of a hill in a semi-circular shape and curving toward a large stage building; stone retaining walls may have supported the sides of the structure. The Ennaeans probably entered the theater through two narrow entryways near the stage, taking their places in the large seating area (*cavea*) while city officials stayed down near the orchestra to give orations or call on speakers. On this day, Pinarius came before the assembled people in the theater wearing his civilian garb—Livy mentions his toga—and again explained that he lacked the authority to remove the garrison without permission. As he spun out the debate, Roman soldiers took positions above the building, possibly on the hillside, and in the exits and roads around it.

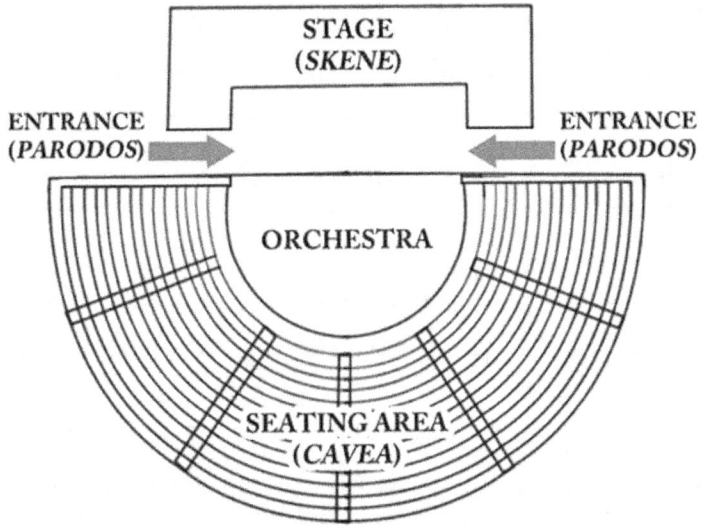

Figure 5.1. Top: Plan of a Hellenistic theater based on surviving examples at Morgantina and Segesta. Bottom: Nineteenth-century illustration "The Theatre at Egesta." (Top: Illustration by author. Bottom: Ward, Lock & Co. 1881–1884.)

Meanwhile the Ennaean assembly became increasingly agitated with Pinarius, and the whole crowd began demanding that he hand over the gate keys. None of our sources say how many angry faces Pinarius faced in the theater that day or even how many people lived in Enna. The geographer Strabo simply says that the city had a "few" inhabitants, but he was writing two centuries later and after Enna had suffered several disasters. For the sake of comparison, scholars estimate between 6,000 and 10,000 inhabitants lived in the nearby town of Morgantina, and medieval Enna had around 10,000 inhabitants, according to thirteenth-century tax rolls. We can cautiously assume something similar for the population of Enna in the late third century, with possibly one or two thousand adult male citizens shouting down at Pinarius from their seats. Livy claims that tensions rose to the point that the Ennaeans were threatening violence. Another source, the ancient military writer Polyaenus, says that the Ennaeans even voted "to revolt," perhaps meaning that they voted to remove the Roman garrison. Whether or not these details are true, eventually Pinarius gave a signal to his troops. They were prepared. Rushing down from above, they announced themselves to the assembled citizens with a shower of missiles. Many must have found their mark, drawing cries of anguish as they lanced into the crowded mass. Any Ennaeans who tried to flee through the exits ran into men with unsheathed swords, who mercilessly slew the unarmed crowd. Livy paints a scene of terrible chaos as the soldiers closed in around the mass of Ennaeans:

> The men of Enna, shut up in the cavea, were slain and piled together not only owing to the slaughter, but also by the panic, since they rushed down over each others' heads, and as the unharmed fell upon the wounded, the living upon the dead, they were lying in heaps. (Livy 24.39.5; trans. Moore, modified)

After slaughtering the political assembly, Livy says, the Romans moved off "like in a captured city" and killed many of the other inhabitants. When the proverbial smoke cleared, most of the Ennaeans were dead or had fled, and the strategic city was secured for Rome.[31]

Marcellus soon learned of Pinarius's bloody deed—and condoned it, believing that "the Sicilians would be deterred by fear from betraying their garrisons." A surviving inscription indicates that he took plunder from the city ("Marcus Claudius son of Marcus, consul, took this from Enna"), and he may have let Roman soldiers loot it for good measure. But Enna was famous throughout Sicily for its formidable defenses and its connection to Demeter and Persephone, and many saw the killings as a terrible atrocity. If Pinarius had successfully prevented Enna from going over to Carthage, the massacre only pushed more communities into the Carthaginian camp.[32]

In the meantime, Marcellus pressed the siege of Syracuse throughout the winter. Early in 212, he received intelligence that the city would soon celebrate a festival to the goddess Artemis, during which large quantities of wine would be given to a population that was short on food. Marcellus calculated that many Syracusans would be inebriated as they drank copiously on empty stomachs, and he prepared to take full advantage. After night had fallen over a day of celebration in the city, Marcellus sent about a thousand men—courageous individuals whom he and his officers had carefully chosen—to a stretch of wall on the city's north side, near the Hexapylon gate (see figure 5.2). These soldiers quietly scaled the wall, killed sleeping and drunken sentries in two nearby towers, then covertly moved to the gate and opened it. While Marcellus moved his other forces up to the open gate and undefended section of wall, the Roman infiltrators deliberately announced themselves with a trumpet blast. The blaring signal sheared through the night air and sent a jolt of terror through the inhabitants and defenders nearby. Many Syracusan guards assumed that the whole city had been lost and "some fled along the wall, others leapt off of it, while others were pushed off by the terrified crowd." By dawn most of the other legionaries had entered through

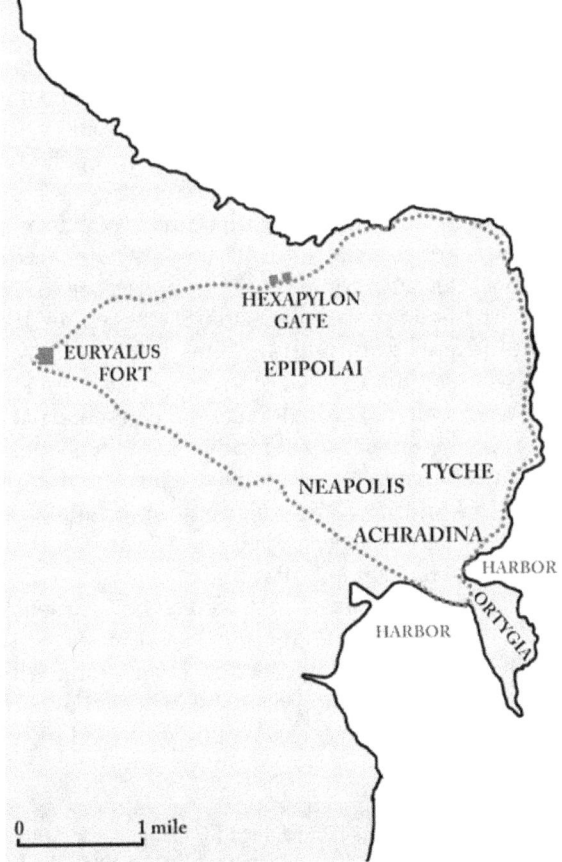

Figure 5.2. Plan of ancient Syracuse. (Illustration by author, after maps in Goldsworthy 2000 and Lazenby 1978.)

the Hexapylon gate, and the Roman invaders quickly captured a large area of the city called Epipolai.* But the siege was far from finished. Epicydes, one of the two brothers who led the defenders, still held the fortified Achradina quarter, the harbor, and Ortygia on the southeastern side; another Syracusan commander held the fortress Euryalus on the northwestern end.[33]

Marcellus failed to persuade the defenders of Euryalus to surrender, so he pitched camp between two quarters of the city called Neapolis and Tyche and let some of his soldiers plunder these neighborhoods. Livy insists that Marcellus forbade bloodshed and only permitted his men to plunder the property. And perhaps it happened that way, since he would have wanted to maintain discipline while much of the city remained under enemy control. Nonetheless, it is difficult to see how Marcellus would have restrained his soldiers once they had scattered off to plunder. Whichever way his troops behaved, the famously wealthy city yielded massive quantities of loot, much of which would later adorn the city of Rome.[34]

For the next several months, the siege went better for the Romans. The defenders at Euryalus soon surrendered their fort, perhaps feeling hopeless as they watched Roman soldiers shredding their way through Neapolis and Tyche. The Syracusans still had firm holdouts in Achradina and Ortygia, but Marcellus just dug in to continue the blockade. Additionally, his army fought off relief attempts from land forces outside the city, repelled counterattacks from within, and received aid in the form of a pestilence that autumn. The sickness afflicted both the Roman and Carthaginian-Syracusan camps outside the walls, but hit the latter especially hard. Many died, including the generals Himilco and Hippocrates, and some soldiers deserted as the terrible disease spread. (One modern historian speculates that it was typhus.) A Carthaginian fleet arrived at the eleventh hour with supplies, but its admiral lost his nerve and sailed away when Roman vessels moved to intercept. After that, Epicydes abandoned the city, too. Before long a Spanish mercenary commander, who was fighting for the Syracusans, decided to switch sides. He opened a gate for the Romans at Ortygia, and the last defenders there were killed or ran. The remaining holdouts in Achradina sent representatives offering their surrender, asking only assurances for their lives. Marcellus berated them for their betrayal but accepted their submission.[35]

Marcellus gave terms to Syracuse that were relatively lenient, given that the siege had been ferociously difficult and lasted months (it was now late 212). He left the Syracusans their liberty and laws, along with the territory around the city. But this moderation must be balanced against the fact that he allowed his men to loot the place. Marcellus protected the houses of Syra-

---

* From the elevated plateau of Epipolai, Marcellus was able to look down over the rest of the city. Our sources claim that he wept as he imagined his soldiers tearing the place apart. One may question whether the commander who approved of the massacre at Enna and carried out the destruction of Megara felt this swell of emotion. Still, he probably understood the significance of this moment, as his forces infiltrated a city that long seemed invincible to invaders. For Marcellus's tears, see Livy 25.24.11–14; Plut. *Marc.* 19.1–2.

cusans who had stayed loyal to Rome, but otherwise the sack was as violent as any other; Livy says there were "many shameful examples of anger and greed," and he and other authors particularly fixate on the death of Archimedes. The elderly mathematician, whose clever siege machines had so frustrated the Roman attackers, was callously cut down by a rampaging soldier while he drew mathematical figures in the dust (figure 5.3). And what the Syracusans did not suffer at the hands of legionaries, they suffered through the impoverishment of war, siege, and plundering. Diodorus claims that some of the inhabitants were unable to procure food after the city fell, so they sold themselves as slaves in hopes that their new masters would feed them.[36]

Marcellus was unrepentant. Two years later, after he had returned to Rome, an embassy of Sicilian Greeks visited the Senate to protest the treatment of Syracuse. Marcellus responded that "the question is not what I should have done—for the norms of war defend anything I did to my enemies—but what they *deserve to have suffered*" (my emphasis).* The city had

---

* Livy and Plutarch both suggest that the complaints against Marcellus were motivated by politics rather than moral considerations. Plutarch says that Marcellus's political rivals encour-

Figure 5.3. Nineteenth-century illustration "The Death of Archimedes." (Ward, Lock & Co. 1881–1884.)

defended itself vigorously in a long, costly siege. While Plutarch and Livy aver that the Roman troops could not be restrained (a plausible claim, since the Romans must have experienced grievous hardship during the siege), we need not doubt that Marcellus willingly channeled their rage and greed in order to punish the city's recalcitrance. The Senate obviously accepted his actions and his logic, because Marcellus was not penalized or censured. In fact, he received an ovation (a lesser version of a triumph) in recognition of his achievements in Sicily.[37]

Although the loss of Syracuse dealt a major blow to Carthage, the war in Sicily continued. The Carthaginians still had an army on the island and held several towns, and they were confident enough to offer battle to Marcellus (though they lost) shortly before he returned to Rome in early 211. Carthaginian reinforcements also landed that year, pulling more Sicilian communities over to their side or pulling back communities that had already rejoined Rome. Meanwhile, Numidian horsemen ranged around the island and ravaged the territory of Rome's allies. Regardless of these new difficulties, a Roman commander named Marcus Cornelius Cethegus was able to recapture the strategic town Morgantina. Archaeological evidence suggests that he violently destroyed it: Morgantina's sanctuaries ceased to operate, and several dwellings in the city show signs of damage, abandonment, or both; human remains were also found in some structures, as were coin hordes that were never collected; finally, the inhabited area of the city shrank enormously after 211, implying a sudden and massive loss of population to perhaps half or a third of the pre-211 size. At the Senate's instructions, Cethegus gave Morgantina's territory to the mercenaries who had helped capture Syracuse, suggesting that the original community was killed or dispersed. By devastating the city and repopulating it with allies, Cethegus retaliated against their betrayal a few years prior. Moreover, he ensured that enemy forces would not possess this important stronghold.[38]

Next, the consul Marcus Valerius Laevinus arrived in Sicily in 210 and took charge of the effort there. By then, Agrigentum was the last major enemy stronghold on the island. And Laevinus was fortunate when he approached the city because there was rivalry between leaders in the Carthaginian camp. Soldiers from one of the squabbling groups decided to open Agrigentum's gates for the Romans, who rapidly captured the city, killing many Sicilians and Carthaginians in the process. The Punic commander managed to escape with a few others and abandoned Sicily to Rome.

Laevinus had Agrigentum's leaders scourged and beheaded and sold the rest of the population into slavery. This exercise in mass violence was surely punitive, as the city's defection had provided the Carthaginians with an essential base to challenge Roman positions in Sicily. Moreover, this display terrified other communities, and forty Sicilian cities surrendered without a

---

aged the Syracusans to come forward, and Livy mentions the hostility of other senators. See Plut. *Marc.* 23.1; Livy 26.29.5, 26.32.5.

fight when they learned of Agrigentum's fate. Laevinus then mopped up the rest of the island and captured twenty-six more towns. Livy states that the consul "rewarded and punished" the leading men of each community as each "deserved," implying that other leaders were executed. Finally, he encouraged the Sicilians to return to their fields and begin the long process of recovery, and with that the Romans secured Sicily for the rest of the war.[39]

When faced with defection and rebellion, Roman commanders in Sicily had largely used the same tactics as their colleagues in Italy, using mass violence and terror as tools to reestablish Roman control, encourage surrender, and discourage further drift toward Carthage. Livy is explicit that Marcellus destroyed Megara to terrorize other communities into surrender, and that he condoned the killings at Enna because he thought it would deter other betrayals. Similarly, Pinarius's massacre at Enna was preventative, since it kept the strategically important town out of enemy hands.

As in Italy, this approach could potentially blow up in the Romans' faces, eliciting horror and disgust that drove more communities into the arms of the enemy. Mass violence was especially counterproductive in Sicily before 212, when the Roman position looked most precarious. When Marcellus destroyed Megara, Syracuse was holding its own, and Carthaginian help was on the way. Likewise, when Pinarius massacred the Ennaeans, Syrcause still held tight, a large Carthaginian army had arrived, and many Sicilian towns had changed sides. In both cases, Roman acts of mass violence were followed by more defections to Carthage. On the other hand, by 210, Roman victory looked increasingly likely and Roman threats looked more credible, since Carthaginian forces were gone and Syracuse had fallen. At that point, the mass executions and mass enslavement at Agrigentum quickly frightened many Sicilian cities into surrender. Violence and terror were effective—when Rome was winning.

"HANNIBAL ACCOMPLISHED NOTHING FURTHER": ROME VICTORIOUS IN ITALY AND SPAIN

The fall of Capua and Syracuse signaled that the fortunes of war were shifting. The Senate pressed its advantage by consistently keeping over ten legions in action, fielding half or more of these in Italy to hem in Hannibal and limit his movements. At the same time, a few Roman commanders became much more brazen in their approach to Hannibal and more willing to confront him directly. Marcellus, the conqueror of Syracuse, especially tried to make life hell for the invader by perpetually stalking the enemy army, fighting numerous skirmishes and stalemate battles to keep them scrambling. Roman armies also continued to retaliate against rebels. Most notably, Fabius Maximus nabbed Hannibal's other great prize, Tarentum, in 209, putting many of the inhabitants to the sword and selling 30,000 survivors. The cumulative effect of all this Roman pressure was that Hannibal was slowly bled of allies and soldiers in Italy.[40]

Even so, Hannibal still had friends in the south, and he could still win victories. He annihilated Roman armies near the town of Herdonea in 212 and then again in 210. He also ambushed Marcellus and his co-consul T. Quinctius Crispinus when they were reconnoitering; Marcellus was killed outright, while Quinctius later died of his wounds. Meanwhile in the East, Philip V of Macedonia remained a potential threat. Although the Senate sent forces to keep him in check, they would not resolve that struggle until 205 (see chapter 6). Additionally, Rome encountered new complications with its still-loyal allies. In 209, almost half of the Latin colonies declared that they were too exhausted to send more troops—at a time when the Senate was in no position to press them—and there were rumblings of rebellious sentiment in Etruria. The dispatch of Roman soldiers kept the Etruscans in line, but the threat of further rebellion remained potent.[41]

And through it all, Hannibal could still expect reinforcements from Spain or North Africa, which might strike at Roman morale, bolster the confidence of his Italian collaborators, and swing the momentum back in his direction. As it happened, forces led by his two brothers eventually arrived in Italy. In 208 and 207, Hasdrubal Barca marched soldiers from Spain into Italy, and in 205 Mago Barca landed seaborne troops in the northwest of Italy. Hasdrubal's triumphant arrival was quickly undone when two Roman armies, led by the consuls Gaius Claudius Nero and Marcus Livius Salinator, converged on him in Picenum (northeast Italy). The Romans wiped out his forces near the river Metaurus, and afterward threw his severed head into Hannibal's camp. Mago fared better at first, raising recruits among the Gauls and Ligurians and defeating a Roman army. Even so, these early successes were soon reversed. In 203 Mago was wounded in battle and routed, and he died while sailing back to Carthage.[42]

"Hannibal accomplished nothing more in Italy" (as Livy bluntly puts it). He was effectively cornered in the far south, the last place where he had any significant supporters. If Hasdrubal or Mago had been able to muster more forces, or if they had left a strong Punic Spain in their rear, Hannibal might have held out. Unfortunately for him, Carthaginian power in Spain was crumbling even before Hasdrubal Barca departed for Italy.[43]

The Romans were active in Spain since the beginning of the Second Punic War, and we must return briefly to the year 218 to catch up on Spanish affairs. When war was first declared, the Senate had no inkling of Hannibal's planned invasion; they expected to fight the war in Spain and North Africa and began to marshal their forces accordingly. So just as Hannibal's army was steadily marching from Spain to Italy, one of the consuls for 218, Publius Cornelius Scipio, was sailing to Spain with his legions. The two armies nearly crossed paths in southern Gaul, and Scipio tried to intercept the Carthaginian force then and there. When he failed, he sent his army ahead to Spain under the command of his brother, Gnaeus Scipio, while he rushed back to northern Italy to gather local troops and meet Hannibal's invasion there. In Spain, Gnaeus Scipio campaigned successfully against the Carthaginians, and later in 217 he was rejoined by his brother Publius. The two achieved notable successes for the next several years, defeating enemy armies and allying with local tribes. In 211, however, they launched a doomed offensive into Carthaginian territory. Dividing their forces and marching separately to attack different targets, the brothers were outmaneuvered, overwhelmed, and killed along with many of their troops. Although some survivors made it to safety, and the Senate sent reinforcements in 210, most of the Scipios' gains in Spain were lost. The Romans were reduced to a small Spanish enclave north of the Ebro River (see map 5.1).[44]

Soon after the deaths of the Scipios, an unlikely Roman general would radically revise this situation. He was the son of Publius Cornelius Scipio and shared his name, but he is better known to history as Scipio Africanus.* We already met the younger Scipio after the Battle of Cannae, when he supposedly prevented his fellow Romans from abandoning the Republic. In 210, even though he was only twenty-six years old and had never held a military command, one of Rome's citizen assemblies voted him a special command in Spain as proconsul. It is not clear why the voters made this unusual choice, though surely they acted with senatorial approval. Possibly they knew that the tribes in Spain respected personal bonds of loyalty, and they wanted to capitalize on the relationships and alliances made by Scipio's father and uncle. Or perhaps the Spanish command was simply unpopular, or there was a shortage of senior commanders available to take the post. Whatever the reasoning, the young Scipio turned out to be an extraordinarily gifted commander and quickly took charge of the situation.[45]

Late in 210, Scipio arrived with reinforcements at the city Tarraco, in northeastern Spain, where he spent the winter investigating local conditions and planning. Since his first campaign in Spain culminated in a well-known episode of mass killing, it will be helpful to examine Scipio's reconnaissance, planning, and choices with very close scrutiny.† In his investigations that winter, Scipio discovered that there were three Carthaginian armies in Spain, which together outnumbered his own forces. Fortunately, they were widely scattered and were not watching his position. At least one of them was busy fighting local tribes, and perhaps the others did not see the young Roman general as a threat. Scipio decided that it would be unwise to engage

---

* "Africanus" was an honorific that Scipio received after the war.
† We are fortunate to have insight into Scipio's thinking during his first Spanish campaign, chiefly because Polybius had exceptionally good sources and provides a detailed account. Polybius claims that he read a letter written by Scipio that described the capture of New Carthage (his main achievement in the campaign of 209). Polybius also personally observed the topography and fortifications at New Carthage, and likely spoke with Laelius, Scipio's second-in-command who led the Roman fleet during the attack. Polybius probably also referenced the contemporary texts of the Roman historian Fabius Pictor and Hannibal's historian Silenus. For Polybius's sources, see Polyb. 10.9.1–3, 10.11.4; Livy 26.49; Walbank 1967, 191–96, 204, 212; 1985a, 125–31.

any of the Punic armies separately, since they might converge and destroy him as they had his father and uncle. Instead, he focused his attention on the chief city of Carthaginian Spain, New Carthage.* He learned that the city was the main supply station and communication hub for Punic forces in Spain; its capture would deprive Carthage of a secure base for its army and fleet, cut off the most convenient point of provision and reinforcement, and ultimately threaten Carthaginian control of the whole Spanish coast. As the best harbor in the region, the city could also provide a new point of contact with Rome and support further incursions into Punic territory. Additionally, he discovered that the Carthaginians held many Iberian hostages at New Carthage, and used these to maintain the loyalty of local communities. If the hostages fell into Roman hands, it might undermine the Iberians' allegiance to Carthage. Finally, he found out that the city had a largely civilian population, and since the Carthaginians did not expect an attack so far south, it only possessed a small garrison of a thousand regular soldiers. In short, New Carthage was utterly vital to the enemy, it was vulnerable, and its capture would be a coup.

Regardless of New Carthage's tempting vulnerability, Scipio would have to cross over 250 miles to reach the city, driving deep into enemy territory, and he would have to strike the city swiftly. He could not get bogged down in a long siege, as any delay would give Carthaginian armies time to relieve the city and perhaps an opportunity to trap and destroy his forces. He had a very small window of opportunity.[46]

In early spring of 209, Scipio seized the moment, marching his army south while his second-in-command, Gaius Laelius, shadowed it along the coast with a fleet of thirty-five vessels; the presence of the fleet would allow them to beat a hasty retreat if necessary, underlining the riskiness of Scipio's plan. Their forces totaled about 25,000 foot soldiers and 2,500 cavalry. After arriving and encamping to the east, the Romans confronted a city that was virtually unassailable due to its formidable natural and artificial defenses. New Carthage was nearly surrounded by water, with a lagoon to the north, a sea harbor to the south, and a water channel to the west, and it was only approachable by a narrow isthmus on the eastern side (figure 5.4). Additionally, the city was nestled into a group of five hills. Livy calls one of these the "citadel of Hasdrubal" (*arx Hasdrubalis*, where a fortress was indeed located), while Polybius names the rest of the hills as "Asclepius," "Hephaestus," "Aletes," and "Kronos." (These five eminences seem to correspond to the modern hills Mt. Molinete, Mt. Concepcion, Castillo de Despeñaperros, Mt. San José, and Mt. Sacro, respectively.) Together the hills formed a protective perimeter around the city, and any assailants who entered New Carthage would find themselves surrounded by steep terrain on every side.

---

* New Carthage is modern Cartagena, Spain. The founder of the city was Hamilcar Barca's son-in-law Hasdrubal, who actually just called it *Qart Hadasht*—Carthage—but the Romans took to calling it *Carthago Nova*, "New Carthage."

In addition to the topography, Polybius says that New Carthage had tall defensive walls topped with battlements that extended for twenty stades (about 12,100 feet). Traces of the well-made Carthaginian defenses have been found in the area of the isthmus (figure 5.5), revealing double-facing "casemate" fortification with internal compartments for storing weapons and other equipment. As noted, the city also had a citadel on one of its hills, where traces of similar fortifications have been found. The hill of Asclepius (modern Mt. Concepcion) offered good natural defense as well, since it was

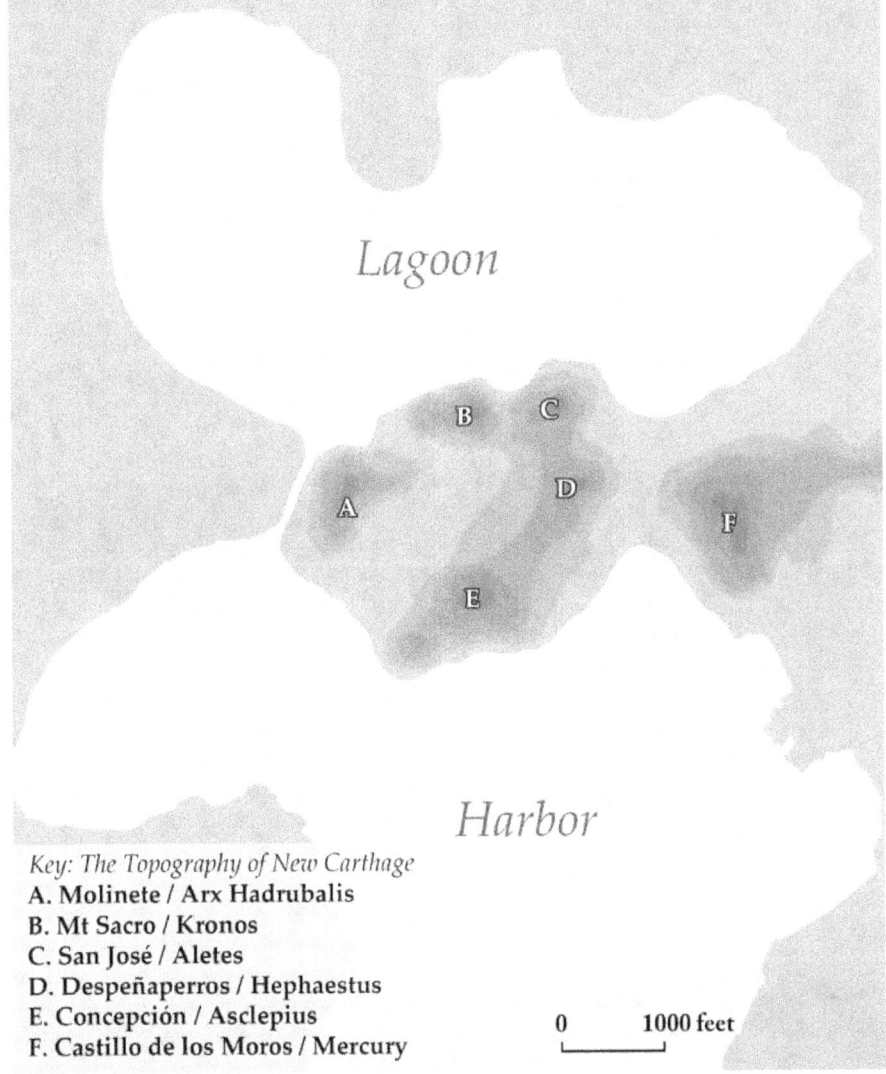

Figure 5.4. Plan of New Carthage. (Illustration by author, after maps by Julio Mas in Ruiz Valderas and Balanza 2002, and Lazenby 1978.)

the highest point in the city, with a modern elevation of about 160 feet. It may have hosted fortifications, too: Livy says that there was a "fortress [*castellum*] sited on a high hill" *and* a "strongly-fortified citadel [*arx*]" in New Carthage.[47]

The Carthaginian garrison commander, Mago (*not* Mago Barca, but a different man with the same name), utilized the defenses, topography, and population to maximize his small force. When Scipio's army arrived outside the city, Mago stationed half of his regulars on the hill of Asclepius, where they could repel seaward attacks, and put the other half on the citadel. The men in either spot would have had good views of the

Figure 5.5. Ruins of the Punic fortifications in Cartagena, Spain. (Photos by author.)

city and could respond rapidly if the outer defenses were breached.* They could also dig in and maintain a foothold if the Romans broke through the walls, and relief armies could come in a matter of days. Finally, he armed 2,000 townspeople and stationed them near the isthmus, telling the rest of the population to defend the walls.[48]

As Scipio's soldiers settled into their encampment, he called his troops to an assembly and laid out his plan. After arguing that it was possible to capture the city, he promised gold crowns to the first man over the wall as well as other rewards for conspicuous courage. Then, to fire their spirits further, he explained that the god Neptune had given him the battle plan in a dream and would provide divine aid at just the right time. This sort of personal myth making was not new for Scipio. According to one story, he used to go to the Capitoline Temple in Rome by night and commune with Jupiter before making any major decisions, and he developed a reputation for divine inspiration and good fortune. Whether this behavior was borne of cynical calculation or genuine belief, at New Carthage it surely helped inspire his men.[49]

The attack began the next morning. Scipio first sent Laelius with the fleet to encircle the city by sea, then drew up 2,000 of his best men on the isthmus with ladders. Mago responded by sending the 2,000 armed townspeople against them, hoping to throw the attackers off balance and hinder their assault. The townspeople dashed out of the gates and resolutely ran into Scipio's troops. A fierce struggle ensued, but the Romans were able to bring more reinforcements into the fight, and the Carthaginians were thrown back through the gates. The defenders on the wall began to panic at this early defeat, and some fled from their posts.

Scipio directed the unfolding action from a nearby hill that Livy calls the "Hill of Mercury," probably the modern Castillo de los Moros, from which the city and its five hills would have been clearly visible (as they are today). Scipio could plainly see that the defenders were abandoning the walls, and he ordered a general assault. His troops rushed forward, following close behind the fleeing Carthaginians and placing their scaling ladders on the walls. Due to the great height of the fortifications, though, some of the Roman ladders broke under the weight of the escalading soldiers. Other Roman attackers became disoriented as they climbed to the top of the tall walls, and they were easily thrown off when they met the slightest resistance. The Carthaginian defenders began to rally and lobbed wooden beams or other debris down on the Romans, sweeping ladders clean of men and hurling them to the earth below. Unable to break through and suffering casualties, Scipio sounded the retreat.

---

* The distance from the modern Mt. Concepcion to the area of the ancient isthmus is about 2,000 feet, and it is about 2,600 feet from the citadel (Mt. Molinete) to the ancient isthmus (according to my on-site measurements using the GPS Essentials application for Android). Walking at a leisurely speed of about 3.5 mph, I crossed to the area of the ancient isthmus from the foot of Concepcion in seven minutes, and from the foot of the Molinete in eleven minutes.

The fighting had consumed the better part of the day, and it was approaching evening. If the Carthaginians felt any sense of triumph for repelling the first Roman attack, their elation did not last. Scipio renewed the assault, sending forward more men and placing ladders in more places. Some soldiers, overlapping their shields above their heads in the famous *testudo** formation, reached the gates and began to hack at them with axes. The Roman fleet also assailed the walls on the seaward side of the city. The outnumbered defenders kept up their stubborn resistance, but their attention was fully occupied by the reinvigorated attack. In the middle of this confusion, Scipio played his ace. The previous winter, he learned from local fisherman that the lagoon to the north of the city was relatively shallow and fordable in places; moreover, a natural phenomenon lowered the water level further around evening.† While his other soldiers threw themselves against the walls, Scipio sent a sent a flanking party of five hundred legionaries through the lagoon, and they were able to wade across and mount the north wall without opposition. These infiltrators moved along the walls toward the isthmus, using their swords to cut down enemies in their path. They quickly reached the gates and opened them for their compatriots just as the ladder-bearing troops swarmed over the battlements. The defenders on the wall must have been shocked as the flood of invaders broke the levy of their fortifications. They were rapidly overpowered by attacks from front and rear, and survivors ran back into the city streets.[50]

Scipio's army had forced its way in, but it still had not secured New Carthage. As they entered the city, they would have found themselves on unfamiliar terrain, surrounded by steeply rising hills and ever-darkening streets as the sun sank toward the horizon. To make matters worse, many possible threats waited within: there were Carthaginian regulars holding the citadel as well as the southern hill (Mt. Concepcion), in addition to the inhabitants armed by Mago and the others who fought on the walls. These defenders had given a good account of themselves, resisting determinedly and, once, even throwing the Romans back. Some of the locals had died in the fighting, of course, but others escaped back into the town. From the perspective of Scipio and his officers, this meant that there were armed individuals mingling with the rest of the population, whose whereabouts they could only guess. They beheld a city brimming with risk, and Livy is probably right that the Romans entered the city "in battle order with their officers," maintaining protective cohesion as they passed inside.[51]

---

* When they formed the *testudo* ("tortoise"), groups of Roman soldiers locked their shields together over their heads, creating the appearance of a tortoise shell or a tiled roof.

† Polybius says that a "tide" changed the water level, but there is no significant tidal action in this part of the Mediterranean. On the other hand, Livy (26.45.6–8) claims that wind could change the water level. And, lo and behold, wind changes the water level in the harbor of modern Cartagena by as much as a meter and a half (United States National Geospatial-Intelligence Agency 2014, 43). Polybius probably mistook something like this for a tide. In any event, the lagoon was relatively shallow to begin with, and due to some natural phenomenon it was even shallower on this occasion.

Scipio's entire strategy required that he capture New Carthage quickly. If his army got hung up besieging the citadel or fighting house to house, Carthaginian relief forces would arrive before he had subdued the city. His entire plan reflected this as, apparently, he did not contemplate a siege, and he had the fleet ready for a quick retreat. Thus, once he and his soldiers got inside the walls, they acted quickly. One group of Romans attacked the southern hill (Asclepius/Mt. Concepcion) and dislodged the defenders. Scipio led another group to attack the citadel. But before he moved against the citadel, "when [he] thought that a sufficient number of troops had entered he sent most of them . . . against the inhabitants of the city with orders to kill all they encountered and to spare no one, and not to start pillaging until the signal was given."

Polybius interpreted this massacre as calculated violence designed to terrorize the population ("they do this, I think, to inspire terror"), with bloodthirsty soldiers killing beyond what was militarily necessary for maximum shock ("when towns are taken by the Romans one may often see not only the corpses of human beings, but dogs cut in half, and the dismembered limbs of other animals, and on this occasion such scenes were very many owing to the numbers of those in the place"). And as Scipio and his troops pushed into New Carthage, this massive display of violence certainly cowed anyone who was still willing to fight, vitally quickening the pace of surrender when Scipio's whole plan hinged on rapidity. In the longer term, moreover, this sharp brutality might persuade neighboring Spanish communities to yield at once.[52]

Modern scholars have more or less followed Polybius here, contending that the killing at New Carthage was meant to terrorize or shock the populace. Yet "terrorism" is not the only way to interpret Scipio's orders, and it may be an oversimplification. Scipio marshaled his troops carefully and deliberately after breaching the walls, sending them to respond to specific threats and perhaps following a prearranged plan. Some Romans attacked and captured the southern hill, clearing it of enemies. Scipio personally led another group against the citadel, where the enemy leader held out with five hundred soldiers. It follows that the other Romans, sent to kill anyone they met, were also responding to a specific danger in the tense moments after the breach, when everything hung in the balance. Accordingly, the order to "kill all they encountered" is best regarded as a large-scale mop-up operation—an effort to seek out and destroy any remaining threats, actual or potential, throughout the still-dangerous city. After all, the inhabitants were not simply bystanders killed for maximum shock, but proven enemies who remained threatening as long as they might recover or counterattack.[53]

When Scipio marched against the citadel, Mago resisted for a time, but soon saw that the city was lost, and surrendered. At that juncture, satisfied that New Carthage was secure, Scipio passed orders to stop the killing—further indicating that the violence was a pointed and purposeful response to a specific situation—and allowed his troops to loot the town (though additional violence may have been committed by his plundering soldiers).[54]

New Carthage had fallen. On the following day, Scipio's men gathered together "little less than ten thousand prisoners" and sorted them, separating the citizens and their families from about 2,000 people that Polybius calls the "workmen"—*cheirotechnai*—who were possibly poorer members of the community, lacking citizenship. The citizens were spared and sent back to their homes while the workmen were made public slaves of Rome, with the promise of future freedom if they performed well. Other young- and strong-looking prisoners were incorporated into his naval crews, probably as oarsmen. Scipio showed particular generosity to three hundred Spanish hostages found in the city, giving them gifts and promising to protect their young women from assaults by his soldiers. He also asked the hostages to send messages to their relatives, to inform them "first that they were safe and well, and second that the Romans were willing to restore them all to their homes unharmed"—if they allied with Rome, that is.[55]

The bold strategic moves, exemplary violence, and ostentatious mercy at New Carthage would come to define Scipio's conduct in Spain. For the next three years, he launched yearly offenses into Carthaginian-held territory, capturing important towns and routing enemy armies. During these campaigns he showed himself every bit the tactical master that Hannibal was, executing complex maneuvers with a well-trained army. He went on the offensive in 208, defeating Hadrubal Barca at the Battle of Baecula. Hasdrubal escaped to make his ill-fated expedition to Italy, but Scipio was back at it in 207, striking into southern Spain. Then in 206, the Carthaginians made a stand near a town called Ilipa, where Scipio shattered their forces. Most of the Carthaginian leaders made their escape to North Africa, and the last Punic stronghold at Gades surrendered later that year.[56]

These battlefield victories were only one element of Scipio's larger strategy. Reminiscent of Hannibal in Italy, he tried to peel away Carthage's Spanish subjects using diplomacy, generosity, and force. And he was often successful. For example, Scipio's merciful treatment of the survivors at New Carthage ensured their good behavior for the rest of the war—and when the Carthaginians tried to recapture the city a few years later, the inhabitants offered no help to their former masters. Scipio's kind treatment of the Spanish hostages bore fruit, too. Iberian communities started going over to Rome shortly after the conquest of New Carthage, and they continued to join him in subsequent years following his battlefield victories. These new allies sent warriors to fight alongside the Romans, allowing Scipio to achieve greater parity with the enemy and causing great consternation among Carthaginian leaders.[57]

Yet Scipio's use of leniency and diplomacy should be balanced against his use of ruthless force, which he used to shock stubborn foes into compliance. When he lunged deep into Punic-held territory in 207, the Carthaginian commander resolved to avoid battle and instead scattered his troops in various towns. Scipio realized that it would be a long slog to capture each city individually, so he sent his brother Lucius Scipio to make an exemplary raid on a single community, Orongis. Some of the town's defenders feared that the Romans would break inside and massacre them all indiscriminately (perhaps they had heard about New Carthage), so they decided to surrender. Exiting the city from one of the gates, they raised their right hands to show that they were unarmed but still held their shields on the left. The Romans either misunderstood the gesture or did not care, and killed the lot of them. Then they marched into the open gate, captured the marketplace, and scattered throughout the town to kill any remaining defenders. In the end, they captured the Carthaginian garrison and about three hundred leaders but spared the other survivors. Having demonstrated that resistance was futile, Lucius rejoined Scipio's main force before the whole army went into winter quarters.[58]

After driving Punic armies out of Spain, Scipio also consolidated his position by attacking and violently subduing defiant holdouts. Thus in 206, he marched against Ilurgia (or Iliturgi) and Castulo, two Iberian towns that supposedly betrayed his father and uncle earlier in the war. Though the story of their treachery has been doubted, certainly these places were among the last centers of resistance in Spain. Scipio reached Ilurgia first and found that the town was prepared for a fight. The Romans went ahead and launched an attack, but the Ilurgians fought vigorously and flung back several assaults. Scipio personally advanced to the walls to direct the action up close, but was wounded in the neck in the midst of the fighting. Incensed by the harm done to their general, the legionaries renewed their attack and soon burst inside the city. They slaughtered the inhabitants indiscriminately—Livy says they spared neither women nor children—and destroyed the place, setting fire to some structures and demolishing those that would not burn.

For the ancient authors Livy and Appian, the uncontrollable fury of Scipio's soldiers led to the destruction of Ilurgia. Livy goes so far as to claim that the massacre was driven by "anger and hatred," and that the legionaries sought "to extinguish every vestige of the city and to wipe out the memory of their enemies." The town's alleged treachery, along with Scipio's wound, may indeed have enraged the soldiers. On the other hand, Livy and Appian both enliven this event with high drama, stressing the unrestrained havoc of vengeful legionaries and suspiciously deflecting responsibility away from their hero, Scipio. Though the rank-and-file Romans may well have been angry, we should not doubt that Scipio deliberately unleashed his troops against this city. Ilurgia was one of the last holdouts of Punic Spain, and it had closed its gates in defiance. Moreover, most of the inhabitants participated in the defense, including women and children who carried weapons and munitions or shored up the walls with stones. Such defiance would have provided Scipio with sufficient cause to punish their stubbornness and send a message to other Spanish communities; he likely encouraged, or at least condoned, the massacre of the population. And of course, the destruction of the urban fabric would have required organization and labor, and cannot be easily chalked up to angry soldiers acting on their own initiative.

In the end, if Scipio sought to make an example of Ilurgia, it apparently worked, since the inhabitants of nearby Castulo feared a similar fate and surrendered without struggle. They were treated mildly.[59]

The fight in Spain was not quite over. Before the year was out, Scipio fell ill, faced a mutiny of troops who were resentful about a lack of back pay, and had to deal with a revolt by some of his new Spanish allies. These difficulties were each handled in turn. Scipio recovered, the troops were placated (and the leaders of the mutiny executed), and the Iberians were brought back into line. Spain was hardly "conquered" in the sense that Rome unequivocally controlled it, but it was quiet for the time being and emptied of Carthaginian armies. The scene was set for the war's final clash.[60]

"AFRICA BURNS WITH WAR": THE ROMAN COUNTER-INVASION

Scipio sailed home in late 206 to great fanfare and stood for the consular elections. He had not held the normal sequence of magistracies leading up to the consulship, but his prestige was so great that it did not really matter. Elected consul for 205, Scipio announced in the Senate his intention to bring the war to Africa. He met some resistance from the elderly Fabius Maximus, who argued that Rome should direct its energies against Hannibal in Italy, but most senators eventually sided with Scipio. The young consul then spent most of the year training his army in Sicily, got his command extended by the Senate, and invaded Africa in 204.[61]

The Roman expeditionary force landed near Utica, a large coastal city that was subject to Carthage, and occupied the high ground nearby, with their fleet pulled up on the adjacent beach. For the remainder of 204 Scipio deflected probing attacks by the enemy and ravaged the countryside but fought no major actions. He also made a crucial alliance with Massinissa, a Numidian prince. The Numidians lived in North Africa to the west of Carthage and provided the Carthaginians with their excellent light cavalry. The Romans had played on Numidia's internal politics throughout the conflict, courting various factions in order to drive a wedge between the Numidians and Carthaginians. Now those efforts bore fruit, as Massinissa and a small band of his followers joined up with Scipio's invading legions. Although most Numidians stuck to their alliance with Carthage, Scipio's new friend brought crucial local knowledge along with additional manpower. The Romans would need the help because two enemy armies—one Carthaginian and one Numidian—marched up and established camps nearby as winter set in.[62]

In spring of 203, after protracted talks with the commanders in both camps got nowhere, Scipio launched an audacious raid against the hostile encampments. His men approached the enemy camps stealthily under cover of night, then set their wooden huts on fire. Caught unawares or asleep, many of the Carthaginian and Numidian occupants were burned to death. Others raced out of the bright blaze only to be hacked to bits by Roman swords waiting in the darkness. Any survivors fled.

Polybius, though praising the skillfulness of the Roman raid, remarks that nothing could exceed this attack in its "horribleness" (*deinotēs*)—and many nearby towns were thoroughly terrified. Afterward, the Roman army approached a city close by, which surrendered immediately without a fight. Scipio spared this community but allowed his soldiers to sack two other towns in the neighborhood. The reasoning behind this violent outburst is unclear. It is possible that Scipio sacked these places because they had resisted, and he wanted to punish their recalcitrance and warn others. He probably needed to reward his men, too, buoying their morale in the middle of a high-stakes invasion and avoiding another mutiny like the one he had faced in Spain. According to Appian, in fact, Scipio briefly lost control of his men in 204, during an attack on a Punic city called Locha, indicating that the young commander was having problems with discipline. It was essential that he keep his troops satisfied and under control, since a disordered army could put all his gains at risk, which may explain the city-sackings.[63]

In any case, the Carthaginian government had a little fight left, resolving to raise a new army and continue the struggle. Later that year, Scipio met another combined Carthaginian and Numidian force at a place called the Great Plains and eradicated it in battle, racking up another field victory. That same year, Scipio sent some troops to help Massinissa against his rivals in Numidia and warded off a Carthaginian naval attack on his coastal encampment. Carthage's resolve then melted away, and they opened peace talks. Scipio offered the following terms: the Carthaginians must evacuate Italy and all other territories outside Africa, hand over Roman prisoners and deserters, and pay an indemnity of 5,000 talents. Carthaginian leaders apparently accepted the terms, and both they and Scipio sent envoys to Rome to confirm the peace. The Roman Senate approved it, and the long war neared a conclusion at last.[64]

But then things got complicated. It appears that while the Carthaginians were negotiating with Scipio, they also recalled Hannibal and his army from Italy. These forces arrived in Africa in late 203. As Scipio waited to hear from Rome about the ratified peace, some Roman supply ships got blown off course and ended up in a bay close to Carthage itself. The Carthaginians sent out fifty warships to seize the vessels, grabbing most and towing them back home. Scipio dispatched delegates to complain about this violation of the armistice, but these men were met with hostility and accomplished nothing. As the ambassadors took their leave from Carthage, their vessel was set upon by three smaller ships and deluged with missiles, killing many onboard. In a relatively short time, the war in Africa quickly reignited.[65]

It is difficult to explain this breach of armistice. Perhaps the Carthaginian authorities could not pass up the opportunity offered by the helpless cargo ships. It is also possible that some Carthaginians wanted to renew the fight, since Hannibal and his veteran army had returned to Africa. Whatever the case, the armistice was off, and Scipio was furious. After all, he had been on the verge of ending a terrible war that had lasted over fifteen years, cost tens of thousands of lives, and pushed Rome to the brink of defeat. His own glory and reputation were on the line. Worse still, he might be recalled and replaced by another commander if the conflict continued. The Senate had extended Scipio's command in 204 and 203, but they could always change their minds, and some leading senators had been agitating to get the African command for themselves. With the politics of the moment pressing in on him, Scipio went back on the warpath, going

> round the towns, no longer receiving the submission of those which offered to surrender, but taking them all by assault and selling the inhabitants as slaves, to manifest the anger he felt against the enemy owing to the treacherous behaviour of the Carthaginians. (Polybius *Histories* 15.4.2; trans. Paton)

By reopening hostilities and brutally devastating Punic towns, he surely gave shape to his anger. Yet there must have been calculation here, too, given the trouble he took to lead these potentially dangerous assaults, not to mention the effort he made to round up and sell whole populations into slavery. Rather than giving purposeless expression to his rage, Scipio probably sought to pressure the Carthaginians back to the negotiating table or force them into another battlefield encounter—ensuring that *Scipio* reaped the magnificent glory of ending the war.[66]

He got what he wanted. In 202 Scipio's expeditionary force met Hannibal's army to the southwest of Carthage, possibly in the countryside near modern Sidi Youssef. There, Hannibal fought his last field encounter against the Romans, the Battle of Zama, in order to defend a homeland he had not seen in over thirty years. For that task, his surviving veterans were reinforced by new drafts of mercenaries, Carthaginian levies, and eighty war elephants. Hannibal opened the battle with a charge of his elephants. The sight of so many lumbering beasts stampeding across the field would have been fearsome, but the Romans showered them with missiles and drove off many. They also herded some of the creatures through lanes in their formation that Scipio had deliberately left open for this very purpose. Most of the elephants were quickly neutralized. Others ran back into the Carthaginian lines, throwing them into disorder. Scipio's cavalry, led by his new ally Massinissa on one flank and his lieutenant Laelius on the other, soon charged and successfully chased Hannibal's horsemen from the field. In the center, Scipio's and Hannibal's infantrymen engaged in a long, gruesome struggle, but eventually Massinissa and Laelius returned and attacked Hannibal's line in the rear. The Carthaginian army collapsed into a rout. Hannibal escaped and urged Carthage's government to sue for peace.[67]

Scipio held a war council to determine his next move, and some of his more wrathful advisers pressed him to attack and destroy Carthage completely. Roman politics intervened. According to Livy, Scipio realized that besieging and capturing such a large city would take a very long time. If he settled on retribution or sought to eliminate Carthage once and for all, then the Senate might send someone to replace him before he could finish the job, snatching away credit for the ultimate victory. With that in mind, he proposed peace terms that were harsher than his previous offer, but still arguably lenient given the difficulty and duration of the war. His terms dictated that the Carthaginians would lose their overseas empire, though they would maintain control of their domains in Africa; they would give up their elephants and most of their fleet, could no longer wage war outside of Africa, and could only fight inside Africa with Rome's permission; they would pay Rome an indemnity of 10,000 silver talents over fifty years, but keep their property and live under their own laws, without a Roman garrison or governor. The Carthaginian government accepted his proposal. Political considerations had prevented the destruction of Carthage for the moment, leaving that task to Scipio's grandson by adoption, Scipio Aemilianus, fifty-six years later.[68]

CONCLUSION

The Second Punic War saw the Republic brought within a razor's edge of defeat. Many Romans must have felt that they were living their most desperate hour during Hannibal's invasion and crushing victories. Their predicament was immeasurably worsened when their allies began to defect after the Battle of Cannae in 216. At that point, the Senate and its generals adopted a new strategy in Italy and Sicily. Rather than try to defeat Hannibal in battle, they mostly avoided him and deployed their still-staggering manpower everywhere else. Their aim was to undermine Carthaginian support, prevent further defections, and strike at rebel communities.

Mass violence was an essential component of the new strategy in Italy and Sicily. When Roman generals reconquered rebel communities, they often beheaded the leaders responsible for rebellion, and in some cases they sold all or part of the population. Roman commanders also destroyed some rebel towns, physically or symbolically, as when Marcellus demolished Megara in Sicily or when the Senate deprived Capua of its independent civic existence. These same tactics were used preemptively against communities suspected of traitorous designs, like Nola and Enna. Making dreadful examples of rebels (real or suspected) punished their betrayal, discouraged future defections, and encouraged other rebels to surrender.

At times this ruthless approach was self-defeating in both Italy and Sicily, prompting indignation and further rebellion—most notably when the Romans executed the hostages from Thurii and Tarentum, or when Pinarius killed the Ennaeans, and more communities just went over to the Carthaginians. Nonetheless, it is significant that these atrocities took place between 214 and 212, when Roman victory was far from inevitable. In Italy, Hannibal was still making gains and major defectors like Capua still held out. In Sicily, a large Carthaginian army had arrived to shore up resistance to Rome and Syracuse was handily fighting off Roman attacks. Brutal violence only became an effective tactic once the military momentum had swung in Rome's favor. For instance, the rest of the Campanian defectors rapidly yielded after Hannibal abandoned them, Capua fell, and the Romans executed Capuan leaders. Likewise, the recapture and harsh punishment of Agrigentum, the last major Carthaginian base in Sicily, prompted forty other Sicilian towns to surrender.

The dynamics of Roman violence were somewhat different in Spain, where Scipio was not dealing with defecting allies but operating in foreign territory. Sometimes he used mass violence to terrify the enemy, neutralize threats, and hurry along surrender, as with his indiscriminate massacre at New Carthage. Scipio's army also destroyed communities like Ilurgia when they refused to knuckle under, thereby demonstrating the price of continued resistance. On the other hand, Scipio apparently coupled ruthlessness with mercy more than his counterparts in Italy and Sicily, treating some prisoners mildly and generally sparing towns that yielded without a fight. Greater moderation was essential, since the Carthaginians controlled more manpower and resources in the Spanish peninsula when Scipio arrived in late 210. The young Roman commander needed to win over local tribes and raise rebellions if he was going to win.

In Africa, Scipio continued to use a combination of brutality and mercy, but political considerations may have loomed especially large in his calculus. After the armistice broke down in late 203/early 202, he sacked and enslaved several Punic towns, refusing to accept their surrender. He was angry that the Carthaginians had violated a truce, but also probably feared he would be replaced by another commander and deprived of the victory. His uncompromising attacks put pressure on the Carthaginians and expedited a conclusion to the war. And later, after his victory at Zama, he ultimately decided not to attack and destroy Carthage. Capturing the city would take too long, and again, someone might replace him in the African command before he could finish the job. The stakes of this final campaign were simply too high, and Scipio's political ambitions heavily influenced his use of mass violence.

We must remember that these displays of violence in Italy, Sicily, Spain, and Africa often required coordination, effort, and the organization of resources. Roman commanders had to decide which enemy leaders to punish, separate them from larger groups of prisoners, put them under guard, take them to public locations, and task soldiers with their execution. When the condemned prisoners numbered in the dozens or hundreds, these killings would take time and labor. Mass enslavement likewise required sorting, guarding, and selling off crowds of prisoners. Very large-scale killings, as at Enna, required significant advanced planning and coordination. None of this was easy or quick. However, Roman commanders made these deliberate choices throughout the war because they believed mass violence would resolve their military and political problems, advance Roman strategy, and achieve their goals.

NOTES

1. Cannae: see esp. Polyb. 3.113.1–3.118.10; Livy 22.45.5–22.52.3; Daly 2002; Hanson 1992. Five/six square miles: Goldsworthy 2000, 208. Army sizes: Hoyos 2015, 120; Goldsworthy 2000, 200; Daly 2002, 26–32; Lazenby 1978, 75–76, 79–81. Casualties: Polyb. 3.117.3 says 70,000 dead; Livy 22.49.13 says 45,500 dead; most modern scholars err toward the lower total, e.g., Goldsworthy 2000, 213; Hoyos 2015, 123; Daly 2002, 199; Lazenby 1978, 84–85.

2. Escape to Canusium: Livy 22.52.4, 22.53.1–3; App. *Hann.* 26; Dio fr. 57; Zon. 9.2. "Clinging": Livy 22.53.4–5. "Acknowledged reputation": Polyb. 10.3.7; cf. Livy 21.47.7–10.

3. Scipio's intervention: Livy 22.53.6–13; Dio fr. 57; Oros 4.16.6; Val. Max. 2.9.8, 5.6.7; Front. *Strat.* 4.7.39. Story doubted: Ridley 1978, 161–65; Evans 1989, 117–21; Daly 2002, 223 n.18.

4. Barcid conquest of Spain: Polyb. 2.1.5–9, 2.13.1–7, 2.36.1–7, 3.9.6–3.15.13, 3.17.1–11, 3.20.1–3.21.8, 3.29.1–3.30.4, 3.33.1–18; Livy 21.1.1–21.16.2, 21.18.1–14, 21.21.1–21.22.5; App. *Hisp.* 7, *Hann.* 2, *Pun.* 6; Dio fr. 48. Outbreak, causes of war: Hoyos 1998, 196–259, 2003, 55–97, 2015, 80–94; Goldsworthy 2000, 143–50; Lazenby 1978, 1–28; Beck 2011b; Rich 1996; Eckstein 2006, 174–75.

5. Hannibal's march: Polyb. 3.34.1–3.56.6; Livy 21.23.1–21.24.5, 21.26.4–21.31.12, 21.32.6–21.38.9. Ticinus and Trebia: Polyb. 3.65.1–3.74.11; Livy 21.39.10–21.48.10, 21.52.1–21.56.9. Lake Trasimene: Polyb. 3.82.9–3.84.15; Livy 22.2.1–22.7.3. Dictatorship of Fabius Maximus: Polyb. 3.87.6–3.94.10, 3.101.1–3.105.11; Livy 22.8.6, 22.11.1–22.18.10, 22.23.1–22.31.11. Hannibal's invasion down to Cannae: Hoyos 2015, 101–25; Goldsworthy 2000, 158–221; Lazenby 1978, 29–85.

6. "Not waging a war": Livy 22.58.3. Prisoners: Polyb. 3.77.4–7, 3.85.1–4; Livy 22.7.5, 22.13.2. Hannibal's strategy: Fronda 2010, 34–37; Hoyos 2015, 100–101, 2003, 101–2; Goldsworthy 2000, 149–50, 155–56; Lazenby 1978, 29–32; Briscoe 1989, 46.

7. Revolts in Etruria: Zon. 8.7, 8.18. "Outside force": Fronda 2010, 29. Rome and Italy: Fronda 2010, 13–34.

8. Apulians/Samnites: e.g., Polyb. 3.118.3, Livy 22.61.11, 23.1.1–4, 23.11.10–11, 24.45.1, 27.1.3. Capua: Livy 23.6.1–23.10.13. Other Campanians: e.g., Polyb. 3.118.3; Livy 22.61.11, 23.20.1, 26.34.6–7; Zon. 9.2. Bruttians: e.g., Livy 23.11.10–11, 23.20.4, 25.1.2, 29.38.1, 30.19.10. Lucanians: e.g., Livy 23.11.10–11, 25.16.5–8; App. *Hann.* 35. Tarentum: Polyb. 8.24–34; Livy 25.7.10–25.11.20. Other Greek cities: e.g., Livy 22.61.12, 23.30.6–8, 24.1.2–13, 24.2.1–24.3.15, 25.1.1, 25.15.5–17; App. *Hann.* 35. See also Fronda 2010; Lomas 1993, 54–70.

9. Macedonian-Carthaginian treaty: Polyb. 7.9.1–17; Livy 23.33.1–12. "Second front": Walbank 1967, 42–44. See also Gruen 1984a, 373–77.

10. Hiero's support: Polyb. 3.75.7; Livy 22.37.1, 22.37.10, 23.21.5, 23.38.12–13. Hieronymus intrigues: Polyb. 7.2.1–7.5.8; Livy 24.4–7. Hieronymus's death: Livy 24.21–29; Plut. *Marc.* 13.1. See also Eckstein 1987, 112–31, 135–55; Hoyos 2015, 37–38, 71–72, 135–36, 145, 159; Goldsworthy 2000, 74–75, 129–30, 260–62.

11. "Any other people": Livy 22.54.10.

12. "One in three": Hoyos 2015, 150. Recruitment, commanders: Fronda 2010, 38; Goldsworthy 2000, 226–28; Hoyos 2015, 136–39; Rosenstein 2012, 147, 149–50.

13. Legions in Campania: e.g., Livy 23.39.6, 23.46.9–10, 25.13.1–3. "More than seventy": Livy 23.17.2. Massacres and maneuvering at Casilinum: Livy 23.17.7–23.18.9, 23.19.1–16. Casilinum's significance: Fronda 2010, 100.

14. Samnite towns: Livy 23.37.12–13. "Past disasters": Livy 23.41.13–14. 25,000 killed/captured: Livy 24.20.5–6. Arpi: App. *Hann.* 31. Cf. Livy 24.47.9–11; Fronda 2010, 257–58.

15. "Harshness": Livy 25.8.1. "Their anger": Livy 25.15.7. See also Fronda 2010, 230–31.

16. Ultimatum, closing the ring: Livy 25.22.12–16. Capua as priority: Livy 26.1.1–4. Fighting at Capua: Polyb. 9.3.1–9.7.10, Livy 26.4.1–26.12.2.

17. Despair in Capua, Numidian "deserters": Livy 26.12.10–19. "Broke the spirit": Livy 26.13.1. Capuans decide to surrender, suicides: Livy 26.13.1–26.14.5.

18. "How pitiable": Aesch. 2.182. Capuan executions: Livy 26.14.6–26.16.4; App. *Hann.* 43; Zon. 9.6; Val. Max. 3.8.1, 3.2.ext1; Oros 4.17.12.

19. "Powerful, nearby," Capua's fate: Livy 26.16.7–13, 26.34.1–13. Cicero on Capua: Cic. *Off.* 1.35; *Agr.* 2.32–33. Fighting in Italy to 211: Hoyos 2015, 143–54; Goldsworthy 2000, 222–35; Fronda 2010, 243–58, 260–61; Lazenby 1978, 87–100, 110–15, 121–24.

20. Advantages controlling Italian towns: Rawlings 2011, 303–4. Punitive strategy: e.g., Zimmermann 2011, 288–89; Briscoe 1989, 77–78; Fronda 2010, 234–79 *passim*, 307–10, 313–14.

21. Hannibal punishes waverers: e.g., Livy 27.1.14, App. *Hann.* 54. Cf. Polyb. 3.86.9–11; Livy 23.15.6. Italian armies defeated by Rome: e.g., Livy 23.35.13–19, 24.14.1–24.16.5. "Forced": Livy 26.16.13. Weakness of Hannibal's strategy: Fronda 2010; Hoyos 2015, 128–31; Rosenstein 2012, 152–56; Goldsworthy 2000, 222–26.

22. Leontini raids: Livy 24.29.1–12.

23. Marcellus's duel: Plut. *Marc.* 7.1–3; Livy *Per.* 20; Val. Max. 3.2.5; Front. *Strat.* 4.5.4.

24. Attack on Leontini: Livy 24.30.1–6. Cf. Livy 26.30.4; Plut. *Marc.* 14.1.

25. Syracuse defects: Livy 24.30.6–24.33.9; Polyb. 8.3.1; Plut. *Marc.* 14.2; Zon. 9.4. Syracuse and trade: Casson 1991, 72. Syracuse's defenses: Beste 2016, 195–96; Lazenby 1978, 106; Hoyos 2015, 159. "Did not reckon": Polyb. 8.3.3 (Paton). Cf. Livy 24.34.2; Plut. *Marc.* 14.3–9.

26. Spring 213, legions: Lazenby 1978, 105, 107–8; Clark 1992, 182–91. "Against the gods": Plut. *Marc.* 16.2. Assaults on Syracuse: Polyb. 8.3.2–8.7.12; Livy 24.34; Plut. *Marc.* 15.1–17.3; Zon. 9.4; Polyaen. 8.11.

27. "Sacked and destroyed," "terrify the others": Livy 24.35.1–2.

28. Carthaginian and Roman reinforcements, Syracusan army, defections: Livy 24.35.3–24.37.1. Agrigentum's strategic significance: Livy 25.23.3; Morton 2014, 20–38. Morgantina's strategic significance: Cerchiai et al. 2007, 236; Bell 1981, 5–6; Stone 2014, 14 n.46.

29. "Lofty": Livy 24.37.2. "Navel": Cic. *Verr.* 2.4.106. Temple to Demeter: Strabo 6.2.6. Enna's role in slave revolt: Morton 2014, 20–38. Byzantines, Arabs: Metcalf 2009, 14–15, 25; Citter et al. 2017, 308.

30. 2,000 soldiers: this is a guess based on the garrison at Nola in Livy 24.19.4. "Might be known": Livy 24.37.11. Pinarius's plan: Livy 24.37.5–24.38.9; Front. *Strat.* 4.7.22; Polyaen. 8.21.1. "Raise a shout": Livy 24.38.7.

31. Ennaeans assemble, Roman positions: Livy 24.39.1–4; Polyaen. 8.21; Front *Strat.* 4.7.22. Theaters as Greek assembly places: Johnstone 1996, 106–9. Contemporary theaters in Sicily: Sear 2006, 48–49. A "few" Ennaeans: Strabo 6.2.6. Morgantina population: Stone 2014, 12. Medieval Enna (Castrogiovanni): Epstein 1992, 54. Massacre: Polyaen. 8.21; Livy 24.39.1–6; Front. *Strat.* 4.7.22. Cf. Barrandon 2018, who argues for the reliability of Livy's account here.

32. Marcellus's approval: Livy 24.39.7. Plunder, inscription: CIL 1².608. More defections: Livy 24.39.8–9.
33. "Some fled": Livy 25.24.5. Infiltration: Polyb. 8.37.1–13; Livy 25.23.1–25.24.10; Plut. *Marc.* 18.3–4.
34. Plundering: Livy 25.25.5–9; Plut. *Marc.* 19.2–3, 21.1–5.
35. Typhus: Hoyos 2015, 162. Marcellus continues siege, relief efforts fail: Livy 25.26.2–25.27.13. Fall of Syracuse: Livy 25.30.1–25.31.7.
36. Terms: Livy 25.28.3; Eckstein 1987, 159–60; cf. Marc. *Plut.* 20; Cic. *Ver.* 2.2.4, 2.4.120, 2.4.130; Val. Max. 5.1.4. "Shameful examples": Livy 25.31.9. Capture of city, death of Archimedes: Livy 25.31.1–11; Plut. *Marc.* 19.3–6; Cic. *Fin.* 5.50; Val. Max. 8.7.ext7; Pliny *NH* 7.125; Flor. 1.22.33–34; Oros. 4.17; Zon. 9.5. Thoroughly plundered: Polyb. 9.10.1–13; Livy 25.31.11–12, 26.30.9; Plut. *Marc.* 21. Starvation, slavery: Diod. 20.26.2. Brutal sack: Ziolkowski 1993, 81–83; Eckstein 1987, 162–63; Walbank 1967, 134–35; Champion 2004, 50.
37. "What they deserve": Livy 26.31.2; cf. Plut. *Marc.* 23.1–5. Marcellus's ovation: Livy 26.21.1–13; Plut. *Marc.* 22.
38. Offering battle to Marcellus: Livy 25.40.1–25.41.8. Cethegus's activities: Livy 26.21.14–17. Destruction at Morgantina: Stone 2014, 11–13; White 1964, 273–77.
39. Capture of Agrigentum, mop up: Livy 26.40.1–18. Laevinus's motives: cf. Eckstein 1987, 180–81. Roman reconquest of Sicily: Hoyos 2015, 159–63; Goldsworthy 2000, 262–68; Lazenby 1978, 103–8, 115–19, 172.
40. Marcellus hounds Hannibal: Livy 27.2.1–12, 27.12.7–27.14.14, 27.27.1–10; Plut. *Marc.* 24, 25.2–26.4, 29.1–8. Fall of Tarentum: Livy 27.16.5–8; Plut. *Fab.* 22.4, *Mor.* 195F; App. *Hann.* 49; Polyaen. 8.14.3; Zon. 9.8; Cic. *De Or.* 2.273.
41. Hannibal's battles: Rawlings 2011, 301–2. Battle(s) of Herdonea: Livy 25.21.1–10, 27.1.3–15; App. *Hann.* 48; Front. *Strat.* 2.5.21; Polyaen. 6.38.7; Rosenstein 1990, 207–8. Marcellus's death: Polyb. 10.32.1–6; Livy 27.27.1–14; Plut. *Marc.* 29.1–9; App. *Hann.* 50; Val. Max. 1.6.9; Nepos 22.5; Oros. 4.18; Zon 9.9. Latin colonies: Livy 27.9.2–6. Etruria rumblings: Livy 27.21.6–8, 27.24.9.
42. Hasdrubal's invasion: Polyb. 11.1–3; Livy 27.19.1, 27.38.1–14, 27.43.1–27.49.9, 27.51.11; App. *Hann.* 52–53. Mago's invasion: Livy 28.46.4–6, 29.4.5–6, 29.5.1–9, 30.18.1–19.5; Front. *Strat.* 3.6.5; App. *Hann.* 54.
43. "Nothing more": Livy 30.19.12. War in Italy down to 203: Hoyos 2015, 186–200, 209–10; Goldsworthy 2000, 235–44; Lazenby 1978, 170–92, 196, 199, 214–15.
44. Scipio brothers in Spain, aftermath: Polyb. 3.76.1–13, 3.97.1–3.99.9; Livy 22.19.1–22.22.21, 23.26.1–23.29.17, 24.41.1–24.42.11, 25.32.1–25.39.19; App. *Hisp.* 14–15, 16–17, 32; Hoyos 2015, 165–71; Goldsworthy 2000, 246–53; Lazenby 1978, 125–32.
45. Scipio's command: Livy 26.18.1–26.19.9; Goldsworthy 2000, 270–71; Hoyos 2015, 172–74; Lazenby 1978, 132–33.
46. Scipio not seen as threat: Hoyos 2005, 144. Carthaginian commanders, Scipio's recon: Polyb. 10.7.1–10.8.10. Strategic significance of New Carthage: Ziolkowski 1993, 88; Lazenby 1978, 134.
47. Scipio's army, fleet, camp: Polyb. 10.8.8, 10.9.4–7, 10.11.1–3, 10.17.13; Walbank 1967, 204–5, 211–12; Scullard 1930, 67 n.1, 82; Huertas 2012, 67–69. Polybius's description of New Carthage, accurate but misoriented by about 45°: Lazenby 1978, 135; Walbank 1967, 205–12; Scullard 1930, 289–99. New Carthage's defenses, topography: Polyb. 10.11.4, 10.13.8; Ruiz Valderas and Madrid Balanza 2002; Ramallo Asensio 2006, 91–92; Martín Camino and Belmonte Marín 1993; Blánquez Pérez and Roldán

Gómez 2009, 93–104; Noguera Celdrán et al. 2011–2012. "Fortress," "citadel": Livy 26.48.4.

48. Mago's troop placement: Polyb. 10.12.2–3; Livy 26.44.2; Lazenby 1978, 137–38; Scullard 1930, 83; Walbank 1967, 213–14; Kern 1999, 29.

49. Scipio's speech: Polyb. 10.11.5–9; Livy 26.43.1–8. Scipio's divine inspiration: Polyb. 10.2.5–13, 10.5.4–7; Livy 26.19.3–9, 26.49.9.

50. Roman assaults: Polyb. 10.12.1–10.15.2; Livy 26.44.6–26.46.6, 26.48.5–14. Hill of Mercury as Castillo de los Moros: Lazenby 1978, 136; Walbank 1967, 214. Lagoon phenomenon: Lazenby 1978, 136; Hoyos 1992, 127–28; Walbank 1967, 194, 1985, 132–33.

51. Defenders flee inward, Roman "battle order": Livy 26.46.7–8.

52. "Spare no one," "inspire terror," "when towns": Polyb. 10.15.4–5 (Paton, modified). Strategic terror in Roman warfare: Gilliver 1996, 234–35; Eckstein 1976, 135. Cf. Front. *Strat.* 3.8. Anger as an aggravating factor: Ziolkowski 1993, 84–86; Levithan 2013, 205–28; cf. Jones 2012, 65. Other communities: e.g., Livy 28.3.10.

53. Modern scholars follow Polybius: e.g., Eckstein 1987, 210; Isaac 2004, 216; Shatzman 2011, 50; Harris 1979, 51; Thornton 2006, 168; Gilliver 1996, 225 (though Gilliver argues that the killing was also meant to encourage surrender; see below). Cf. Gaca 2010, 136. Population a potential threat: Goldsworthy 2000, 275.

54. Mago resists and surrenders, plundering: Livy 26.46.8–10, 26.48.2; Polyb. 10.15.7–9.

55. Treatment of survivors, hostages: Polyb. 10.17.6–10.18.15, 10.19.3–7; Livy 26.47.1–4, 26.49.7–26.50.14. "Workmen": Walbank 1967, 216. Oarsmen: Morrison 1988, 251–53. "Safe and well": Polyb. 10.18.5.

56. Baecula: Polyb. 10.38.7–10.39.9; Livy 27.18.1–27.19.1. Ilipa: Polyb. 11.20.1–11.24.9; Livy 28.12.13–28.15.13. Gades: Livy 28.37.10.

57. New Carthage's population stays loyal: Livy 28.36.4–6. Iberians defect after New Carthage: Polyb. 10.34.1–10.35.3, 10.35.6–10.38.6; Livy 26.18.13–14, 27.17.1–17. Iberians defect after battlefield victories: e.g., Polyb. 10.40.2, 10; Livy 27.19.3–12, 28.15.14–16. Iberians supplement Scipio's forces: e.g., Livy 28.13.1–15.

58. Orongis: Livy 28.2.15–28.3.16.

59. Attack on Ilurgia/Iliturgi: Livy 28.19.1–28.20.7; App. *Hisp.* 32; Zon. 9.10. Cf. Barrandon 2018, who doubts the participation of women and children—but as discussed in chapter 2, this was not uncommon in ancient warfare. Exemplary violence: Eckstein 1987, 223. Location, name of Ilurgia: Richardson 2000, 131; Walbank 1967, 305; Hoyos 2015, 167; Lazenby 1978, 152. Castulo surrenders: Livy 28.20.8–9; cf. App. *Hisp.* 33.

60. Scipio's Spanish campaigns: Hoyos 2015, 172–86; Goldsworthy 2000, 271–85; Lazenby 1978, 133–56.

61. "Africa burns": Livy 28.44.6. Scipio's election, invasion plan: Livy 28.38.12, 28.40.1–28.45.14.

62. Initial invasion, alliance with Massinissa: Livy 29.28.1–29.33.10; App. *Pun.* 13–15; Zon. 9.12, Dio fr. 57.

63. Night attack, assaults on towns: Polyb. 14.4.1–14.6.5; Livy 30.3.8–30.7.2; App. *Pun.* 19–23; Zon. 9.12. Locha: App. *Pun.* 15. Cf. Livy 29.29.2–3.

64. Great Plains, aid to Massinissa, fleet attack: Polyb. 14.8.1–14.10.9; Livy 30.8.1–30.10.21. Terms: Livy 30.16.2–30.16.15.

65. Hannibal's return: Walbank 1967, 440–41; Hoyos 2015, 210. Attacks on supply ships, envoys: Polyb. 15.1.1–15.2.15; Livy 30.24.5–30.25.8.

66. Breach of armistice: Eckstein 1987, 253–54; Hoyos 2015, 208; Goldsworthy 2000, 299; Lazenby 1978, 215–16; Walbank 1967, 441–42. Scipio's anger: Eckstein 1987, 254; but cf. Hoyos 2015, 211–12. Others seek African command: Livy 30.24.1–4, 30.27.1–5, 30.40.6–15, 30.44.3. Risk to Scipio's command: Goldsworthy 2000, 300.

67. Zama: Polyb. 15.9.2–15.15.9; Livy 30.32.1–30.35.3; App. *Pun.* 41–48; Zon. 9.14; Front. *Strat.* 2.3.16; Nepos 23.6. Overviews of war in Africa: Hoyos 2015, 198–219; Goldsworthy 2000, 286–309; Lazenby 1978, 193–232.

68. Political considerations: Livy 30.36.9–10, 30.44.3; App. *Pun.* 52, 56. Peace terms: Polyb. 15.18.1–8; Livy 30.37.2–7; App. *Pun.* 49–55; Dio fr. 57; Zon. 9.14; Nepos 23.7. Politics shape peace terms: Eckstein 1987, 255–60; Rosenstein 2012, 173–75; Goldsworthy 2000, 308 (Goldsworthy calls the terms "harsh"); Lazenby 1978, 228–30.

6

"So Much Destruction and Utter Ruin"

Politics and Pragmatism in the Third Macedonian War

Perseus was the last king of Macedonia and the last ruler of the Antigonid dynasty. The Romans defeated the monarch in a grinding conflict known as the Third Macedonian War, and in 167 BCE* they removed him from power and divided up his realm into four smaller states. Afterward, they hauled Perseus off to Rome, where he was humiliatingly paraded through the streets in the victor's triumph. That would be the end of his storied kingdom, little more than 150 years after its most famous ruler, Alexander the Great, had conquered an empire stretching from Greece to India.

Of course, Perseus had no way of knowing this dire outcome four years earlier, when his war with the Roman Republic first began. At the outset, in fact, he was quite ready to fight. Even though his kingdom was much reduced in size following a previous clash with Rome, years of careful planning allowed Perseus to assemble 43,000 troops. This was the largest force ever mustered by any king of his dynasty—perhaps larger even than Alexander's great conquering army. At its heart was the terrifying Macedonian phalanx. In battle, its soldiers stood in a tightly packed rectangular formation that could stretch over a mile long. Each man in the phalanx wielded a fifteen- to twenty-foot pike (a kind of extra-long spear), which allowed soldiers in the first five ranks to project their weapons far out in front of the battle line. In effect, this created a wall of men bristling with thousands of spear points, which was virtually irresistible from the front. Perseus's phalanx, over 20,000 strong, was supported by a roughly equal number of light troops and cavalry drawn from Macedonia and mercenaries from beyond.

---

* Henceforth, all dates in this chapter are BCE.

With this formidable army, Perseus swept south into northern Greece in the spring of 171. By that time, Rome had already declared war, and the king planned to confront the coming Roman invasion in Thessaly (map 6.1). When he found that the Romans had not yet arrived, he had time to capture a series of key towns and strongholds that controlled mountain passes into Macedonia. While Perseus was happily securing the southern boundaries of his kingdom, Roman legions were also moving toward Thessaly. Though they did not match Perseus in raw numbers, the formidable Roman force possibly numbered 30,000 men, more than enough to put up a fight. However, they were making slow progress: the commanding consul, one Publius Licinius Crassus, had chosen to reach Thessaly by marching southeast from the coastal town of Apollonia, passing through difficult terrain in Epirus and Athamania and running down his army in the process. Livy snidely remarks that the war would have ended then and there if Perseus had been near enough to attack the struggling Romans.[1]

Luckily for Licinius, who may have lacked any significant command experience, Perseus was not near enough to take advantage of their difficulties, and in May the Roman army hobbled into Thessaly and made camp at the river Peneus. There they linked up with Greek contingents from the Achaeans, Aetolians, and Thessalians, as well as a force from Pergamum led personally by its king, Eumenes. These Greeks were part of an allied coalition that the Romans had assembled before the outbreak of war. Unlike Rome's subordinate allies in Italy, the Greek allies were all technically independent partners who made voluntary commitments to fight alongside the Republic against Perseus.[2]

As Licinius and the allied leaders debated their next move, Perseus boldly seized the initiative. For the next several days, he repeatedly brought his forces up to the Roman camp and tried to tempt Licinius into battle, but repeatedly failed to draw the Romans into a major engagement. Taking note of his opponent's lethargy, Perseus then went a step further and moved his own encampment to within five miles of the Romans. At dawn one day, he drew up his heavy infantry phalanx about a mile from the Roman camp and then deployed all of his cavalry and light troops a half mile away on a hill called Callicinus, within sight of his opponents. And his forces would have been quite a sight. On the right wing, the spear-wielding Macedonian cavalry stood in their bronze helmets and armored cuirasses. Intermixed with these were Cretan footmen, famed for their archery. Next came the royal cavalry, heavily equipped and interspersed with units of horsemen recruited from many nations; we should envision Greeks as well as "barbarian" Gauls, their contingents clad in a variety of garments and wielding assorted weapons. On the left wing, Thracian warriors—fierce tribesmen who hailed from the northeast of Macedonia—alternated their units of light infantry with squadrons of cavalry. The ancient author Plutarch immortalizes the Thracians' intimidating appearance at a later battle, describing "men of lofty size, whose black tunics were covered by the gleaming white armor of

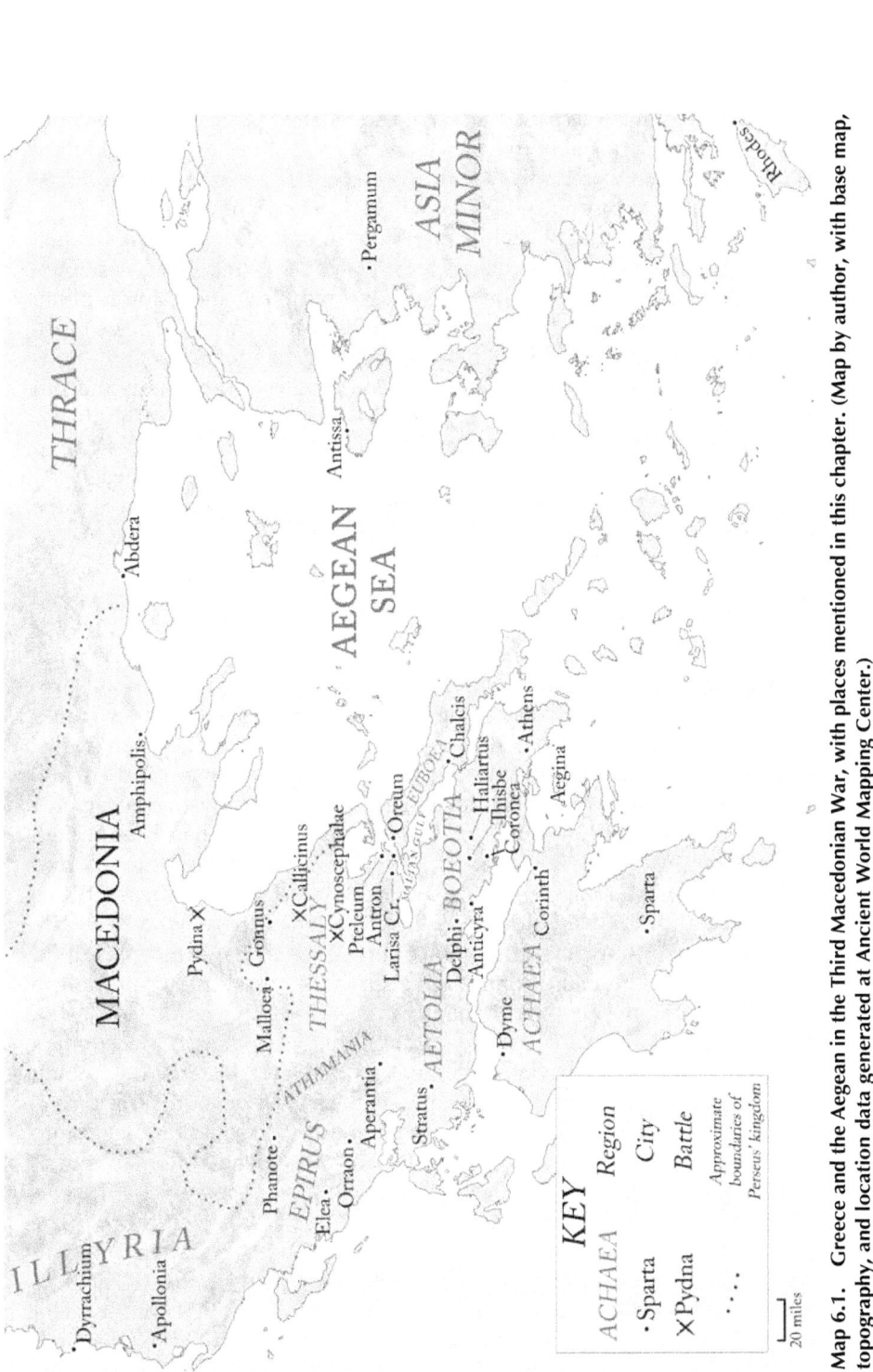

Map 6.1. Greece and the Aegean in the Third Macedonian War, with places mentioned in this chapter. (Map by author, with base map, topography, and location data generated at Ancient World Mapping Center.)

their shields and greaves." In the center, Perseus led the elite cavalry, which Livy calls the "guard" (*agema*) and "sacred squadron" (*sacrae alae*). The variegated peoples, with different dress and languages, glittering arms and bright standards, must have presented a startling picture. Perseus had about 20,000 men on the hill that morning, 4,000 of whom were cavalry.[3]

As the early light crept over the horizon, a Roman sentry reported a dust cloud "larger and nearer than usual." The watchman's report was doubted at first, yet soon more men shouted from the ramparts and came running back into camp. Confusion rippled through the Roman army, officers dashing to headquarters while the soldiery ran to their own tents. As the legions assembled, Licinius led his light infantry and cavalry out of camp and put them into a battle line opposite Perseus. The Macedonians probably had a slight numeric advantage, but the Romans placed some of their Greek allies in reserve behind their main line and formed up their heavy infantry inside the camp. Both precautions provided additional layers of support in case the Macedonians broke through their front ranks.[4]

Finally the moment of battle approached and the opposing forces clashed together. Both sides opened the engagement with volleys of javelins and sling stones. According to one Roman writer, javelins and other missiles made unnerving hissing sounds as they sliced through the air, and we should imagine something similar here, with volleys of fast-moving projectiles whistling back and forth in deadly exchanges. Many missiles would have clanged heavily against shields and armor, punctuated by cries of anguish as a few hit their targets. Soon the heavy horsemen hurtled into one another, and a mass melee developed, which quickly turned against the Romans. The Thracian cavalry, "like wild beasts who had been locked up for a long time," smashed into the Roman right wing and threw it into bloody disorder. Meanwhile, Perseus charged the Roman center with his elite cavalry and swatted aside the Greeks stationed there, who fled with the king's men in hot pursuit. Some of the Greek allies formed a second perimeter behind the main Roman force, crucially providing protection to their fleeing comrades, but the Romans were still nearing complete disaster. Soon Perseus's phalanx showed up on the initiative of its two Macedonian officers, who wanted to drive the momentum into a total rout of the Roman army. Perseus thought better of it. Acting on the advice of another officer, he broke off the engagement, hoping to leverage this triumphant moment into an honorable peace treaty rather than attack a well-defended Roman camp.[5]

The Battle of Callicinus, as it is known, was a Macedonian victory: Licinius lost 2,200 men killed and 600 captured, while Perseus lost only 60. The Romans were in a dark mood after this setback, and the chastened consul pulled his army back to a safer position across the Peneus River. There he held a council in his headquarters tent to take stock of their predicament and determine why the battle had gone so badly. He learned (or decided) that blame for their defeat lay with a contingent of Greek allies: specifically, an Aetolian officer claimed that five of his countrymen had fled early in the battle, spreading panic and precipitating disaster. Licinius put the accused Aetolians under guard and sent them to Rome. This overbearing treatment of ostensibly independent leaders from an ostensibly independent state would be deeply troubling to other Greeks, and would carry consequences in the future.

In contrast to the Romans, the Macedonian army was feeling celebratory. The Thracians strutted through camp singing songs and bearing the severed heads of their enemies spitted on spears. Perseus distributed captured weapons, armor, horses, and prisoners as spoils, and cheered his soldiers on to new victories and new heights of glory. Although the king made notable errors in the battle's aftermath by failing to follow up on his initial victory, he still won the first engagement of the war. It was time to open negotiations. According to the standards of ancient war and diplomacy, it would have been typical for Perseus to offer terms that were unfavorable to the Romans, since he had won and was speaking from a position of strength. Instead, he effectively proposed terms as if *he* had lost the battle, offering to pay a war indemnity and to abandon some of his territories. Clearly Perseus wanted to end to the war as soon as possible. Licinius would have none of it, however:

> It was unanimously decided [by the Romans] to give as severe a reply as possible, it being in all cases the traditional Roman custom to show themselves most imperious and severe in the season of defeat, and most lenient after success. . . . They ordered Perseus to submit absolutely, giving the senate authority to decide as they saw fit about the affairs of Macedonia. (Polybius *Histories* 27.8.7–10; trans. Paton)

The Macedonians were baffled (and perhaps a little frightened) at the Romans' response, and so Perseus proposed terms again, offering to pay an even larger indemnity. Again to no avail. With negotiations stalled, he withdrew his army to a more distant encampment.[6]

Perseus won no new allies from his success at Callicinus, and the Romans' coalition of Greek allies held together. Still, there were warning signs. According to Polybius, "when word of the cavalry battle spread through Greece, the populace's favor for Perseus—which most had concealed before that time—burst forth like fire." Polybius compares this sentiment to a boxing match, when an underdog lands a blow or two against a seemingly invincible champion and immediately wins the sympathy of the crowd. Everyone cheers for the inferior combatant and jeers his stronger opponent because people are "naturally inclined to favor the weaker." And in Perseus, many Greeks found their underdog, "someone who at least appeared to be a capable adversary to the Romans."[7]

Whatever one makes of Polybius's analysis, Greek support for Rome was plainly finite. As we will see, Perseus began his reign by courting Greek opinion, and he managed to win over a handful of his own Greek allies once the war began in earnest. At Callicinus he won a major and unexpected victory, and Greek sympathies started drifting openly in his direction. The Romans recognized the danger, and they spent the rest of the war attempting to undermine Perseus's support in Greece. For some commanders, this meant attacking Greek towns that supported Perseus, or that insufficiently supported Rome, with a singular ruthlessness. Whole cities were left in ruins by Roman armies, their populations put to the sword or sold at auction. There was nothing unusual about this behavior, and we have already seen that the Romans used similar tactics to keep allies in line and punish defectors in the Second Punic War. But when Perseus's fortunes in the war improved, and as Roman behavior became increasingly arbitrary, the Greeks became more disaffected from Rome. A few openly switched sides, while some neutrals threw in for the Macedonians. For three years after the Battle of Callicinus, the Senate would struggle to balance the military conduct of its commanders—and particularly their treatment of the Greeks—against the need to maintain their Greek alliances. The Romans' use of mass violence ebbed and flowed along with these political realities.

ROME AND MACEDONIA IN THE EARLY SECOND CENTURY

The causes of the war between Rome and Perseus are controversial, and the ancient sources are tainted by hindsight and propaganda. However, just as the Second Punic War grew out of the First, the roots of the Third Macedonian War can be found in previous conflicts. Perseus's father, Philip V, was the first Macedonian king to face the Roman legions in battle. A competent general with an eye for opportunity, we have already seen how Philip's gaze turned toward Italy in 215, after Hannibal defeated the Romans in three major battles. Philip wagered, not unreasonably, that Carthage would become the most important power in the western Mediterranean. He also had designs on areas of Illyria that fell within Rome's ambit, and now seemed as good a time as any to snatch them up. So he made an alliance with Hannibal and started campaigning in Illyria. The Romans sent ships and men eastward when they caught wind of the alliance, and in the ensuing First Macedonian War (214–205), their main aim was to occupy Philip in Greece so he could not help Hannibal in Italy. To that end, they made an alliance with Philip's main Greek rival, the Aetolian League; the Aetolians would engage Philip by land while the Roman fleet raided Macedonian-controlled cities along Greece's coasts. The Aetolians eventually realized that they were doing most of the fighting—Roman support was somewhat less than advertised—and made peace with the Macedonian king. The Romans soon struck their own treaty with Philip so that they could focus fully on the war with Carthage.

Shortly after the defeat of Hannibal at the Battle of Zama (202), several Greek states requested Roman help against an aggressively expanding Macedonia. The Romans, who were surely still bitter about Philip's opportunistic pact with Hannibal, responded affirmatively to Greek pleas and went back east for a rematch with the Macedonians. The Second Macedonian War (200–197) ended when the charismatic Roman general, Titus Quinctius Flamininus, decisively defeated Philip at the Battle of Cynoscephalae ("the Dogs' Heads"). In the ensuing peace agreement, the Romans stripped Philip of all his territories in Greece, forced him to surrender his fleet, and saddled him with a war indemnity. On the other hand, he got to keep most of his kingdom as well as his throne, and he was recognized as a "friend" (*amicus*) of the Roman people. Meanwhile, Flamininus declared the "freedom of the Greeks": every community in Greece would henceforth be free from Macedonian domination and live under its own laws. All Roman troops were withdrawn, no garrisons were left behind, no territory was annexed, and a bright new polish was added to Rome's reputation throughout the Greek world.[8]

From the Romans' perspective, Philip's defeat and surrender meant that the Republic was superior to Macedonia, and that their subordinate "friend" would henceforth defer to Rome. Yet Philip spent the rest of his life pushing his independence any way he could and attempting to rebuild his kingdom. He increased revenues from farms and harbor dues, reopened old mines and established new ones, and, to restore the population, he encouraged childbearing and transferred in new settlers from abroad. He also fought wars against tribes to the north and east of his kingdom, keeping his army in fighting shape. And the Romans even granted him some modest territorial gains when he supported them in another eastern war. In the midst of all this recovery, Philip's twilight years were marred by a dynastic crisis: it was rumored that his son Demetrius was scheming to usurp the eldest son, Perseus, as heir to the throne, and Philip had the younger man executed (a decision he later regretted deeply). Still, despite this family tragedy, Philip had successfully revived the strength of his kingdom by the time he died in 179.[9]

When Perseus became king, he went further than his father, stockpiling weapons and supplies to support a large army for years on end. He also worked tirelessly to restore Macedonia's reputation abroad, forgiving debts and recalling political exiles to signal his magnanimity in grand public fashion. Further, he made marriage connections with two eastern kings, wedding his sister, Apame, to King Prusias II of Bithynia and personally marrying Laodice, the daughter of King Seleucus IV of the Seleucid Empire. In Greece, he intervened in economic crises and sent military support to a couple of communities, and many Greeks took a liking to the dashing and dignified young king. Unfortunately for him, Perseus was unable to transform this popularity into many meaningful political relationships, and some of his diplomatic efforts failed. That said, he did manage to make an alliance with the Boeotians, giving him a possible base of support in central Greece. He also conquered some territory in Thrace, to the northeast of his kingdom, and made an alliance with an influential ruler there. Finally, after putting down a rebellion in Greek territories that he had inherited from his father, he marched his full army through the heart of Greece—without harming any lands along the way—traveling to the sanctuary at Delphi and consulting its famous oracle. He looked to be a king on the rise.¹⁰

The steady restoration of Macedonia's strength alarmed some of its neighbors, especially King Eumenes of Pergamum, who delivered the Roman Senate a laundry list of accusations against Perseus. Macedonia's revival alarmed the Romans, too. Perseus was probably not a direct threat to Rome, yet he still appeared to be reasserting himself as a leading power by rebuilding Macedonia's military and diplomatic standing in the Greek world. The newly minted king was acting too much like an equal, all the while gaining popularity and authority in Greece (at Rome's expense).* The Romans decided to give the king, and the rest of the Greek world, a forceful reminder of Macedonia's place. They demanded that Perseus accept responsibility for crimes that he had supposedly committed—including various treaty violations, conspiracies to assassinate foreign leaders, and secret plans for a war of revenge against Rome—submitting himself totally to Rome's judgment.

Many of the Romans' accusations were dubious in the extreme, however, and Perseus and his representatives would not simply roll over. Instead, during diplomatic talks in 172 and early 171, the Macedonian king refuted the accusations one by one. While neither side wanted war, and both sought diplomatic solutions up until the moment of armed conflict, Perseus refused to debase himself. This behavior seemed to confirm that he saw himself as an equal and that Rome's authority was breaking down, so the Romans determined that a military response was necessary in order to cut the king down to size. Negotiations came to an end, and by May 171, the consul Licinius was marching through Greece, where he suffered his humiliating defeat at the Battle of Callicinus.¹¹

"THE WAR WAS FOUGHT IN GREECE WITH CRUELTY AND GREED": LATE 171 AND EARLY 170

Perseus and Licinius spent the rest of the campaign season darting around Thessaly. Perseus tried several times to bring about a decisive action, or at the very least to give the Romans another bloody nose. First he attempted to attack the Roman camp by night (but was spotted and had to retreat). Then he tried making another surprise appearance outside the Roman encampment with his cavalry and light troops, hoping to draw out the Romans and repeat his success at Callicinus (but the Romans would not come out). And

---

* Though the Romans' reputation had improved with their declaration of "Greek freedom," the Greeks expected Rome to take a more active role in their affairs thereafter. For the next several decades, Greek embassies frequently approached the Senate requesting arbitration and intervention in their squabbles. Yet the Romans were not interested in becoming an active hegemon in Greece. They listened to ambassadors and sent legations eastward to "investigate," but rarely took decisive action one way or the other. Their inactivity may have led some Greeks to look around for an alternative, and Perseus was happy to step in. See Gruen 1984a, 101–31, 418, 481–505, with additional analysis by Eckstein 2008, 359–65.

then he tried attacking Roman foragers while they were scattered through local grain fields (but he was driven off by a Roman counterattack). Unable to land any further blows, Perseus turned his army back toward Macedonia. Still, the campaign was not a total loss. Earlier in the season, Perseus had captured several strongholds in northern Thessaly and destroyed at least one town that resisted. Among his gains was Gonnus, a well-fortified settlement that guarded a mountain pass into Macedonia; Perseus left a body of troops here to protect his rear. The acquisition of these strategic centers and the arrival of winter snows sealed off the southern approaches to his kingdom.[12]

But if the entryways to Macedonia were shut, Perseus's meager domains and allies in Greece were exposed to Roman attack. Aside from the towns he had recently seized in Thessaly, Perseus held some territory along the northeastern coast of Greece and around the Malian Gulf. And although he had failed to secure many new Greek friends, a triad of cities in Boeotia, namely Haliartus, Coronea, and Thisbe, decided to stick with Perseus when war broke out. (The other Boeotians declared for Rome.) All these places became immediate Roman targets.

Around the time that Perseus and Licinius were maneuvering in Thessaly, a Roman praetor named C. Lucretius Gallus was operating in Boeotia. Lucretius commanded a Roman fleet as well as 10,000 marines and 2,000 allies; rather than pursue naval operations along the coast, he sent these troops inland to besiege and assault the pro-Macedonian town Haliartus. This was no easy task, as the town was surrounded by thick stone walls and possessed a fortified acropolis. Furthermore, since it was perched on a hill next to Lake Copais, much of the ground around it was swampy and poorly suited to siege operations. Perhaps emboldened by their tough defenses, the inhabitants fought resolutely against Lucretius's attack, repeatedly charging out their gates and disrupting the Roman siege works. When the Romans brought up a battering ram, they dropped down heavy objects from their walls to disable it. And when the Romans still managed to batter down parts of the walls, the defenders worked through the chaos of battle to erect new barriers. Lucretius was not getting anywhere, and perhaps he was growing impatient as this relatively small town deflected his assaults, but he finally spotted an opportunity when a sizeable stretch of Haliartus's walls collapsed. The praetor picked out 2,000 of his men and ordered them into the breach, anticipating that the defenders would concentrate their forces to oppose this new intrusion. Meanwhile, he distributed scaling ladders among the rest of his men and sent them to mount the city walls in multiple places. The defenders responded exactly as Lucretius had predicted, coming down off the walls and gathering at the break in their fortifications to meet the Romans' 2,000-man strike force. Atop the battered ruins of their broken wall, the Haliartans piled bundles of dry wood and prepared to set them alight with blazing torches, hoping to create a large fire that would stop the Roman intruders and buy precious time to rebuild the defenses. This time they were unlucky. Darkening clouds were gathering above, and the heavens chose this moment to rip open and pour heavy rain down upon their heads. The desperate defenders could only get their woodpiles to smolder and smoke, and could not create flames to block the hole in their walls. The Roman invaders simply dragged the smoldering bundles of timber aside as they clambered up through the breach. While this scene unfolded below, the other Roman soldiers scaled and captured the walls above. Soon the invaders were racing inside the town, and the defenders ran for their lives.

In the initial panic of the city's capture, Livy says, the Romans indiscriminately killed "the old men and the young boys whom they encountered." This might indicate that they massacred everyone they met, though old men and boys were probably also helping in the defense of the city, and would have been out in the open as the Romans burst in. Such massacres were common when the Romans captured cities, and in this case the explosion of violence probably suppressed any remaining resistance and gave expression to Roman anger after a short but sharp siege. About 2,500 defenders were able to escape the onslaught and reach the acropolis, where they held out until the next day. Looking down on the carnage in the streets below, they decided that further resistance was pointless and surrendered to Lucretius—who promptly sold them into slavery. The combined killing and enslavement may have eliminated most of the population, and indeed, archaeologists have found signs of a steep reduction in Haliartus's habitation in this period. Aside from their assault on the populace, the Roman soldiers also plundered the city, lugging its valuables to their ships. Finally, Lucretius's men destroyed the city's physical fabric. Livy says Haliartus was "demolished (*diruta*) to the foundations," and the geographer Strabo likewise claims that the city was "demolished" (*kataskapheisa*), noting that it no longer existed in his lifetime; their comments suggest that Lucretius tore down a number of buildings, likely including the walls. The Romans probably torched some of the buildings as well. According to the geographer Pausanias, the city was burned, and the temples still appeared "half-burned" in later centuries. Between the initial massacre, the enslavement, and the destruction, Haliartus was effectively finished.[13]

Lucretius, however, was not finished. He marched his force southwest about sixteen miles to Thisbe, another Boeotian town that had sided with Perseus. The inhabitants likely surrendered at his approach—Livy says that he seized the town without a struggle—and Lucretius proceeded to hand the community over to exiles and pro-Roman leaders. Those who favored Perseus were apparently sold at auction along with their families and property. Soon afterward, the city's new pro-Roman government sent delegates to Rome requesting (among other things) permission to refortify their city and its acropolis. Their request indicates that Lucretius also demolished Thisbe's defenses, in whole or in part, thus rendering the community indefensible. With the city satisfactorily neutralized, Lucretius then returned to the Roman fleet anchored at Chalcis.[14]

Meanwhile, after the Macedonians withdrew behind their mountainous borders, the consul Licinius turned his army against the towns that Perseus had nabbed in northern Thessaly—perhaps hoping to accomplish something, anything, before the end of his consulship. Gonnus was his first objective, since he recognized that it provided both "a secure barrier into Macedonia and a convenient route down from Macedonia into Thessaly." When he found the city utterly impregnable, he moved west against a town called Malloea, capturing it at the first assault and unleashing his army to sack the place. Livy uses the word *direpta* to describe the sack, implying that Roman soldiers tore the place apart and perhaps committed individual acts of vandalism, rape, and murder as well. After this display, the rest of the region fell into Licinius's hands without incident.¹⁵

Next Licinius dispersed some of his men into winter quarters in Thessaly before taking his remaining forces further south toward the Malian Gulf. He was clearly determined to strip Perseus of his small domains there, and attacked the Macedonian-held towns Pteleum, Antron, and Larisa Cremeste. None was very willing to fight. He found Pteleum completely abandoned by its inhabitants, allowing him to capture it easily and, in Livy's words, "raze it to the foundations." Again, this probably means that he demolished the walls and some other public buildings, presumably also looting anything the inhabitants left behind. With the destruction of Pteleum freshly in mind, Antron voluntarily surrendered to Licinius, as did Larisa Cremeste after it was deserted by its Macedonian garrison.

By then it was late summer or perhaps early autumn. Licinius learned that Perseus's last Boeotian ally—the town Coronea—was causing trouble for its pro-Roman neighbors, and moved further south to deal with them. We do not know quite what happened next because large chunks of Livy's text are missing at this point in the narrative (one of the many hazards of ancient history). However, the Roman Senate issued a decree about the city of Coronea in 170, which was inscribed on stone and partly survives. With this inscription, as well as hints in Livy and Polybius, we can reconstruct the likely course of events. When Licinius appeared before their walls, the Coroneans may have resisted at first, and the Romans probably besieged them; Polybius mentions a siege of Coronea that may refer to Licinius's attack. At some point, the Coroneans saw the writing on the wall and surrendered. They likely made a formal *deditio* (ritual unconditional surrender), handing over themselves, their city, and their property to Roman discretion. Licinius then may have sold the pro-Macedonian leaders and perhaps other inhabitants into slavery, much as Lucretius had at Thisbe. To cap off the affair and make the town less serviceable as a potential enemy stronghold, he might have demolished the walls.¹⁶*

---

* The destruction of Coronea's fortifications is difficult to demonstrate positively, but the evidence hints in that direction. In 170, the Senate issued decrees concerning several Greek cities captured by the Romans in 171. A nearly complete decree survives concerning Thisbe, in which the Senate allows the Thisbeans to fortify their acropolis but *not* their city. A similar

With the approach of winter in 171/170, the Romans had taken significant steps to strip Perseus of friendly bases in Greece. Then their conduct took a decidedly indiscriminate turn, sweeping up their own allies in a torrent of violence. Chalcis, a friendly city that hosted Rome's fleet, was one of the first to suffer. Lucretius treated it very much like a captured enemy city rather than a friend and ally. For one, he stripped the temples and shrines of valuables, carting the plunder to his ships and thence back to Italy. His men also enjoyed lax discipline and openly looted the locals' possessions. At least some Romans felt emboldened enough to abduct townspeople and sell them into slavery. With winter approaching, Lucretius quartered his sailors in private Chalcidian dwellings, where they were free to assault local women and children. Again they acted as if Chalcis was a prize of war and not an ally, leaving the citizens terrified and vulnerable in their own homes.[17]

Lucretius was succeeded as fleet commander in 170 by the praetor Lucius Hortensius, who attacked several pro-Macedonian cities on the north Aegean coast. Each one shut its gates and repelled his raids. Eventually he sailed to Abdera, a friendly city in the same neighborhood, and demanded a large quantity of supplies. Other Greek cities had been subject to requisitions as well, but the Abderans showed a flicker of defiance: rather than submit immediately to Roman demands, they asked for time to send ambassadors to the consul and to the Roman Senate. Their ambassadors "had barely come to the consul when they heard that their city had been stormed, their leading men beheaded, and the rest of the population sold at auction." With Abdera's population eliminated, and his Aegean campaign coming to nothing, Hortensius returned to Chalcis. Like Lucretius, he treated the town as a treasure house for himself and a discipline-free playground for his troops, with dire consequences for the inhabitants.[18]

Communities on the periphery of Greece also saw unprovoked Roman aggression early in the war. Before hostilities began, tension had been building between the Romans and the Illyrian ruler Genthius, whose kingdom was west of Macedonia. In 172, the Romans had sternly rebuked Genthius for launching raids against one of his neighbors, and they had tried and failed to secure his support in the fight against Macedonia. At the start of the war with Perseus, Lucretius's brother also commandeered fifty-four of Genthius's warships for the Roman navy. Livy writes that he "came upon" Genthius's ships in port at Dyrrachium and then, "pretending to believe that these had been mustered for Roman use," he took them.[19]

Later in 171, at about the time that Perseus won the battle of Callicinus, Gaius Cassius Longinus brought fire and sword into Illyria. Cassius had been elected consul alongside Licinius, and he was apparently bitter because

---

decree survives for Coronea, and, though the inscription is fragmentary, it seems to contain the same language about fortifications (*RDGE* 2, lines 10–11). Notably, Coronea possessed strong fortifications (Diod. 16.58.1). If the Coroneans requested the right to (re)fortify their city after the Roman attack, despite previously possessing strong walls, then the Romans had probably destroyed their walls during the campaign of 171.

he had not gotten command of the war with Macedonia (which had legally fallen to his consular colleague). A brazen man with an inflated sense of his own importance, Cassius was not one to be thwarted by rules in his pursuit of glory. He took matters into his own hands, leaving his assigned area of operations in northern Italy and taking his army by land toward Macedonia. Evidently, he wanted in on the action there, but instead he brought trouble to communities that had nothing to do with the war. First he passed through Alpine country, where he ravaged fields and abducted thousands of people for sale as slaves. His army then moved into the territory of several tribes along the northern Adriatic coast. These contacts were initially friendly, and Cassius only requested guides who could lead his legions to Macedonia. Yet at some juncture, he gave up on the road to Macedonia, turned back around, and attacked the tribes he had met along the way. According to Livy, his army "made slaughter, pillage, and fire everywhere," and his victims "did not know . . . the consul's reasons for treating them as enemies." Cassius also sent a legate (i.e., a lieutenant) with soldiers against two wealthy Illyrian towns, Ceremia and Carnuns. The former town surrendered to the Roman legate, who treated the inhabitants well. He hoped that this display of kindness would encourage Carnuns to surrender, too, but that city stubbornly refused to yield, and the Romans were unable to capture it. So, just to be sure that "his soldiers had not been exhausted from two sieges for nothing, he sacked the city [Ceremia] which he had previously left unharmed."[20]

It is worth pausing here to consider why these Roman commanders fought with such ruthlessness in Greece and neighboring regions. Livy asserts plainly that their campaigns had been conducted with "cruelty and greed" (*crudelius avariusque*), and modern observers largely agree, commenting on the "rapacity and cruelty," "burgeoning cruelty and greed," and "savagery" of Licinius, Lucretius, Cassius, and Hortensius. These characterizations are fair enough, but there must have been complex motives behind their decision-making. After all, the physical urban destruction wrought at Pteleum and Haliartus (and perhaps at Thisbe and Coronea) would have required organization, labor, and resources on the part of the Romans. Likewise for the whole or partial enslavement of pro-Macedonian populations. These were not easy or cheap choices and, as noted previously, even mass enslavement was less lucrative than often imagined.

Certainly, the attacks on Perseus's allies in Boeotia and his holdings in the Malian Gulf made some military sense. Modest as these places were, they furnished Perseus with a sprinkling of useful strongholds. The towns around the Malian Gulf controlled important sea-lanes and provided a toehold in central Greece. And of the Boeotian towns, Haliartus was essential: situated in a pass between Lake Copais and Mount Helicon, it watched over a main marching route through central Greece. Altogether these places gave Perseus the potential to disrupt Roman supply lines into the interior or even to extend Macedonian control further into Greece, at the very least, these towns could harass pro-Roman settlements in the region, as the Coroneans did. So when the Roman destroyed these communities, or removed their pro-Macedonian inhabitants, they pointedly deprived Perseus of fortresses and supporters in Greece, secured Roman communication routes, and protected Rome's friends.[21]

The brutal treatment of pro-Macedonian towns also mirrors the strategy employed by Roman commanders in the First and Second Macedonian Wars. In both of those conflicts, the Romans attacked Macedonian-controlled or Macedonia-friendly cities in Greece. When these communities resisted, the Romans treated them with great violence in order to terrorize others into quick submission. In other words, exemplary violence had a point. None of our sources claim that the Romans were following a similar playbook in 171 or 170, but the logic is implicit. When Licinius sacked Malloea, neighboring communities fell into his hands. With the destruction of Pteleum, Antron "surrendered at once," while the Macedonian garrison at Larisa Cremeste deserted "out of fear." In similar fashion, Lucretius's merciless handling of Haliartus was followed by the quick surrender of Thisbe. In each instance, the Romans' frightful methods discouraged enemy resistance and precipitated surrender.[22]

Nevertheless, such military logic can only explain so much, and clearly many Greek and Illyrian communities suffered for the sake of material and political profit. Lucretius personally benefited from spoils taken from Haliartus, Thisbe, and Chalcis. Upon his return to Italy, he used his plunder to fund a water channel for the city of Antium (south of Rome), where he had an estate. He also decorated a temple with precious artwork that he had seized abroad. Like many other commanders, Lucretius leveraged these war profits for political purposes, investing them in visible public places that were likely meant to increase his popularity. The rank-and-file soldiers also hoped to profit from the war. When the consuls conducted the levy for 171, at least some Romans volunteered for the legions because they expected rich plunder. Soldiers had come home rich after previous campaigns in Greece, and the new generation wanted a piece of the pie. At the same time, there are hints that Roman commanders were struggling with morale problems in the army and tried to reward their troops to tamp these down. A desire to placate the soldiery may have influenced the Roman legate in Illyria: he only sacked Ceremia to reward his men, and it may not be a coincidence that he targeted two "wealthy towns." The need to appease the soldiers might explain the lack of discipline at Chalcis as well, where Roman troops were allowed to plunder and commit violence against the populace of an allied town.[23]

The behavior of other commanders is harder to explain, but some combination of the above—strategic needs, terror tactics, greed, and politics—likely shaped their choices. Hortensius's decision to storm allied Abdera, executing its leaders and enslaving the rest, was probably punitive and exemplary. Greek loyalties were already beginning to waver after Callicinus, if not before, and the Romans were sensitive to any signs of disloyalty. When the Abderans did not immediately bend to Hortensius's demands, his brutal response both punished disobedience and sent a sharp message to other cities that had the audacity to resist Roman orders. It probably did not hurt that his soldiers profited from sacking the city, since they had failed to achieve anything otherwise. Last, Cassius's unsanctioned march toward Macedonia clearly stemmed from his desire for military glory and its associated political rewards, but his decision to plunder and enslave neutral communities along the way is less comprehensible. It is possible that he was simply seeking any opening for military success or material reward. When opportunities were not immediately forthcoming, he simply created new ones by launching unprovoked attacks against neutral tribes. The behavior may appear extreme in isolation, but it fits rather well with the larger picture of Roman conduct in these years.

"THE BASEST VIOLENCE": GREEK REACTIONS TO ROMAN WARFARE

Whatever their motives, the Romans' conduct in the first two years of the war sent shockwaves through Greece, and their carefully assembled coalition of allies began to crack. In late 171 and early 170, several embassies appeared before the Roman Senate to express outrage and make pleas for restitution. Ambassadors from Thisbe and Coronea, which were now under the control of pro-Roman leaders, apparently protested the rough treatment of their towns (though a gap in Livy's text makes it impossible to know the substance of their complaints). Abderan envoys "wept before the Senate and lamented that their town was stormed and sacked by Hortensius." An ambassador for Chalcis came to "deplore the disasters of his homeland" and the acts of "arrogance, greed, and cruelty" inflicted upon his city:

> [Their] temples had been stripped of all their adornment . . . ; free persons had been rushed away to slavery; the possessions of allies of the Roman people had been plundered and daily were being plundered [still]. For, according to the precedent set by Gaius Lucretius, Hortensius too was quartering his sailors, in summer no less than in winter, in private houses, and the homes of Chalcis were full of the mob from the [Roman] fleet; at large among the Chalcideans and their wives and their children there were men utterly reckless in word and deed. (Livy 43.7.10–11; trans. Schlesinger, modified)

Ambassadors from other allied cities were less confrontational and emphasized their continued support for Rome. But even friendly representatives from Athens registered a subtle complaint between their effusive promises of cooperation. They explained that Lucretius and Licinius had demanded a huge quantity of grain, and that Athens had dutifully delivered it "even though they tilled barren soil and fed their own farmers with imported grain." Their point was plain enough: they fulfilled the Romans' demands, but doing so was incredibly burdensome.²⁴

More serious warnings came out of rugged northwestern Greece, in a region called Epirus. By the time of the Third Macedonian War, Epirus was inhabited by over a dozen semi-autonomous political groups ("tribes" or *ethnē*), which were politically united in the federal Epirote League. The most important of these tribes were the Chaonians in the north, the Thesprotians further south, and, largest and most powerful, the Molossians in central and eastern Epirus.* The Epirote League had agreed to cooperate with Rome at the beginning of the war, but this may have been a decision of expediency, since the legions would have to cross through Epirus on their way to fight Perseus. No one wanted to find himself on the sharp end of a Roman sword. Furthermore, many Epirote leaders disagreed about where to place their loyalties. Close relations between Epirus and Macedonia went back centuries, and many leading Epirotes had old ties to the Macedonian royal house. At the same time, there were Epirote politicians who wanted to cozy up to Rome to enhance their own status.

The fragility of the Epirotes' support for Rome was exposed for all to see in 170. That year, two Molossian leaders plotted to abduct one of the newly elected consuls, Aulus Hostilius, as he passed through Epirus to take over the Roman army encamped further east. The two Molossian plotters sent off letters to Perseus, informing him of their bold plan and urging him to come to Epirus. Luckily for Hostilius, this all came to nothing, as Perseus was delayed, and the consul caught word of the conspiracy and escaped. It was still a warning sign.

Shortly after the kidnapping debacle, many Epirote communities defected outright to Perseus. Their motives for abandoning Rome will never be known for certain, but our sources imply that the Romans' overbearing treatment of Greek allies pushed the Epirotes into the Macedonian camp. According to Polybius, one Epirote leader feared that he would be arbitrarily arrested—like the hapless Aetolians who were carted off to Rome after the Battle of Callicinus. So "he resolved that he would do everything possible so that he would not be seized without trial and sent to Rome because of the false accusations" of his political rivals. Presumably the Epirotes also knew about the destruction of Abdera, the plundering of Chalcis, and the various other atrocities the Romans had committed against allies, neutrals, and enemies alike. Eventually, for whatever reasons, they reached a breaking point, and most of the Molossians, along with some Thesprotians and Chaonians, switched their allegiance to Macedonia. Most of the other Epirotes stayed with Rome, but the League had been ripped in two.²⁵

Clearly, some Greeks had come to regard the Roman conduct of war as capricious and cruel. And given the Romans' heavy requisitions, arbitrary be-

---

* The war leader Pyrrhus, who invaded Italy and fought the Romans in the early third century, was king of the Molossians and leader of the Epirote League. By the Third Macedonian War, Pyrrhus's dynasty was extinct, and only the federal League government remained.

havior, and overt violence in 171 and 170, complaints and even defections are not so surprising. Nonetheless, we need to put these reactions in perspective. Greek warfare was hardly gentle before the Romans arrived on the scene. In the fifth and early fourth centuries, Greece was dominated by competing city-states like Athens, Sparta, and Thebes. War was virtually unremitting in this "Classical" period of Greek history, and inter-Greek conflict was regularly punctuated by atrocious displays of violence. Some Greek writers in this period insisted on the rights of the victor to treat the vanquished as he pleased, without a hint that the victims deserved anything better. The fourth-century author Xenophon, an experienced Greek soldier, wrote that "it is an eternal law among all men that whenever a city is captured in war, the bodies and property of the inhabitants belong to the conqueror."[26]

With the rise of Macedonia and the conquests of Alexander the Great in the latter half of the fourth century, warfare remained a harsh and uncompromising affair. Alexander himself destroyed cities, executed prisoners, and used terror tactics to subdue his foes. According to one modern scholar, Alexander "massacred on such a large scale [in India] that [he] must have come close to wiping out entire tribes." Alexander's death in 323 did little to bring peace to the Greek world or the eastern Mediterranean, and the subsequent "Hellenistic Age" of Greek history was born in violence. Alexander's Successors painfully and laboriously carved their own kingdoms out of his sprawling empire, waging a series of destructive wars to do so. By the early third century, three large Hellenistic kingdoms had emerged, centered on Macedonia (under the Antigonid dynasty), Egypt (under the Ptolemies), and Syria and Mesopotamia (under the Seleucids). These kingdoms continued to struggle for the scraps of Alexander's long-disintegrated conquests, and war between city-states, federal leagues, and great powers remained a near-constant in the Greek east. Around the time that the Romans began to intervene in the eastern Mediterranean, expansionist monarchs like Philip V were meting out their own shares of destruction and mass violence (see appendix 2). Polybius, writing over two centuries after Xenophon, still describes warfare as a sort of ruthless calculus: it is necessary to seize and destroy the enemy's cities, people, and resources, he says, in order to weaken them and strengthen your own people.[27]

Yet if Hellenistic warfare remained a pitiless exercise in might makes right, it also developed certain conventions—both pragmatic mechanisms and expectations around normal military conduct—that moderated the treatment of captives and conquered communities. First, Hellenistic monarchs and larger Greek states were generally more interested in *ruling* cities and territories intact than in destroying them. Polybius states this explicitly, writing that Macedonian rulers and Alexander's Successors had once "run all risks against each other in the field and did everything to prevail over one another by force of arms, but spared cities in order that they might rule over those they conquered and be honored by their subjects." This is not empty rhetoric. As economic hubs, administrative centers, and strongholds, Greek cities were valuable assets; they were generally not targets to be destroyed, but prizes to be won by arms or through mercy, gifts, and generosity. Warlord-kings like Demetrius "the Besieger" (Perseus's great-great-grandfather) frequently spared Greek communities that they captured in war, thereby earning praise and at least temporarily bolstering their reputations in Greece. Similarly, when the Seleucid King Antiochus III invaded Greece in 192, he dealt moderately with captured towns in order to cultivate a positive reputation, and some cities voluntarily joined him due to his well-known benevolence. Evidently, war leaders like these (and there were many others) could reap practical benefits from more restrained conduct.[28]

Although victorious Greeks and Macedonians might enslave their captives, there were incentives for treating prisoners more mildly. For instance, it may have been typical to ransom prisoners in Hellenistic warfare. Polybius even says it was "customary" for the Greeks to ransom prisoners, suggesting that the practice was relatively common—though only prisoners of higher status could normally afford this option. Additionally, it may have been more lucrative for captors to ransom their prisoners than to enslave them, and third parties could ransom prisoners in order to bolster their prestige and good repute.[29]

Finally, there were special incentives for treating enemy soldiers moderately. Hellenistic states drew on a large portion of their total manpower when they mobilized for war and relied significantly on mercenaries. This created a situation where armies were both expensive and fragile, and where a single defeat could have a generational impact on a state's military and economic potential. Furthermore, some Hellenistic soldiers were quite willing to switch sides and join a rival army, especially to improve their prospects for pay or survival. As a consequence, quality recruits were in high demand, and rulers took steps to acquire them whenever possible. We often hear of kings and warleaders sparing captured soldiers or defectors, then incorporating them into their own military forces. Commanders who treated captives generously could also gain a reputation for benevolent mercy, which in turn might attract soldiers to their banner. To give just one example, Plutarch reports that the Macedonian troops of Demetrius "the Besieger" deserted to Pyrrhus of Epirus (a non-Macedonian invader!) because the latter reputedly treated prisoners well. Such fluid loyalties were probably most common in the immediate years after Alexander the Great's death, when the political situation remained unsettled. Even so, Hellenistic soldiers were still switching sides, getting spared, and getting enrolled into rival armies over a hundred years after Alexander perished. Macedonian troops even developed a special gesture to signal their surrender or their intention to switch sides: they would raise their pikes upward toward the sky instead of pointing them forward toward the enemy. They were still using this signal, which Polybius calls "customary," in the early second century.[30]

Hellenistic rulers and generals were not lambs, and they could be brutal when it suited their purposes—but they also had a vested interest in maintaining a positive reputation, since cities "bounced around like footballs," and soldiers might change allegiances. (Reputation went both ways, too: very ruthless leaders could find themselves diplomatically alienated, as happened to Philip V during his violent Aegean campaigns in the late third century.) Overall, moderate conduct seems to have been as or more usual than harsher alternatives, and it is poignant that no city in mainland Greece was destroyed for one hundred years after Alexander's death, just as it is poignant that the destruction of Mantinea—which broke this silence—aroused hostility and bitterness from Greek observers.[31]

When the Romans first waged war in Greece in the late third century, they too used moderation and diplomacy to cultivate relationships and win allies. On the other hand, Roman military practices had been forged in a different environment, and there was less incentive for restraint on the part of Roman armies than there was for their Hellenistic counterparts. Indeed, our sources hint that Roman methods ran counter to the conventions and expectations of the Greeks. The first thing to note is that the Romans, with their vast manpower reserves, had no need or desire to absorb Greco-Macedonian troops into their legions or to treat surrendering soldiers with restraint. This fatal disconnect between Hellenistic convention and Roman practice led to a bloodbath at the Battle of Cynoscephalae (197). At a critical moment in the battle, the Romans flanked the Macedonian phalanx, turning the tide and putting most of the Macedonians to flight. As victorious Roman legionaries furiously pursued the enemy, some of the Macedonians halted and pointed their pikes upward instead of fleeing for their lives. The Romans did not know that this was a signal for surrender or defection, and they butchered the Macedonian soldiers. Other survivors were later sold as slaves. A similar clash between Roman practice and Hellenistic expectations occurred in 207, when a Roman commander captured the Greek city Aegina. After the commander collected the inhabitants on his ships with the intention of selling them, the captives begged permission to send messengers to their kin to obtain ransom money. Although the Romans occasionally ransomed prisoners, apparently this was not their normal practice. So the commander

> at first refused very sharply, saying that they ought to have sent envoys to their betters to come and save them while they were still their own masters, not now [that] they were slaves. . . . So at the time he dismissed [them] . . . but the next day, summoning all the prisoners of war, he said he was under no obligation to be lenient to [them], but for the sake of the rest of the Greeks he would allow them to send envoys to get ransom, as such was their custom. (Polybius *Histories* 9.42.6–8; trans. Paton, modified)

One modern scholar sardonically says that "the outcome was not impressive," as much or all of the population was still sold into slavery.[32]

The Romans' lack of territorial ambitions in Greece also loosened restraints on their conduct. In their eastern wars before 146, the Romans might seek to defeat rival states like Macedonia, but they were not out to annex or garrison conquered territories like an expansionist Hellenistic king. When they captured Greek cities, the *long-term* benefits of restraint may have been less evident, or less important, to Roman generals on the ground; and in some circumstances, the immediate benefits of mass violence would have been more appealing. Whatever their reasons, between 214 and 146, the Romans destroyed, burned, or sacked several dozen Greek cities, and subjected many further Greek populations to massacre and enslavement (see appendix 1). This meant that the scale of Roman depredations was simply greater than anything the Greeks had experienced in recent memory.

Taken together, Roman military conduct frequently ran counter to Greek military conventions. Consequently, almost from the moment the legions cut their way into the eastern Mediterranean, many Greeks decried the Romans as "the most ruthless barbarians." During the first war between Rome and Macedonia, several Greek diplomats delivered speeches criticizing the Romans for their harsh military conduct. In Polybius's *Histories*, one ambassador venomously condemns his fellow Greeks, the Aetolians, for making an alliance with the "barbarian" Romans. His powerful words are worth quoting here:

> You [Aetolians] say that you are waging war against Philip [the Macedonian king] on behalf of the Greeks . . . but you are fighting for the enslavement and destruction of Greece. That is the story your treaty with the Romans tells, which before had existed in writing, but which now can be seen in action. . . . You yourselves, when capturing a city, would not allow yourselves to commit outrages against free men or to burn their towns, since you believe such conduct to be cruel and barbarous. Yet you have made this treaty, by which you have handed to the barbarians all the rest of the Greeks to be exposed to the basest violence and lawlessness. (Polybius *Histories* 11.5.1–3, 5–7)

To reinforce his point, he mentions Oreum and Aegina, two cities the Romans sacked and enslaved. The Romans are a foreign blight spreading from the west, he argues, and they would use the conflict with Macedonia as a pretext to subjugate Greece.[33]

Other Greeks shared this anger and indignation, and Polybius recounts a similar speech delivered at Sparta in 210. One ambassador, who was trying to convince the Spartans to ally with Macedonia, heaps heavy blame on the Aetolians for making an alliance with the Romans. His accusations specifically highlight Roman violence: "It was just the other day that the Aetolians, together with the Romans, captured the unfortunate city Anticyra, enslaving its inhabitants. So the Romans are carrying off the women and children, clearly to suffer the things which must be suffered by those who fall into the power of aliens." Note too the repeated characterization of Roman behavior as "barbarian" or "alien"—that is, un-Greek.[34]

These same sentiments were expressed during the second war between Rome and Macedonia. In the first years of the conflict, Roman ambassadors made an offer of alliance to the Achaean League in southern Greece, which was then allied with Macedonia. However, the League reportedly rejected them "because of certain atrocities that the [Roman] consul Sulpicius committed against the Greeks." According to another source, the Achaeans were reluctant to ally with Rome because one Roman commander had destroyed two Greek cities, while another had "brutally handled ancient Greek cities which had done no wrong to the Romans, and were ruled by the Macedonians against their will." The Achaeans did eventually join with Rome, but they were partly motivated by fear. One Achaean politician argued that the threat of Roman violence *compelled* the League to abandon Macedonia. The Macedonians could not protect them, he asserts, and Rome's fleet had already preyed on numerous Greek cities. For extra emphasis, he raises the specter of Dyme, a city that the Romans had recently sacked. This warning must have had special weight with Achaean leaders, who "shuddered at Roman arms."[35]

Negative reactions to Roman conduct persisted in these and later wars. During a war between the Romans and their former friends, the Aetolians (191–189), some Greek envoys went to the Senate bewailing "slaughter, burning, and destruction," and "wives and children carried away into slavery" by the Roman legions. There is also evidence that Roman armies earned a menacing reputation throughout the Greek east, to the point that they unnerved battle-hardened soldiers. According to Livy, Macedonian troops were struck with "fear and apprehension" when they saw the bodies of comrades cut to pieces by Roman swords, and "trembling, they realized the sort of weapons and the sort of men against whom they would have to fight." In a battle between Roman and Seleucid forces, the Seleucids panicked and broke at the mere sight of a Roman flanking force. In Appian's words, the frightened soldiers had "heard for some time about the Romans' terrifying way of fighting."[36]

These speeches and moments of terror, though not verbatim historical records, reflect anti-Roman feelings in Greece. The reason for this hostility is undoubtedly complex, and we should only generalize with caution, since surely there were many different Greek opinions about Rome. Nonetheless, it does appear that Roman military conduct shaped Greek opinion for the worse—and that the Greeks regarded the Romans as *exceeding* the threshold of acceptable military behavior, both in their methods of waging war and in their treatment of prisoners, cities, and populations. This impression of Roman warfare was confirmed anew in the Third Macedonian War, further threatening Rome's coalition of allies in Greece.[37]

"THE SENATE WAS DISPLEASED": THE NEW POLICY OF RESTRAINT

The Romans had shown little desire to bow to Greek complaints in the past. When Greek embassies appeared in Rome to protest the conduct of Roman commanders, the Senate's *modus operandi* was to give them a hearing and then send them on their way. Usually the senators neither offered redress nor punished the alleged wrongdoers. (Recall Marcellus's ominous response to the complaints of the Syracusans: "The question is not what I should have done—for the norms of war defend anything I did to my enemies—but what they deserve to have suffered.") From the Roman perspective, there was no great incentive to restrain their commanders or punish alleged excesses just to appease the Greeks. After all, mass violence and terrorism had been useful in their prior campaigns in Greece, especially when these tactics had been paired with active diplomacy and Roman battlefield victories.[38]

But if the stark Greek reactions to Roman conduct in 171 and 170 were nothing new, there were significant differences in both the *reasons for* and the *context of* their complaints. In past wars in Greece, the Romans primarily assaulted the allies and subjects of their chief enemy: Macedonia. In the Third Macedonian War, by contrast, they also brought violence and hardship to their own Greek friends, committing atrocities against friendly towns, demanding heavy requisitions, and arresting the leading citizens of allied states on flimsy pretexts. The Romans were treating friends like foes, which must have appeared cruel and arbitrary.

At the same time, Roman failures on the battlefield did not inspire confidence in their inevitable victory. Perseus won the first major engagement of the war, and he continued to strike successful blows. In 170, he made a surprise attack on the Roman fleet while it was stationed on the island Euboea, seizing twenty supply vessels (and sinking the rest) as well as four warships. The same year, he defeated the consul Aulus Hostilius in battle when the latter attempted to invade Macedonia. The experience seems to have cowed the consul, for he refused to meet Perseus in battle again. The tireless Macedonian king next campaigned in northern Greece where he grabbed more cities in Thessaly, then dashed further north to defeat and slaughter 10,000 Dardanians (a neighboring people and a frequent threat). In late 170 or early 169 he also moved west into Illyria to urge the king Genthius to join his cause. Along the way, he captured several forts and towns and drove out Roman garrisons. Later that winter, he pushed into central Greece to seize a strategic Aetolian city called Stratus. He ultimately failed to secure it, but he did occupy another Aetolian town, Aperantia, which gave him a new base in Greece. This streak of successes continued to chip away at Greek support for Rome. When he attacked Stratus, in fact, Perseus had expected the help of an Aetolian leader, a clear sign that more Greeks were wavering.[39]

There were also troubling military consequences for the defections in Epirus. After switching sides, the Epirote defectors allowed Perseus to garrison an important fortress town called Phanote, and they may have given him other strongholds as well. They also lent troops to Perseus and let him march through their territory, allowing him to dive much deeper into Greece (as he did in 170/169 to attack Stratus). At the same time, the Romans were suddenly prevented from passing through Epirus. In the territory of the Molossians, the main Epirote tribe to defect, archaeologists have identified fifty-eight fortified sites. These towns, fortresses, guard towers, and fortlets all possessed heavy stone walls and other defenses.* These sentinel-fortresses controlled every access point into the Molossian heartland (the modern Ioannina Basin), and effectively shut the Romans out of three major routes from the west coast into the Greek interior, through the modern Drino, Kalamas, and Acheron valleys.[40]

This was no small loss since these were essential pathways into northern Greece. In the Second Macedonian War, the consul Flamininus took steps to secure marching routes through Epirus into Thessaly. During the first year of the Third Macedonian War—before the Epirote defections—the consul Licinius had reached Thessaly by passing through the heart of Molossian territory in east-central Epirus. Hostilius, consul in 170, traveled from the coast into central Epirus and was about to enter Molossian territory when he caught wind of the kidnapping plot. This threat stopped him at the fortress town Phanote, forcing him to "abandon the plan of journeying through Epirus" (and to sail much further south before marching overland). The consul of 169 was also unable to take his predecessors' paths through Epirus. He had to use "the only short route to Thessaly which was now open to him," moving further south and then cutting up through central Greece.[41]

The Romans were not blind to these festering problems, and they diverted manpower to attack key Epirote strongholds. In 170, a Roman legate attacked Phanote, which guarded a major pass into Molossian territory; given its strategic significance, it was probably not a random target, and it points to larger Roman goals in Epirus. The attempt on Phanote ended in failure, but Polybius says that the same Roman commander was operating in Epirus the next year, in 169. Evidently Epirus was important enough to warrant the ongoing diversion of Roman forces.[42]

In short, the strategic situation was notably affected when the Molossians and other Epirotes switched sides, at a time when Perseus was enjoying some military success.† The Romans needed to prevent further defections, a fact that gave new urgency to Greek complaints. This time, rather than ignore the Greeks, the Senate issued a flurry of decrees on behalf of Abdera,

---

* For readers interested in ancient military architecture, Epirote fortifications commonly used two types of stonework: "polygonal" and "ashlar" masonry. Polygonal masonry is composed of large, many-sided stone blocks, finely dressed so that they interlock tightly together (usually without mortar). Ashlar masonry is of more regular, rectangular stone blocks, often laid at right angles and in horizontal rows and sometimes bonded with mortar.

† Even the support of more peripheral Greek communities could have strategic consequences. Later in the war, for example, the city Antissa provided a friendly harbor and supplies for Macedonian ships operating in the Aegean. See Livy 44.28.1–44.29.4, 45.31.13.

Coronea, and Thisbe. These decrees granted privileges to pro-Roman politicians in each community but also confirmed the inhabitants' rights over their property. The Senate further decreed that an "improper war" had been waged against Abdera and sent two ambassadors to find and free all the Abderans who had been wrongly enslaved. They repeated these instructions to the consul Hostilius and the praetor Hortensius in Greece, effectively censuring their behavior and the behavior of their predecessors.

The senators also turned their attention to Chalcis, the allied city that had been ravaged by Rome's naval forces. The Senate first expressed appreciation for the city's many services to the Republic, then declared that the praetors Lucretius and Hortensius had acted without senatorial approval. They also insisted that they were fighting Perseus—and before him, his father Philip—for the sake of Greek freedom, not to harm Rome's friends. Last, they sent a second letter to Hortensius. In it they expressed their firm displeasure, ordered him to find and restore any citizens of Chalcis that he and his men had enslaved, and barred him from quartering any of his regular sailors in private homes.[43]

Lucretius faced an even darker storm of criticism, which may have destroyed his political career. While he was absent from the city (busily putting his war plunder to good use in a neighboring town), the tribunes of the plebs* repeatedly attacked him in public meetings called *contiones*. We should picture crowds of citizens in the Forum gathering around the monumental speakers' platform, raucously booing and cheering as the tribunes made speeches against Lucretius and ridiculed his misdeeds.† Lucretius soon received a summons to appear before the Senate, where an audience full of hostile colleagues confronted him. They jumped on the bandwagon and verbally tore him to bits. He was then dragged to another public meeting where the tribunes lobbed more criticisms at him and set a date for his trial in a formal assembly. In the trial, the citizen jury voted to punish him with a fine of one million *asses*,‡ a substantial sum that would have inflicted financial hardship or ruin, even for a senator.[44]

The Senate's representatives abroad also sought to repair relations in Greece. Roman ambassadors "tried to convince people of the mildness and humanity of the Senate," affirming loyalties and expressing goodwill to the Greeks. They also announced the recent decrees and declared that no state

---

* Tribunes of the plebs (not to be confused with military tribunes) were Roman officials charged with protecting the mass of Roman citizens against the arbitrary power of other magistrates. They possessed legislative and veto powers.

† The speaker's platform was called the Rostra. In this period, the raised stone platform set off a traditional assembly area called the Comitium, where citizens met to vote and probably gathered for *contiones* as well.

‡ A bronze *as* (plural *asses*) weighed about two ounces. The minimum wealth expected of a Roman senator in this period was probably one million *asses*. To put this in perspective, the annual salary for a Roman soldier at this time was about 1,800 bronze *asses*. So Lucretius had to pay a fine equivalent to about 555 annual legionary salaries, quite possibly a substantial portion of his total wealth. For legionary pay in this period, see Southern 2006, 106. For expectations of senatorial wealth in the Middle Republic, see Livy 24.11.7–9; Lintott 1999, 71 n.24.

should provide supplies, money, or troops to Roman commanders without express instructions from the Senate itself. Yet the ambassadors were not completely conciliatory. During their visit to the Achaean League, they made it clear that they knew who supported Rome in a not-so-subtle warning, and when they checked in with the Aetolian League, they initially asked for hostages (but then walked back the request). Still, damage control was the overwhelming goal of Rome's diplomatic activity in 170 and into 169.[45]

THE FALL OF PERSEUS

One of the new consuls for 169, Quintus Marcius Philippus, got the command in Greece. To our modern eyes, Philippus might not seem the best candidate to bring the war to a successful conclusion. He was "sixty years old and fat," in Appian's words, and during his previous consulship, he was ambushed and defeated by a Ligurian tribe in northern Italy. However, his father had enjoyed a relationship of "guest-friendship" with Perseus's father, and Philippus had been to Greece in both diplomatic and military roles. He knew the physical and political terrain well and put that knowledge to good use. Concentrating on the task of invading Macedonia, Philippus pushed his army through a treacherous mountain pass that brought him to the kingdom's southeastern coastal plains.* There he established a small but secure foothold by the end of the campaigning season. Further, not only did Philippus finally move the fighting out of Greece and into Macedonia, but he also fully embodied the Senate's new policy of restraint. He responded positively to Greek representatives who approached him and avoided burdensome requests of their governments. He also treated captured cities mildly, very much in contrast to his predecessors. When one Macedonian town surrendered, "he was contented with taking hostages, and he promised the inhabitants that he would leave the city without a garrison, that they would be free from tribute and live under their own laws." He did this "in order to win the hearts of the remaining Macedonians," indicating that he sought to wipe out the reputation for cruelty that sullied previous Roman generals.[46]

Even so, the war was not yet over. Philippus could still be dislodged from his precarious beachhead, and the Illyrian king Genthius finally decided to enter the conflict on the Macedonian side. Additionally, Perseus was diplomatically engaged with the island republic of Rhodes and with Eumenes, the king of Pergamum. Rhodes had cautiously supported the Romans throughout the war, but by 169, its representatives were trying to mediate between

---

* Perseus was bathing when he learned that a Roman army made it through the mountains into Macedonia. He reportedly leapt out of the bath in a panic, smacking his thigh furiously and exclaiming that he had been beaten without a battle. His next move was rash: he pulled his men back from *all* the mountain passes and withdrew further north, a decision that left almost every door to his kingdom wide open and allowed Philippus to secure his position and supply chain. It was a decision Perseus would regret, and was not his only strategic error in the war.

Rome and Perseus. For his part, Eumenes was a close Roman ally and had provided energetic support from the beginning of the conflict. The fact that he and Perseus were allegedly talking at all, and that some Rhodian politicians were becoming openly pro-Macedonian, did not bode well for the Romans.[47]

The next Roman commander needed to achieve something significant in order to move the war forward at last. As it happens, Philippus's successor was up to the challenge. Lucius Aemilius Paullus was also in his sixties in 168, when he took up the Macedonian command. Unlike his predecessor, however, he had an impressive record of military accomplishments, with major victories in both Spain and Liguria to his name. He was also a thoroughly blue-blooded member of an old aristocratic family and refused to court popularity in public life. When he was elected to his second consulship in 168, rather than give the traditional speech expressing gratitude to the Roman people, he asserted that he was under no obligation to them; after all, *they* wanted *him* for a general, or they would not have elected him. And he did not want or need their gossip-driven advice. As he put it gruffly, "if [you] try to command the commander, [your] campaigns will be even more absurd than they are now." This attitude extended into the military camp, where he was a strict disciplinarian and rarely gave in to the complaints or desires of his soldiers just to win their favor.[48]

Paullus brought that strictness of character with him when he arrived in Macedonia that spring. By that time, Perseus had planted his army along the banks of a dry riverbed near Mount Olympus. It was a highly defensible position, but Paullus had a plan. For two days, he launched diversionary attacks against Perseus's main army while a Roman detachment made its way around the enemy's back. On the third day, this secondary force seized a strategic pass in the Macedonians' rear. When Perseus discovered that he had been outmaneuvered, he withdrew further north to the town of Pydna and staked out some flat terrain that was suitable for his forces. Paullus followed with his legions and found the Macedonian army ready for combat, on ground that was dangerously favorable to Perseus's phalanx. The consul decided to delay battle that day, despite the eagerness of his men, but the climactic clash was creeping closer.[49]

That night—it was June 21, 168—there was a lunar eclipse. Soldiers in both armies took this as an omen that Perseus's kingdom would fall, which Polybius says was simultaneously joyous for the Romans and terrifying for the Macedonians. Yet even if the heavenly phenomenon had buoyed Roman spirits, the consul was in no hurry to meet Perseus on the battlefield. Paullus's father had died fighting Hannibal at Cannae, and he would have known that it was risky to fight on ground of the enemy's choosing. He tried to delay again the next day, holding war councils, sacrificing animal after animal to secure the gods' blessings, and waiting for some strategic advantage to appear. Soon, however, circumstances spiraled out of his hands. Our sources disagree about why the battle began, but it seems that a minor skirmish escalated until both generals drew up their full armies.[50]

As Paullus beheld the assembled Macedonian army, he was presented with an awesome sight. It consisted of 40,000 infantry and 4,000 cavalry, Macedonians as well as auxiliary and mercenary troops from Thrace, Gaul, and Greece. At its center was a 21,000-man phalanx, sixteen ranks deep and nearly a mile long, with weapons and armor that "filled the plain with the gleam of iron and the glitter of bronze." Paullus was reportedly terrified at what he saw:

> As the attack began, Aemilius [Paullus] came up and found that the Macedonian battalions had already planted the tips of their long spears in the shields of the Romans, who were thus prevented from reaching them with their swords. And when he saw that the rest of the Macedonian troops were drawing their shields from their shoulders round in front of them, and with long spears set at one level were withstanding his shield-bearing troops, and saw too the strength of their interlocked shields and the fierceness of their onset, amazement and fear took possession of him, and he felt that he had never seen a sight more fearful; often in later times he used to speak of his emotions at that moment and of what he saw. (Plutarch *Life of Aemilius* 19.1–2; trans. Perrin, modified)

The Roman army may have numbered between 35,000 and 40,000 in total, but as the battle began, their slight disadvantage in numbers was less important than their asymmetric equipment. As the phalanx advanced like a juggernaut across the plain with pikes thrust forward, the sword-wielding Romans were unable to come to grips with the enemy. Some soldiers tried to batter the Macedonian pikes aside with shields, blades, or even their bare hands, but many were struck dead as enemy spear points pushed through their shields and armor. The Romans did not flee, but they fell back as the seemingly impenetrable mass bore down on them.[51]

Fickle Fortune soon favored the Romans. While the phalanx advanced, it slowly lost its integrity. Different parts of the formation met greater or lesser resistance from the opposing Roman infantry, and the Macedonian front became uneven. When gaps began to appear, Paullus saw his opportunity. After sending his war elephants (a loan from Rome's Numidian allies) and Latin troops to attack the Macedonian left, he ordered his flexible Roman units to infiltrate the openings in the enemy phalanx. The pikemen were helpless as the legionaries slipped into their lines, and their unwieldy pikes and small daggers were no use against Roman swords in close combat. The gaps in the phalanx ballooned into holes, and soon Perseus's army collapsed into a rout.

What followed was an extraordinarily bloody massacre. Such was typical in ancient battles, as victorious troops surged forward to release their fear and rage against foes that only moments ago had threatened their lives. The Romans killed some of the Macedonians where they stood, slew others who attempted to surrender, and chased fugitives for miles (see figure 6.1). Paullus's own son, the seventeen-year-old Scipio Aemilianus, is said to have returned to camp after dark covered in the blood of enemies he killed in the rout, and he was just one of many who participated in this hours-long

Figure 6.1. Reliefs from the Aemilius Paullus Monument at Delphi. Paullus discovered that a pillar was being erected at Delphi to honor Perseus, so he had it repurposed as a monument to his victory at Pydna. The scenes may depict "the massacre that ensued after Perseus's phalanx had broken," according to Roman historian Michael Taylor (2016, 574). (Photos courtesy Foto Marburg/Art Resource, NY.)

slaughter. Livy and Plutarch tell us that over 20,000 of the enemy were killed and 11,000 captured, compared to around a hundred Roman troops who mostly fell in the early phases of the battle. It is difficult to say whether these figures are accurate, but an imbalance in casualties between victor and vanquished was also typical in ancient warfare, and here the scales may have tipped heavily toward the triumphant Romans. Livy bluntly states that they had "never killed so many Macedonians in a single battle."[52]

Perseus escaped and went on the run, but was rapidly abandoned by his remaining friends and attendants and soon surrendered unconditionally. He and his family would become the jewel in the crown of Paullus's triumphal procession back in Rome, and a few years later, he died in captivity in Italy. It was an anticlimax for the king who had seemed, for a time, a worthy enemy of Rome. Meanwhile, a Roman praetor named Lucius Anicius defeated Genthius's Illyrian kingdom in only thirty days. He then marched into Epirus and attacked enemy strongholds, including Phanote and four other fortified centers. He subdued them all rapidly, and with that the war was over. Next came the Roman reckoning.[53]

"THE AXE FELL": AFTERMATH AND THE SCOURING OF EPIRUS

The historian Peter Derow writes that after Perseus's defeat, "the need for moderation was over, and the axe fell." In the immediate aftermath of the battle, Roman armies unleashed torrential fury throughout Greece and Macedonia. The nearby city of Pydna surrendered, but Paullus "gave" it to his troops to sack. Paullus also sent soldiers to sack three towns in Macedonia and one in Thessaly that had not shown sufficient loyalty during the war or had continued to resist. Down in central Greece, a Roman garrison helped in the massacre of 550 Aetolian statesmen and in the exile of others, acting in support of pro-Roman leaders in Aetolia. Paullus reprimanded the garrison commander for participating in the massacre, perhaps because he had acted without orders, but ultimately condoned the killings and the expulsions.[54]

After this initial outburst of violence, Paullus worked with ten commissioners from the Roman Senate (the *decem legati*) to hammer out a postwar settlement for Greece and Macedonia. Sometime in the summer of 167, Paullus and the commissioners ordered dozens of Macedonian leaders to go to the city Amphipolis. There the Macedonians assembled in a large group around a public tribunal, where Paullus and the ten commissioners sat in judgment. A herald called for silence, cutting short the murmur of the crowd, and Paullus solemnly announced the Senate's decisions in Latin (even though he was fully conversant in Greek) while another man translated. He declared that the Macedonians would be free to enjoy their own laws, to keep their cities and lands, and to elect their own leaders. However, the Romans would abolish the monarchy, divide the kingdom into four independent republics, and collect half the taxes that the Macedonians formerly paid to their king. They imposed a similar settlement on Genthius's Illyrian kingdom.[55]

The Romans next summoned Greek leaders before the tribunal in order to determine "which side had favored the Romans and which side had favored the king, more than who had done wrong or been wronged." Two Roman commissioners were also dispatched to the Achaean League to make the same determination. Opportunistic Greek politicians (or pro-Roman sycophants) seized the moment to feast like carrion birds on the soon-to-be-dead careers of their political rivals. They filled Paullus's and the commissioners' ears with accusations and slander, claiming that so-and-so had openly bragged about his friendship with Perseus or covertly favored the Macedonians. The Romans did not hesitate to take advantage of these claims, however dubious, and ordered hundreds of men from Achaea, Aetolia, Boeotia, Acarnania, and Epirus, as well as many prominent Macedonians, to go to Rome at once. These political prisoners were supposedly sent to stand trial, but just languished in exile in Italy for the better part of twenty years.* Paullus additionally ordered the beheading of an Aetolian and a Theban leader who had supported Perseus, and sent a force to destroy the city Antissa and forcibly remove its population because it had given aid to Perseus's fleet.[56]

The peace settlement was designed to ensure that Rome would never again face such a challenge from Macedonia. Thus Macedonia was carved up, its vigorous and warlike kings removed entirely. As for the Greeks, the Romans would secure their future faithfulness through aggressive political engineering, rooting out the leaders who demonstrated (or were suspected of) insufficient support and leaving loyalists and sycophants to govern in their stead. There was exemplary punishment here, too. The massacres, beheadings, forced migrations, and deportations demonstrated vividly that there was a very high price to pay for supporting Rome's enemies. The lesson, the Romans hoped, would not soon be forgotten.

After Paullus and the ten senatorial commissioners finished settling affairs in Greece, Macedonia, and Illyria to their liking, Epirus still waited for the victor's judgment. The Epirotes who defected to Perseus had materially aided the Macedonians with men and fortresses and closed off several marching routes into northern Greece. In a postwar climate such as this, where a whiff of pro-Macedonian sentiment could lead to exile or death, the most horrifying demonstration of Roman fury soon came down on Epirote heads.[57]

By mid-summer of 167, Paullus marched his army west through Epirus to board ships and return home. At this point, the Epirotes had long since surrendered and received Roman garrisons, but Paullus had specific instructions from the Senate regarding the communities "that had defected to Perseus." As he passed through the region, he contacted Epirote leaders and told them that he had come to remove Roman garrisons, since the Epirotes would be "freed" like the recently "freed" Macedonians. Paullus then instructed them to gather their gold, silver, and other treasures in each of their towns and villages, where it would be collected by his troops—a relatively peaceful, albeit extortive, exchange. He then dispatched soldiers to dozens of Epirote communities that had supposedly switched sides during the war (most of which were Molossian), timing the detachments so that they all arrived on the same day.† Our sources claim that Paullus sent detachments to around seventy "cities," but this poses a problem, as there were not seventy cities in Epirus in this period. However, the region was densely settled with fortified sites, including towns, villages, fortresses, and fortlets. Though many of these settlements were relatively small, there were certainly over seventy walled communities in the region, and this must be

---

* One of these exiles was Polybius, who began writing his *Histories* while in exile.

† Livy says that Paullus sent "cohorts" to each community. In the mid-Republican army, cohorts were maneuvering units, possibly ad hoc, numbering several hundred legionaries but not of standard size (as they would be in the later army). See Lendon 2005, 228; Polybius 11.23.1.

what our sources mean by "cities."* The point to emphasize is that this was a massive operation, with Paullus directing the movements of dozens of detachments marching to multiple locations. He probably deployed all or most of his army, perhaps apportioning his soldiers according to the size of the various targets.[58]

When they arrived at their destinations early in the morning, these troops pretended that they had come to collect the treasure. Towns and villages all over Epirus opened their gates to the Romans, either because they did not suspect anything foul or felt they had no choice, and legionaries moved freely inside. Perhaps nervous crowds gathered to watch as foreign soldiers marched into public spaces to find the waiting piles of gold and silver. Once they were in, Roman officers gave a signal, and at a stroke, the legionaries scattered to sack each of the unsuspecting communities. The orgy of violence that followed must have been horrendous, as soldiers were given license over people and property. Some Romans may have torched homes after they carried out the plunder or dragged out the inhabitants. Anyone who offered resistance was likely struck dead.[59]

While all of this was happening, the Romans would have been forcibly corralling the shocked survivors together into marketplaces or other open spaces and putting them under guard. Perhaps quaestors or other officers made head counts, or even inspected and documented the prisoners, while soldiers prepared the whole lot for sale or transport. Our sources agree that the Romans carried off a staggering 150,000 people that day.† This number, like the number of Epirote "cities," poses problems. On the one hand, as noted previously, the legions could handle very large numbers of captives. Merchants followed Roman armies on campaign, ready to buy up prisoners on the spot, and the Romans could use force, intimidation, and the like to cow their human plunder into submission. Paullus also had an exceptionally large force, perhaps 30,000 to 40,000 men, which could have handled a very large crowd of prisoners indeed. On the other hand, there is no way to check the figure without solid population data, and the number is suspiciously round. Therefore, we should not regard "150,000" as a precise count and perhaps not even as a broad estimate of the captives taken. Instead, this is best seen as the *impression* created by the sheer scale of the mass abduction.[60]

---

* Most of our sources for this episode probably go back to the contemporary account of Polybius. Interestingly, Polybius was criticized by another Greek historian for identifying certain "forts" (*purgous*) as "cities" (*poleis*) (Strabo 3.4.13). In this case, Polybius probably wrote that Paullus attacked seventy *poleis*, but he was using the term loosely and meant something like "settlements." When later authors like Livy followed Polybius's account, they took his wording literally, perhaps ignorant of the fact that Polybius's "cities" could be forts, walled villages, or other small communities. For *purgos* as a fort or fortified village, see Walbank 1979, 270.

† Livy uses the word *abducere*, "to carry off by force," to describe this mass kidnapping. Strabo and Plutarch use the word *exandrapodizein*, a variation of *andrapodizein*, "to andrapodize." According to Gaca (2010, 123), when our sources say that a whole community was "andrapodized," this is probably "a compressed way to signify the same twofold military practice of killing fighting-age males of the attacked city, village, or rural area, and then abducting andrapodized survivors from among the defeated populace."

Whatever the real numbers, the number "150,000" was credible enough for the sober Polybius to believe it and record it, and we can be confident that the Romans tore tens of thousands of people away from their homes that day. Moreover, there are archaeological indications of disruption in Epirus in this period. At Dodona, the most significant political and religious center of the Molossians, thousands of "lamellae" were produced in the fourth and third centuries; these were questions written on small strips of lead and directed to the sacred oracle of Dodona that have been found in large numbers by modern archaeologists. In 167, these questions to the oracle largely dry up, which indicates either a significant change in religious practice or, perhaps, a sudden lack of people visiting the oracle. Further, the Molossians posted many public inscriptions at Dodona in previous centuries, but only a single surviving inscription mentions "Molossians" after 167. The production of Epirote coinage also virtually ceases for a time, pointing to severe economic disturbances, and archaeologists have detected signs of a reduced population, with habitation shrinking or disappearing altogether in some settlements. The ancient geographer Strabo underlines this population decline, stating that Epirus was once "well peopled, even though the land is rough and full of mountains," but was much reduced after Paullus's attack.[61]

The mass sacking and mass enslavement alone would have effectively eliminated these communities, but Livy says that Paullus's troops also destroyed the fortifications before making an exit. There are suggestive hints of this destruction in the archaeological record.[*] At a city called Elea, for example, the archaeology suggests that the northern and most accessible fortification wall was demolished in this period, and at the same time there was a break in habitation (see figure 6.2). At a site called Doliani, which was probably ancient Phanote, archaeologists uncovered destruction layers at the eastern and northern walls of the citadel, noted that the town's outer wall was largely leveled, and identified a reduction in the settlement's inhabited area. More extensive demolition is evidenced at a town called Orraon (or Horreum), where one investigator observed the following:

> The wall throughout its circuit has been thoroughly and deliberately destroyed; many blocks have been shattered and others broadcast, in a manner which cannot be attributed to the conditions of weather or the passage of time on a site where other walls stand to a considerable height. (Hammond 1953, 135–36)

---

[*] There is always some danger in associating archaeological finds with specific historical events, given the ambiguous nature of material evidence. With the Epirote destructions in particular, the picture provided by archaeology is incomplete: archaeologists have only excavated a handful of larger sites in Epirus, and they have done less work in Molossia than other regions. And not every sign of destruction and abandonment necessarily resulted from Paullus's activities in 167. Nevertheless, archaeologists have noted many destruction events in Epirus that are chronologically proximate, a coincidence that must be significant. Above all, destroyed walls coupled with burning and disruption in habitation, within the likely zone of Paullus's attack, are probable signs of Roman actions—the lingering traces of massive, organized violence.

Figure 6.2. Top: The ruined northeast wall, gate, and tower, at Elea, Epirus. Bottom: The shattered blocks of the eastern wall at Orraon, Epirus. (Top: Photo by author. Bottom: Photo by Mandy Peacock.)

This intensive destruction has been attributed to the Romans, indicating that some of Paullus's soldiers exerted great effort to dismantle the walls.[62]

To achieve what Plutarch calls "so much destruction and utter ruin," Paullus had to marshal thousands of soldiers in a highly coordinated surprise attack. These troops, while allowed free rein to plunder, also had to subdue and corral tens of thousands of people and physically demolish heavy stone fortifications. In other words, this was a complex military operation that required planning, organization, labor, and decision-making at the highest levels of command. What had motivated the Romans to carry out their "greatest post-war atrocity"?[63]

The sources strongly suggest that the Romans hoped to profit from the attacks. This is indicated by the coordinated sack of the Epirote communities, the extortive request for gold and silver, and the massive culling of captives. Indeed, Livy and Plutarch put profit front and center in their accounts, emphasizing that Paullus's men were "given" these communities for the sake of plunder, and tallying profits gained. Moreover, such naked profiteering did not arise out of nothing. Paullus was notoriously reluctant to share the Macedonian royal treasures with the rank-and-file soldiers, insisting that the kingly loot should go to the Roman treasury—to the great displeasure of his men. But when the Senate directed Paullus to devastate Epirus, he was presented with an opportunity to address his soldiers' complaints without giving up any of the royal treasures. As Livy puts it, Paullus had hoped the looting of Epirus would "satisfy the expectations of his soldiers."[64]

But the soldiers' greed is not an adequate explanation on its own. There were easier ways to gain loot, ones that did not involve a massively coordinated attack or the destruction of fortifications. Notably, the Senate ordered Paullus to attack the Epirote communities "that had defected" or "rebelled." When placed in the larger context of the postwar settlement of Greece and Macedonia, when the Romans punished far lesser offenses with violence and deportation, it is fairly obvious that the devastation of Epirus was punitive. The Molossians and the other defectors were the only Roman allies who joined the enemy outright. Furthermore, by adhering to Perseus, they had aided the Macedonians with troops and strongholds, and frustratingly closed off important routes through northern Greece. The defections called for a response that would punish the Epirotes and send a powerful message to other states. Alongside other violence that followed the defeat of Perseus, the Roman treatment of Epirus seems less isolated and unusual.[65]

It is also significant that Paullus's soldiers tore down the walls of so many fortified sites in Epirus. As shown in chapter 3, such demolition would have been laborious, particularly on this scale, and required the targeted and deliberate expenditure of manpower. I am aware of only one comparably large-scale demolition of fortifications in this period, which offers an important parallel. In 195, the consul Marcus Porcius Cato reportedly destroyed the walls of every town east of the Baetis (Guadalquivir) River in Spain. Like Epirus, archaeological evidence shows that this region enjoyed "a relatively dense concentration of indigenous sites which might have been costly to capture by storm." According to several ancient authors, this mass demolition was meant to prevent future defiance from these cities and, Appian adds, removing their walls made them "easier to attack" (or "more accessible"). Region-wide defortification followed a strategic logic.[66]

In Epirus, the defectors' control of fortified sites had allowed them to shut down marching and communication routes during the war, and the Romans were repulsed when they tried to attack one of these strongholds, Phanote, in 169. This must have produced considerable frustration for Rome, particularly as the war dragged on beyond expectations. By stripping the "rebellious" Epirotes of their walls, Paullus ensured that they would be "easier to attack"—much like the rebel cities in Spain—and made these strongholds useless for the foreseeable future. The Romans would not face this threat again if they marched back east.[67]

CONCLUSION

The clash with Perseus had not been easy. The Macedonian king proved a surprisingly doughty enemy in the first two years of the conflict, defeating the Romans in battle and tirelessly ranging across Greece and Macedonia to seize strategic sites or thwart challenges. Roman commanders spectacularly failed to achieve battlefield victories or build up any significant military momentum against Perseus in 171 and 170. However, they could and did turn against the Greeks, ravaging some communities, destroying others, and bringing death and enslavement to their populations. Cities like Haliartus, Coronea, and Thisbe were obvious targets, since they had joined with Perseus and harassed Roman allies in central Greece, but friendly towns like Abdera and Chalcis also became victims of Roman predation. Meanwhile, the Romans were demanding (or outright commandeering) supplies from allies such as Athens and neutrals like the Illyrian king Genthius, and they arbitrarily arrested several Aetolians on thin pretexts.

These behaviors stemmed from several overlapping pressures, including strategic necessity, greed, and Roman politics. Regardless of the motives, the Greek allies became increasingly disenchanted, particularly when Rome's ruthlessness was coupled with an overall lack of military success against Perseus. A series of Greek embassies shuffled westward, bringing complaint after complaint about the demands, violence, and indignities they had suffered. By 168 the Molossians and other Epirotes had defected, Genthius had entered the war on the side of Macedonia, and support for Perseus was more openly expressed in Aetolia, Rhodes, and even peripheral communities like Antissa. All of this had substantial consequences, increasing Perseus's striking distance, closing off marching and communication routes, and giving the enemy more bodies and strongholds with which to prosecute the war. And presumably, Rome's remaining Greek allies became less enthusiastic about participating in a struggle where they themselves might suffer Roman attacks.

It is little wonder that the Senate sought to smooth over relations in Greece beginning in late 170, censuring their commanders, offering redress, and launching a diplomatic counter-offensive. This new moderation was a matter of political and strategic necessity. Curbing Rome's reputation for capriciousness and cruelty counteracted pro-Macedonian sympathies and headed off any further defections. Furthermore, by offering remedies to the towns that had been brutalized *after* their surrender, the Senate signaled that the Romans would deal fairly with enemy communities that did not resist, expediting surrender in the future.

It must be stressed that the Senate was not attempting to curtail the use of mass violence as such, nor were they adopting a new ethical stance toward massacre, mass enslavement, urban destruction, and the like. At Thisbe and Coronea, which had been roughly handled after they surrendered, the senators' remedial actions solely benefited pro-Roman members of the community. Apparently, they did not allow either town to rebuild its walls, and at Thisbe, the new pro-Roman government was authorized to arrest anyone opposed to the Republic. In other words, the Senate tacitly approved of measures that harmed Macedonian sympathizers or served a clear military purpose. The people of Haliartus, whose city had been dealt a full measure of Roman atrocity, also learned the limits of senatorial restraint. At the end of the war, the Haliartans requested (through a third party) that the Senate spare their surviving inhabitants, and perhaps allow them to rebuild their community. But Haliartus had joined Perseus, fought against Rome, and was captured by force; it was neither an ally, like Chalcis or Abdera, nor a community that had surrendered without a fight, like Thisbe. Unlike these other cities, which received redress, the Senate rejected the Haliartans' request. Then the Senate went a step further and gave Haliartus's territory away, ensuring its death as an independent political community. The senators' uncompromising approach here illustrates the pragmatic nature of Roman restraint: massacre, enslavement, and destruction were still perfectly acceptable military options in the case of a community that had supported the enemy and resisted to the end.[68]

The Romans' victory at Pydna exposed the sheer pragmatism behind all of this moderation and conciliatory diplomacy. When Perseus was crushed at last, Paullus unleashed massive violence throughout Greece. In doing so, he punished the disloyalty that had threatened the war effort, and deterred, preempted, and neutralized future challenges that might arise. Though Paullus directed most of the deportations, destructions, enslavements, and executions, he acted in cooperation with senatorial commissioners and carried out directives from the Senate—the very political body that restrained Roman military conduct earlier in the war. The Republic's leaders had no serious qualms about employing massive violence, so long as it was useful.

NOTES

1. Perseus's army, invasion of Thessaly: Livy 42.51.3–11, 42.53.5–42.54.11, 42.56.8–10. Phalanx: Polyb. 18.29. Roman numbers: Livy 42.31.1–2, 42.32.6; Burton 2017, 99, 114, 125–26. Licinius's route: Livy 42.55.1–4.

2. As for Licinius's possible lack of command experience, religious duties had prevented him from taking up a praetorship in Spain: Livy 41.15.10; cf. 42.32.1–5. Licinius enters Thessaly, confers with allies: Livy 42.55.3–10, 42.57.1–3. The month: Walbank 1979, 305.

3. Perseus approaches: Livy 42.57.4–12; "lofty size": Plut. Aem. 18.3. Perseus's force: Livy 42.58.5–9.

4. "Larger and nearer": Livy 42.58.3. Roman deployment, numbers: Livy 42.57.12, 42.58.12–14. But cf. 42.59.1.

5. Sound of missiles: Amm. 16.12.43, 24.2.14–16, 25.3.13. "Wild beasts": Livy 42.59.2. The battle: Livy 42.59, Plut. Aem. 9.2; App. Mac. 12; Just 33.1.4; Eutrop. 4.6.3; Oros 4.20.37; Zon. 9.22.

6. Roman casualties: Livy 42.60.1; Plut. Aem. 9.2. Retreat and blame: Livy 42.60.3–9; Polyb. 27.15.14, 28.4.6. Aetolian independence vis-à-vis Rome: Gruen 1984a, 29–32; Burton 2011, 274–75. Mood in Macedonian camp: Livy 42.60.2, 42.61.1–10. Perseus's errors: Livy 42.60.4–7; Burton 2017, 132–33. Perseus offers terms, withdraws: Polyb. 27.8.1–15; Livy 42.62.3–15.

7. "When word," "naturally inclined," "capable adversary": Polyb 27.9.1–27.10.4. Cf. Livy 42.63.1–2.

8. First, Second Macedonian Wars: e.g., Eckstein 2008, 77–305; Burton 2017, 18–38; Gruen 1984a, 373–98; Rosenstein 2012, 179–89.

9. Roman international "friendship" (*amicitia*): Burton 2011, esp. ch. 3; 2017, 39–47, 55. Roman view of Macedonia's status: e.g., Livy 42.39.7. Philip's wars, territorial gains, revival: e.g., Polyb 22.14.12; Livy 36.33.7, 38.1.11, 38.2.2, 39.23.10–13, 39.25.17, 39.29.1–4, 39.35.4, 40.21.1, 42.42.1, 42.56.7, 42.67.9–10; Plut. Aem. 8.4–5. Rome and Philip V post-197: Burton 2017, 39–55; Gruen 1984a, 399–402; Derow 1989, 293–95.

10. Perseus's pre-war military, diplomatic activities: Polyb. 22.18.2–5, 25.3.1–8, 25.4.8–9, 27.1.8–9, 27.5.1–8; Livy 41.22.4–8, 41.23.1–41.24.20, 42.12.1–10, 42.13.4–10, 42.29.12, 42.40.6–7, 42.41.10–14, 42.42.1–4, 42.44.3, 42.46.7–10; Diod. 29.33; Paus. 7.10.6–7; App. Mac. 11.

11. Eumenes's laundry list: Livy 42.11.4–42.13.12. Cf. *RDGE* 40. Perseus's rise, Rome's response: Burton 2017, 56–77; Gruen 1984a, 403–19; Derow 1989, 301–3; Rosenstein 2012, 212–18; Eckstein 2013, 88–90. Causes of the Third Macedonian War: Burton 2017, 78–123 (with nuance beyond the scope of this chapter).

12. Campaign post-Callicinus: Livy 42.64.1–42.67.5. Gonnus: Livy 42.54.7–8. Winter snows block Thessalian passes: Livy 43.18.2 (referring to winter 170/169, though the same surely applied in 171/170).

13. Siege, destruction of Haliartus: Livy 42.63.3–12; Strabo 9.2.30; Paus. 9.32.5, 10.35.2. Cf. Fossey 1979, 564. Reduced habitation: Catling 1985/1986, 41; Blackman 1999/2000, 56; Bintliff and Snodgrass 1988, 62–64.

14. Thisbe: Livy 42.63.12. Thisbe's fortifications: Hansen and Nielsen 2004, 458–59. Request to refortify: *RDGE* 3, lines 27–31.

15. "Secure barrier": Livy 42.67.6. Malloea: Livy 42.67.7–8. *Direpta*: Ziolkowski 1993.

16. Pteleum: Livy 42.67.9. Cf. Pliny *NH* 4.29. Antronae and Larisa: Livy 42.67.9–11. The date/season: Burton 2017, 207–8 n.4. Licinius and Coronea: Livy 42.67.11–12, 43.4.9–11; *RDGE* 2; Polyb. 29.12.7. Cf. Livy *Per.* 43; Zon. 9.22; Walbank 1979, 374–75.

17. Livy 43.7.9–11.

18. Abdera/Hortensius's campaign: Livy 43.7.8–11, 43.4.8–10. Cf. Diod. 30.6. Greek cities subject to requisitions: Livy 43.6.2–4, 9, 43.17.1–3. "Barely come": Livy 43.4.10.

19. Rome/Genthius in 172: Livy 42.26.1–6, 42.29.11, 42.37.2. "Came upon": Livy 42.48.8.

20. Cassius tries to get Macedonia: Livy 42.32.1–4. Cassius's ravages: Livy 43.1.1–11, 43.5.2–4. "Two sieges for nothing": Livy 43.1.3.

21. "Cruelty and greed": Livy 43.4.5. "Rapacity": Derow 1989, 311. "Burgeoning cruelty": Burton 2017, 139. "Savagery": Waterfield 2014, 183. Significance of Malian Gulf: Graninger 2011, 36–37; cf. Diod. 20.110.105. Haliartus's significance: Buckler 2006, 129; Austin 1931–1932, 206; Westlake 1985, 123. Roman strategy (and greed) in 171/170: Burton 2017, 139.

22. Terroristic violence in previous Macedonian Wars: Eckstein 1976. Surrender of Malloea's neighbors: Livy 42.67.7–8. "Surrendered at once," "out of fear": Livy 42.67.9, 11. Thisbe: Livy 42.63.12.

23. Lucretius's plunder: Livy 43.4.6–7, 43.7.10. Volunteers: Livy 42.32.6. Cf. Polyb. 1.11.2; Rosenstein 2011a, 142–45. Hints of morale problems: Livy 43.11.10, 43.14.7, 43.15.7–8. "Wealthy": Livy 43.1.1. Cf. Briscoe 2012, 388.

24. Thisbe and Coronea: Livy 43.4.11 (implied); *RDGE* 2–3. "Wept": Livy 43.4.7. "Barren soil": Livy 43.6.3.

25. Epirote dilemma: Polyb. 27.15.1–12; Diod. 30.5. Epirotes support Rome: Livy 42.38.1. Cf. Gruen 1984a, 512. Kidnapping attempt: Polyb. 27.16; Diod 30.5a; Zon. 9.22. "He resolved": Polyb. 27.15.15. Molossians defect: Polyb. 30.7.2; Livy 43.18.2. It is not clear what proportion of Thesprotians and Chaonians stuck with Rome: Meyer 2013, 112 n.319; Cabanes 1976, 293–97; Hammond 1967, 686–88.

26. Harsh warfare, international politics in Classical Greece: e.g., Eckstein 2006, 37–78; Lanni 2008, 469–89. "Eternal law": Xen. *Cyrop.* 7.5.73. Cf. Thuc. 5.89.

27. "Large scale": Van Wees 2010, 244. Establishment of Hellenistic kingdoms: e.g., Shipley 2000, 33–58, 108–52, 192–234, 271–325; Errington 2008, 13–157; Green 1993, 1–200; Bosworth 1996. Philip V's depredations: Polyb. 5.9.1, 5.19.5, 5.100.6–8, 15.22.1–3, 15.23.3, 15.23.9, 15.24.1. Polybius on warfare: Polyb. 5.11.3.

28. "Run all risks": Polyb. 18.3.4–7. Cities as assets: Billows 2007, 305; Shipley 2000, 59, 77. Kings' benefactions to cities, reciprocity: Chaniotis 2005, 68–70; Shipley 2000, 84–85; Billows 2007, 305–6; Ma 2000, 360. Demetrius Poliorcetes's moderation: e.g., Diod. 20.45.3–4, 20.46.3, 20.102.2, 20.103.4, 20.103.5–7 (but eighty leaders were executed), 20.110.2, 20.110.6, 21.14.1–2 (ten leaders executed); Plut. *Dem.* 9.2, 9.8–10, 25.1. Antiochus III's moderation: Livy 36.9.13–15, 36.12.6. Other kings' moderate treatment of cities: e.g., Diod. 19.78.2–5, 19.87.3, 20.107.2, 20.107.4; Polyb. 2.54.5–7, 2.58.13, 2.70.1, 4.63.1–3, 4.71.13–4.72.9, 4.80.10–12, 5.60.10–5.61.2, 5.70.3–6, 8.23.3–5.

29. Ransoming prisoners: e.g., Diod. 19.73.9–10, 20.89.4–5; Livy 32.17.1–2. Ransom customary: Polyb. 9.42.5–8; cf. Garlan 1988, 51–52; Ducrey 1968, 54–55; Pritchett 1991, 245–312. Higher status and ransom: Chaniotis 2005, 113. Ransom lucrative: Braund 2011, 117–18; Chaniotis 2005, 113; Pritchett 1991, 245–312. Third party ransom: e.g., Livy 32.22.10–12.

30. Fragility, expense of Hellenistic armies: Sekunda and de Souza 2007, 336; Serrati 2013, 181; Goldsworthy 2000, 70. Soldiers' economic motivations: Austin 1986, 463–65. Switching sides: Grainger 2016, 22, 58; Martin 2013, 680–84. Sparing enemy troops/defectors: e.g., Diod. 19.73.9–10, 19.85.3–4, 20.53.1, 20.103.5–7, 20.113.3, 22.12.1, 22.13.1; Plut. *Pyrrh.* 26.4–5, 34.6. Demetrius's troops desert: Plut. *Dem.* 44.4–5; *Pyrrh.* 11.4. Cf. Dion. Hal. *Ant. Rom.* 19.88.8, 20.6.2. Defections a century later: e.g., Polyb. 5.70.10–11; Livy 36.14.11–12. "Customary" signal: Polyb. 18.26.10.

31. "Bounced around": Gruen 1984a, 137. Philip V's reputation: Polyb. 9.30.1–2, 16.1.3–6, 15.22.3–15.23.6; Diod. 28.7.1. No city in mainland Greece: Polyb. 18.3.4–8; Roth 2007, 396. Reactions to Mantinea's destruction: Polyb. 2.56–2.58; Plut. *Arat.* 25.4; cf. Polyb. 28.14.1–4.

32. Roman moderation, ruthlessness: Gilliver 1996 (generally); Eckstein 1976 (in Greece). Cynoscephalae: Polyb. 18.26.9–12; Livy 33.10.3–4, 33.11.2. Romans ransoming prisoners: Diod. 23.18.4–5. Attitudes to ransom: e.g., Livy 22.50–61; Plut. *Fab.* 7.4.5; Seneca *Controv.* 5.7. "Not impressive": Walbank 1967, 186.

33. "Ruthless barbarians": Champion 2004, 54. Oreum/Aegina: Polyb. 11.5.8, 9.42.5–8, 22.8.9–10; Livy 28.6.1–7.

34. Speech at Sparta: Polyb. 9.32–39. Cf. Polyb. 5.104.1–11, 10.25.1–5. "Just the other day": Polyb. 9.39.1–3.

35. "Certain atrocities": App. *Mac.* 7. Cf. App. *Mac.* 3; Livy 32.22.8–12. "Brutally handled": Paus. 7.8.2. Roman violence compelled alliance: Livy 32.21.7–28, 32.22.10–11. "Shuddered": Livy 32.19.7.

36. "Slaughter, burning": Livy 38.43.4–5. Cf. Eckstein 2006, 201–2. "Fear and apprehension": Livy 31.34.4–5. "Terrifying way": App. *Syr.* 19. The Greeks of southern Italy and Sicily reacted similarly (see chapter 5).

37. Speeches in ancient sources, especially Polybius: e.g., Walbank 1985b, esp. 258–60; Champion 2004, 436–37, 1996, 321–25, 1997, 111–28; Gruen 1984a, 323–25. Romans exceed threshold: Champion 2004, 428–29. Different Greek opinions: e.g., Polyb. 36.9.1–17.

38. "Displeased": Livy 42.8.7. Unconcerned with Greek opinion: Champion 2004, 50–54. Terroristic violence in previous Macedonian Wars: Eckstein 1976.

39. Perseus's activities in 170/169: Plut. *Aem.* 9.3–6; Polyb. 28.8.8–11, 29.19.7; Livy *Per.* 43; 43.18.1–43.19.11, 43.20.4, 43.21.5–43.22.11, 45.3.7; Diod. 30.4.1. An excellent modern summary, with sources: Burton 2017, 140–41, 144–45. Greek support wavering: e.g., Polyb. 28.3–6.

40. Phanote: Livy 43.21.4–5, 43.23.2–6; Hammond 1967, 630–32. Epirus open to Perseus: Livy 43.22.5; Hammond 1967, 630. Molossian fortifications: Dausse 2007, 205–6, 214–16. Thesprotian fortifications: Lazari and Kanta-Kitsou 2010, 38; Kanta-Kitsou 2008, 11; Suha 2011, 203, 216. Inland routes via Epirus: Hammond 1967, 284–85.

41. Flamininus secures Epirote routes: Livy 32.14.5–6; Zon. 9.16. Licinius's route: Livy 42.55.1; Hammond 1967, 284. Hostilius's route: Polyb. 27.16.1–6; Hammond 1967, 676. "Abandon": Polyb. 27.16.6. Philippus's route: Livy 44.1.3–4. "Only short route": Hammond 1967, 631.

42. Roman legate in Epirus: Livy 43.21.4–5, 43.23.1–3; Polyb. 28.13.7, 9. Strategic importance of Phanote (Doliani): Lazari and Kanta-Kitsou 2010, 37 n.9; Cabanes 1976, 269; Kanta-Kitsou and Lambrou 2008, 15; Hammond 1967, 630–32; Kanta-Kitsou 2008, 21. Roman legate's goals in Epirus, importance of Epirote routes: Oost 1954, 80, 86; Larsen 1968, 482. *Contra* Ziolkowski 1986, 70–71. For Ziolkowski, Epirus lost its strategic significance later in the war. But on the contrary, a Roman legate (Appius

Claudius Centho) was active in Epirus in 170/169, and the praetor Anicius would return to the region the next year.

43. Responses to Thisbe, Coronea, Abdera, Chalcis: Livy 43.4.10–12, 43.8.4–7; *RDGE* 2–3.

44. Lucretius attacked by tribunes: Livy 43.4.6–7. *Contiones*: Morstein-Marx 2004. Potential raucousness at *contiones*: e.g., Plut. *Pomp.* 25.3–6; Val. Max. 3.7.3; Livy 43.16.8; Cic. *Rab. perd.*18. Lucretius dragged before crowd, fined by assembly: Livy 43.8.1–3, 9.

45. "Tried to convince": Polyb. 28.3.3. Embassies: Polyb. 28.3–5, 28.13.11, 28.16.2; Livy 43.8.4–10, 43.17.2–3.

46. "Sixty": App. *Mac.* 14. Philippus's Ligurian campaign: Livy 39.20.1–10; Oros. 4.20.26. Philippus's father and Philip V: Livy 42.38.8–9, 42.39.1, 42.40.11, 42.42.4. Previous missions in Greece: Polyb. 23.4.16, 27.1.1–27.2.10; Livy 39.48.6, 40.3.1–7, 42.37.1–42.47.9, 42.47.10, 42.56.7. Campaign of 169: Livy 44.6.1–17, 44.7.1–6, 44.8.1–44.9.11; Livy 44.10.5–44.13.11; Polyb. 28.10–11; App. *Mac.* 15; Diod. 30.10.1–2. Treatment of envoys: Polyb. 28.13.1–5, 28.17.1–15; Derow 1989, 314. "Contented with hostages": Livy 44.7.5. Cf. 44.9.1.

47. Genthius enters war: Polyb. 29.3.1–29.4.7. Rhodes/Pergamum: Polyb. 29.11.1–6, 29.6.1–29.9.13; Livy 44.18.9, 44.23.4–10, 44.24.1–44.25.12, 44.29.5–8.

48. Paullus's experience: Plut. *Aem.* 4.1–4; Livy 40.27.1–40.28.7. Refusal to court popularity: Plut. *Aem.* 2.6. Paullus's speech: Livy 44.22.1–14; Plut. *Aem.* 11.1–3; Polyb. 29.1. "More absurd": Plut. *Aem.* 11.2. Disciplinarian: Plut. *Aem.* 3.6–7, 13.6–7; Livy 44.33.7–44.34.9. Struggles with soldiers: Lendon 2005, 198–99, 208.

49. Arrival, late spring 168: Burton 2017, 161. Clashes at river, meeting at Pydna: Livy 44.35.11–44.37.4; Plut. *Aem.* 15.1–17.6.

50. Eclipse: Polyb. 29.16; Cic. *Rep.* 1.23; Livy 44.37.5–9; Zon. 9.22; Plut. *Aem.* 17.7–10; Val. Max. 8.11.1; Front. *Strat.* 1.12.8. Probable attempts to delay: Livy 44.37.12–44.39.9; Plut. *Aem.* 17.11–13; Lendon 2005, 199, 204–8. Battle begins: Livy 44.40.3–10; Plut. *Aem.* 18.1–4; Zon. 9.22; Hammond 1984, 44–45; Burton 2017, 167, 218.

51. Opposing armies: Plut. *Aem.* 13.4, 16.6, 18.5–9; Livy 44.38.4–5, 44.40.5–6. Size of Roman army (estimated): Brunt 1971, 675–76 (27,000, plus allies); Rosenstein 2004, 111 (29,400); Hammond and Walbank 1988, 540 (26,000 plus 13,000 allies). "Filled": Plut. *Aem.* 18.8. Italians fight, Romans fall back: Plut. *Aem.* 20.2–3.

52. Gaps in phalanx, rout, massacre: Plut. *Aem.* 20.7–22.9; Livy 44.41–42, 44.44.1–3; Diod. 30.22.1; Zon. 9.23. Imbalanced casualties: Koon 2011, 88–93; Sabin 2000, 5–6; Krentz 1985, 13–20.

53. Perseus's escape, capture, end: Livy 44.43.1–8, 44.45.1–15, 45.4.2–45.8.8, 45.28.9–11, 45.35.1, 45.40.6, 45.41.9, 45.42.4; Plut. *Aem.* 23.1–11, 26.1–27.6, 33.8–34.4, 37.2–4; App. *Mac.* 19; Diod. 30.23.1–2, 31.8.12, 31.9.1–5; Flor. 1.28.10–13; Zon. 9.23. Illyrian War: Livy 44.31.1–44.32.5; Plut. *Aem.* 13.3; App. *Ill.* 9; Zon. 9.24; Flor. 1.29.2. Epirus subjugated: Livy 45.26.3–11.

54. "Need for moderation": Derow 1989, 316. Sacking Pydna: Livy 44.44.6–7. Sacking cities in Macedonia, Thessaly: Livy 44.46.3, 45.27.1–4. Massacre of Aetolians: Livy 45.28.7–8. Paullus condones: Livy 45.31.2. Paullus chiefly concerned with insubordination: Barrandon 2018.

55. Senatorial commissioners: Yarrow 2012, 170–71; Eckstein 1987, 264–65, 294–303. Paullus and commissioners: Livy 45.29.3, 45.31.1, 45.16.2; Briscoe 2012, 658. Paullus's pronouncements: Livy 45.29.1–14. Cf. 45.26.11–15 (for Illyria). Paullus's Greek fluency: e.g., Livy 45.8.6.

56. "Which side favored": Livy 45.31.1. Greeks summoned, accusations, deportations: Livy 45.31.1–11, 45.32.1–9; Polyb. 30.7.5–8, 30.29.1, 30.32.1–9, 33.1.3–8, 33.3.1–2, 33.14.1; Paus. 7.10.7–11; Just. 33.2.8; Zon. 9.31. Antissa: Livy 45.31.12–15. Friendly and neutral powers were also snubbed or punished, including Eumenes of Pergamum (Livy *Per*. 46; Polyb. 30.19.6–14, 31.6.1–6), Rhodes (Livy 45.25.1–13; Polyb. 30.31.1–20; Gell. *NA* 6.3), and Antiochus IV (Polyb. 29.27.1–7; Livy 45.12.1–6; Just. 34.3.1–4; Cic. *Phil*. 8.12; Vell. Pat. 1.10.1; Val. Max. 6.4.3; Pliny *NH* 34.24; Plut. *Mor*. 202F; Zon. 9.25).

57. Roman settlement: Burton 2017, 172–82; Gruen 1984a, 423–29, 514–18, 569–78; Derow 1989, 316–19.

58. Senate instructs Paullus, preparations: Livy 45.34.1–3; Plut. *Aem*. 29.1–3. Fortified sites in Epirus: Dausse 2007, 205; Hammond 1967, 687; Cabanes 1976, 303.

59. Plutarch says the Romans *porthein* these places, a word he frequently uses to describe armies ravaging territory or *chora*: see e.g., *Ant*. 32.1, 38.1; *Caes*. 20.3, *Luc*. 31.2, 35.5; *Mar*. 16.5; *Dem*. 5.2, 41.2, 48.4; *Arat*. 16.1, 28.1, 47.1, 51.2; *Cam*. 37.2, 40.2; *Pyrr*. 11.1, 30.1; *Crass*. 9.6, 21.5. Archaeological evidence of fires at a few sites: Blackman, Baker, and Hardwick 1997/1998, 68; Blackman 1998/1999, 64; Dakaris 1971a, 24, 62, 70–73; Dakaris 1971b, 67, 121; Catling 1984/1985, 37; Baatz 1982, 213; Dakaris 1962, 92–93.

60. 150,000 people: Livy 45.34.5; Plut. *Aem*. 29.4; Strabo 7.7.3 (= Polyb. 30.15.1). Modern skepticism: e.g., Oost 1954, 84; Verberne 1993, 278; Wickham 2014, 164–66. Cf. Morgan 2007, 68, discussing the inflated figures for Mongol atrocities in Persian sources.

61. Lamellae, public inscriptions: Meyer 2013, 17 n.24, 135 n.445. Coinage: Oost 1954, 86. Hellenistic population flourishing, then declining: Forsén 2011, 15–16; Lazari and Kanta-Kitsou 2010, 48–49; Kanta-Kitsou and Lambrou 2008, 37; Kanta-Kitsou 2008, 23; Moderato 2015, 315; Blackman, Baker, and Hardwick 1997/1998, 58; Karatzeni 2001, 166, 170–71; Isager 2001, 11. See also Dakaris 1971b, 71–76; Hammond 1967, 668; Cabanes and Andréou 1985, 532. "Well peopled": Strabo 7.7.9. Cf. Ps.-Scylax 28–32.

62. Destruction of fortifications: Livy 45.35.6. Destruction at Elea: Lazari and Kanta-Kitsou 2010, 48. Phanote: Karatzeni 2001, 170; Kanta-Kitsou and Lambrou 2008, 37. Orraon: Cabanes and Andréou 1985, 521. Other destruction possibly connected to Romans: Blackman 2001–2002, 50; Preka-Alexandri 1999, 169; Kanta-Kitsou 2008, 48. The devastation is also reported in Trog. pr. 33; App. *Illyr*. 9; Eutr. 4.8; Pliny *NH* 4.39.

63. "So much destruction": Plut. *Aem*. 29.5. "Post-war atrocity": Burton 2017, 176.

64. Epirote plunder: Livy 45.34.5; Plut. *Aem*. 29.5. Plutarch and Livy disagree about yield of loot: Briscoe 2012, 721; Hammond 1967, 635 n.1. Paullus prevents plundering: Livy 45.35.6–7. Soldiers' dissatisfaction: Plut. *Aem*. 30.4–6; Livy 45.35.6–9. "Satisfied": Livy 45.34.7. Paullus's role in attack: cf. Derow 1989, 317; Hammond 1967, 635 n.1; Pittenger 2008, 253–55.

65. "Defected": Livy 45.34.1 (*ad Persea defecissent*). "Rebelled": Eutr.4.8 (*rebellabant*). Destruction punitive: Gruen 1984a, 516–17; Larsen 1968, 482; Hammond 1967, 635 n.1; Oost 1954, 86.

66. Cato in Spain: Polyb. 19.1.1 (= Plut. *Cato* 10.3); Livy 34.17.11; Front. Strat. 1.1.1; App. *Hisp*. 41; Polyaen. 8.17.1; Aur. Vict. *Vir. Ill*. 47.2; Zon. 9.17. "Dense concentration": Curchin 1991, 30. Prevent rebellion: Front. 1.1.1; Livy 34.17.8; Aur. Vict. *Vir. Ill*. 47.2. "Easier": App. *Hisp*. 41.

67. Strategic logic of Epirote destruction: Hammond 1967, 635 n.1.

68. Haliartus: Polyb. 30.20.1–7.

7

"He Soaked Spanish Soil with Blood"

Failure and Frustration in the Lusitanian War

A formidable force of warriors streamed across the frontier into Further Spain, the westernmost province of the Roman Republic. Under a warleader known as Punicus, the invading host struck deep into Roman territory and plundered the fields of local communities. Raids such as these were relatively common in ancient warfare and generally did not inflict lasting damage. Nonetheless, they must have been terrifying for the victims, with the attackers looting farms and putting buildings to the torch. Alongside this theft and destruction of property, certainly there was violence against people—those unable to flee or hide—as some of the raiders assaulted, abducted, or killed those they encountered.[1]

These raiders were called "Lusitanians" by classical authors, and they hailed from an area roughly corresponding to western Castile and León, Extremadura, and central Portugal (see map 7.1). They were one of many Spanish peoples beyond Rome's direct control and were hardly the first to attack the provinces. However, though such incursions had happened before, the Romans almost always regarded them as unforgiveable offenses, worthy of the firmest response. If the Republic failed to protect its subjects and allies, not only would Rome look weak, but its apparent vulnerability would encourage other foes to raid across the frontier. It is thus unsurprising that the Roman governor of Further Spain, a commander named Manlius Manilius, marched out to intercept the enemy warriors.

By this time in his life, Manilius had led a successful political career at Rome, establishing a reputation for legal expertise and immense generosity. According to Cicero, his openhandedness went so far that he would pace up and down the Roman Forum, offering his expert advice to all citizens.

Map 7.1. Spain in the mid-second century BCE, with places mentioned in this chapter. (Map by author, with base map, topography, and location data generated at Ancient World Mapping Center. Possible locations of Obulcula, Escadia, and Gemella from Quesada Sanz et al. 2014. Roman frontier after L. Fraga da Silva/Campo Arqueológico de Tavira 2010.)

Yet Manilius was no mere lawyer. He would have had significant military experience behind him as a leading member of the Roman aristocracy. And as a praetor, a magistrate ranking just below the consuls in Rome's hierarchy of officials, he likely commanded an army of 10,000 or more soldiers as he approached the enemy. We have no record of Manilius's thoughts or feelings that day, but he may have been eager to restore Rome's honor and earn an easy victory over some bothersome barbarians. Whatever his expectations, he and his army engaged the invaders—and were soundly defeated. The Lusitanians seem to have happily continued raiding, for Manilius's successor, the praetor Calpurnius Piso, fought them too, probably in the following year. His army was crushed in turn, and by the end of 154 BCE,* the Romans had lost something like 6,000 men.[2]

As the cloud of defeat spread from one Roman commander to another, the Lusitanians were elated, and very likely swollen with pride and plunder after two years of successful incursions into Roman Spain. Other tribes were drawn to their good fortune, like moths to a flame, and fresh warriors linked up with the original raiders after Piso's defeat. Pressing southward, the growing warband besieged some of the wealthier provincial communities on the southern coast. In the course of these operations, their commander,

---

* Henceforth, all dates in this chapter are BCE.

Punicus, was struck in the head by a missile—a lucky shot launched by some forgotten defender—but he was succeeded by another leader who continued to organize attacks within Roman territory.

With the Lusitanians ranging freely across prosperous southern Spain, and after two humiliating defeats, the Romans redoubled their efforts. The next praetorian governor (and future destroyer of Corinth) was Lucius Mummius, who arrived with a fresh army and, one presumes, with an aggressive mandate to drive out the invaders. His new legionaries disembarked in southern Spain, immediately clashing with the Lusitanians and initially putting them to flight. It was an auspicious start, but Mummius's troops were suddenly robbed of victory: the Romans, perhaps overeager in their moment of triumph, became disorderly as they pursued the fleeing enemy. The Lusitanians rallied and turned on their scattered pursuers, reportedly killing 9,000 Roman soldiers. They followed up this latest success by capturing Mummius's camp, a shameful turn of events that highlighted Roman impotence. Glorying in their victory, the Lusitanians seized hold of the Romans' arms and standards and paraded them through central Spain, where tensions were already rising between several Celtiberian tribes and Rome. According to the ancient authors Appian and Diodorus, the Celtiberians were then emboldened to wage their own war against the Roman Republic. To complicate matters further, another large group of Lusitanians, "hostile to the Romans" and over 15,000 strong, marched into the far southwest of the peninsula. This host captured a town that was subject to Rome, moved down to the Pillars of Hercules (the modern Strait of Gibraltar) and crossed into North Africa. The raids had now spilled across the Mediterranean, threatening the fields and urban communities across the strait.[3]

The Romans had been entirely reactive up to this point, and with each new threat they were rapidly losing control of the situation in Spain. Fortunately for them, Mummius proved to be a capable commander after all. The praetor carefully drilled his surviving 5,000 soldiers in camp and refused to engage the enemy until his troops had regained their confidence. Eventually he determined that his men were sufficiently encouraged and moved boldly to recover the situation. After pouncing on and destroying a group of raiders in Spain, Mummius crossed to Africa and fell upon the scattered Lusitanian warriors like a bolt of lightning. Appian claims that not one Lusitanian made it back to Spain to tell the tale. Glorying in his success, the Roman commander dedicated a portion of the captured booty to the gods of war and burned it, then handsomely rewarded his victorious troops with the rest. From there, the army returned to Rome, where the Senate granted Mummius a triumph in recognition of his achievement. In giving him this great honor, the senators must have believed that Mummius had accomplished something significant: by defeating the troublesome raiders, he seemed to regain the initiative for Rome or bring some kind of resolution to the war after so many shameful defeats.[4]

But whatever their hopes and expectations, Mummius had not subdued the Lusitanians, and the Lusitanian War would continue for over a decade with only brief pauses. Throughout the conflict, Rome's commanders met with defeat—like Manilius and Piso—or, like Mummius, scored victories that were fleeting. As Roman leaders became increasingly frustrated with their inability to win decisive battles, they employed more ruthless methods to achieve their objectives. Ultimately, as we will see, they turned to massacre, mutilation, and destruction in their efforts to end this deadly struggle.

ROME, SPAIN, AND THE LUSITANIANS IN THE EARLY TO MID-SECOND CENTURY

This was not the first Roman war fought in Spain. In fact, Rome's presence here had assumed a predominantly military character from the beginning. Over fifty years before these Lusitanian raids began, the Roman legions arrived in the peninsula during the Second Punic War (see chapter 5). Throughout that titanic clash, Roman commanders like Scipio fought and allied with local peoples and established bases in key coastal towns, setting the pattern for their future occupation. When the Senate decided to remain in Spain after the defeat of Hannibal, Roman control initially depended on their military presence and local relationships they had established during the Second Punic War. Some skeletal administrative structures gradually took shape in 198, when the Senate began sending two magistrates to Spain each year—usually praetors, but also the occasional consul—and established provinces for both of them: Nearer Spain (*Hispania Citerior*), based around the lower Ebro in the northeast, and Further Spain (*Hispania Ulterior*) in the south, focused on the Baetis (Guadalquivir) valley. The establishment of these regular Roman officials and provinces may have provided a measure of stability, but again, they should not distract from the overwhelmingly military character of Rome's occupation; in point of fact, Roman "governors" spent the next two decades campaigning almost constantly inside and outside their provinces. This persistent military activity was partly the result of local circumstances, since subject communities rebelled, and external tribes raided the provinces. But just as significantly, the Senate looked upon Spain as an area of active campaigning rather than an area of strictly administrative control. And the peninsula offered many military opportunities to its Roman governors—and opportunities for glory.[5]

In the course of their ongoing wars in Spain, Roman armies became increasingly entangled with inland peoples beyond their direct control. The tribal communities to the west and north, whom the Romans called "Lusitanians" and "Celtiberians," respectively, were of a different character from the tribes and towns along the southern and eastern coasts. The latter communities were politically sophisticated and situated in rich plains and river basins. In contrast, the inland tribes lived in a diverse collection of largely rural settlements, considerably less urbanized than their coastal counterparts and often seated in more rugged terrain. Common languages and cultures cut across the jumble of some inland settlements, and at times several tribes made alliances that tied them together into larger confederations. Yet the political reality for the northern and western peoples was ultimately one of fragmentation, with many small, distinct communities of essentially local character. It was difficult for the Romans to make sense of this bewildering array of peoples—hence their tendency to assign broad, somewhat artificial names to the populations they encountered—and equally difficult to wrangle submission from the medley of diffuse, warring tribes.[6]

Rome waged war with these new foes in central and western Spain from the mid-190s until 179, when the praetors Tiberius Sempronius Gracchus and Lucius Postumius Albinus arrived to take up their governorships. These two men campaigned widely and successfully, thoroughly suppressing the Celtiberians, Lusitanians, and other hostile groups bordering their provinces. And they did not just win battles. The praetors also carefully negotiated treaties with the Celtiberians in 178, and very likely with other Spanish communities. The primary aim of these agreements was to regularize taxation and tribute, thereby establishing administrative formalities in the Spanish provinces that had long been lacking. As a result of their efforts, a period of much-needed tranquility settled over Spain for the next two decades.[7]

In the middle of this relative quiet, the outbreak of new Lusitanian raids in 155 is something of a mystery. However, archaeological and historical evidence suggests that the tribes of western Spain were heavily militarized, and becoming increasingly so, in precisely this period. Their militarization is indicated by many elaborate indigenous graves containing weapons and other war equipment. Given their ostentation and distinction from other local burials, these graves probably point to the presence of a ruling aristocracy, whose status may have been closely connected to warfare. Larger *oppida* ("hill-towns" or "hill-forts") with walls, towers, and bastions also appear more frequently in the early second century, possibly a product of increased fighting between communities, or a monumental form of military display. In addition to the archaeological evidence, classical authors claim that western tribes routinely raided their neighbors and the provinces in this period. Reports of sizeable raiding parties, numbering in the tens of thousands and under the command of charismatic individuals (like Punicus and later Viriathus), also indicate the emergence of more powerful leaders and larger warbands.[8]

This accelerated centralization and militarization might have stemmed from the development of pastoralism in western Spain, which had been growing in importance since the fourth century. Grazing animals required protection, particularly when herds and flocks had to be moved through difficult inland terrain, and livestock raids would have been a recurring threat. Conversely, for the rulers of these Iron Age societies, livestock raids offered opportunities to gain prestige, as well as booty that could be redistributed to followers. And in addition to the internal dynamics that encouraged militarization, external factors probably also accelerated indigenous patterns of warfare and raiding. After the Second Punic War, Carthage stopped recruiting mercenaries from Spain, as they often had in the past. Therefore, Carthaginian mercenary service was no longer an option for aspiring Spanish warriors, depriving them of an outlet for pay, booty, and reputation, and encouraging local raids as an alternative. Native groups also may have militarized and formed larger coalitions in response to the looming threat of Rome, adapting to intensifying Roman hostility, influence, and expansion.[9]

In sum, Rome had campaigned continuously in Spain since the late third century, and in that time had been unable (or unwilling) to create lasting stability in the peninsula. Admittedly, the treaties hammered out in 178 ushered in a moment of peace, but it is just as likely that this respite stemmed from the exhaustion of Rome's enemies as much as any diplomatic savvy. Simultaneously, the native settlements of western Spain were coalescing into larger, more bellicose communities, whose major mode of warfare was based around raiding. The status of their leaders was also connected to war and may have depended on successful raids. Whatever the exact set of circumstances that sparked the invasions in 155, there were more than enough factors in play to put the Lusitanians on a collision course with Rome.[10]

RETALIATION, FRUSTRATION, AND MASSACRE

Mummius's successful campaign drove the Lusitanian raiders out of Further Spain and North Africa and temporarily swung the military momentum back toward Rome. At the same time, he and his predecessors had suffered a string of embarrassing defeats, and the communities responsible had not been brought to heel. From the Roman perspective, every invasion that went unanswered, every raid that went unpunished might erode Rome's reputation for strength and provoke more attacks. A forceful response, on the other hand, would demonstrate that Rome was prepared to retaliate against any insult. Reprisal was in fact a key feature of Roman warfare, and for the next three years, the legions would employ increasingly violent methods to stem the tide of Lusitanian incursions.[11]

In 152, the governors of Further and Nearer Spain launched retaliatory attacks into Lusitania and the surrounding territories. The consul Marcus Claudius Marcellus,* who was governing Nearer Spain, assaulted and took a Lusitanian town called Nertobriga (in modern Badajoz province). To the west of Nertobriga, archaeologists excavated a small fortified settlement that was destroyed by fire around 150. If this destruction is connected to Marcellus's campaign—which admittedly cannot be said with certainty—then the

---

* This Marcellus was the grandson of the Marcellus who captured Syracuse in the Second Punic War. See chapter 5.

consul may have been particularly ruthless, targeting settlements large and small as he drove northward. Meanwhile, the praetor Marcus Atilius was governing Further Spain and may have been coordinating with Marcellus. He also marched into Lusitania, killed a band of several hundred warriors, and at a stroke, captured their largest settlement, Oxthracae (location unknown). This blow sent shockwaves through the region, causing such fear that "all the nearby tribes" came to Atilius in surrender and agreed to some sort of treaty. The terms of this agreement are nowhere preserved, but the Lusitanians may have agreed to hand over their arms and promised to obey the Romans, perhaps in return for arable land. At the very least, the treaty probably mandated an end to the raids. Yet the effort failed, Appian says, "When Atilius withdrew into winter quarters, at once [the Lusitanians] all changed their minds and besieged some of the Romans' subjects." Nothing else is recorded about Atilius's tenure, and it is doubtful that he accomplished anything further.[12]

With Atilius's abortive treaty, the Romans concluded almost five years of fighting without a decisive battlefield victory or a successful diplomatic solution. This failure was exacerbated by a simultaneous and closely connected war with a Celtiberian people called the Arevaci, in which there were also several Roman defeats and no end in sight. The sources lay considerable stress on Roman frustration during these years of hard fighting in Spain. In his contemporary account, Polybius summarizes Roman attitudes in this "fiery war" (*purinos polemos*):

> For while wars in Greece and Asia are as a rule decided by one battle, or more rarely two, and while the battles themselves are decided in a brief space of time by the result of the first attack and encounter, in this war it was just the opposite. The engagements as a rule were only stopped by darkness, the combatants refusing either to let their courage flag or to yield to bodily fatigue, and ever rallying, recovering confidence and beginning afresh. Winter indeed alone put a certain check on the progress of the whole war and on the continuous character of the regular battles, so that on the whole if we can conceive a war to be fiery it would be this and no other one. (Polybius *Histories* 35.1.2–6; trans. Paton)

Polybius's words are emblematic of Roman difficulties. The war was fiery, to quote one modern scholar, "both in its violence and because it spread like fire and kept breaking out anew when it seemed to have been put out." From the Romans' perspective, the clearly defined enemies of their previous wars—Carthage, Macedonia, and so forth—had been replaced by a baffling collection of hostile peoples, whose identities, territorial limits, and allegiances were not entirely certain. Yet the "fire" swept onward as Rome came into conflict with ever more tribes, and frustrations mounted. And not only the Senate, but also the larger Roman population was growing tired of the fighting in Spain. In a rare bout of war weariness, some young men began avoiding the levy, and Roman commanders had trouble filling the ranks of the legions as rumors spread of innumerable battles, heavy losses, and valorous enemies. All of this set off further alarm bells for Rome's leadership.[13]

The Senate's general attitude toward their foes in Spain was also hardening. In 152, for example, the senators rejected a Celtiberian proposal to renew an earlier treaty, and instead insisted on their unconditional surrender. Polybius characterizes the attitude of hardline senators as follows:

> They supposed that if this enemy were overcome, then all would submit to their orders; but that if the Arevaci [Celtiberians] could avert the present threat [from Rome], not only would they be encouraged to fight, but all the other tribes would be encouraged also. (Polybius *Histories* 35.3.9; trans. Paton, modified)

Attitudes toward the Lusitanians were surely similar, given the Romans' poor performance in that theater. Both the Senate and Roman people had grown tired of and frustrated with the wars in Spain.

With these growing anxieties back in Rome, Atilius's successor in Spain would have been under enormous political pressure to resolve the Lusitanian war. As it happens, the next praetorian governor would take extraordinary measures to end the conflict, and would have an outsized influence on the course of the war and its place in Roman memory. The praetor Servius Sulpicius Galba had already enjoyed a distinguished career before he succeeded Atilius to the governorship of Further Spain. He first appears in our sources as a military tribune serving under the consul Aemilius Paullus in the Third Macedonian War. When the Roman army came home following their victory over the Macedonian king, many soldiers complained that Paullus had been stingy with plunder and severe in discipline (despite the fact that he had allowed them to plunder Epirus). Bitter and grumbling, the soldiers openly opposed the Senate's proposal to grant Paullus a triumph for his victory. Galba sided with the rank and file, exclaiming that Paullus "should not hope to reap the fruit of gratitude that he has not earned!" In a fiery speech before the Roman people, Galba nearly persuaded voters to reject the triumph. Ultimately, however, a grizzled old veteran and ex-consul named Marcus Servilius spoke in Paullus's defense, shaming the crowd and taunting Galba to show off his "pretty, unscarred skin." This knock on Galba's lack of military experience (or lack of personal bravery) was probably unfair—as a military tribune, he would have put in substantial time in the army—but the disgruntled soldiers lost the argument, and Paullus got his triumph. Still, if Galba received a political black eye from the controversy, he had showed himself to be something of a lightning rod. Our sources are spottier for the intervening years of his life, but it is certain that he rose up the ladder of offices and built a reputation for outstanding oratorical ability. Cicero unequivocally names him as the best orator of his age, describing his passionate and forceful manner of speaking. Riding on past successes, Galba was elected praetor in 151 while in his early forties.[14]

When he received Further Spain as his province, Galba would have known that the Lusitanians were a serious concern of the Senate. But he was as career minded as any Roman aristocrat, and he may have been relieved that the Lusitanians were back on the warpath. Before the treaties of Gracchus and Albinus, when war was virtually unending in Roman Spain, praetorian governors found numerous opportunities for glory by battling against Spanish foes; at least a dozen praetors earned triumphs for their efforts in Spain between 200 and 167. Victorious praetors were also more likely to win the consulship, the most important elected office in Rome. There can be little doubt that Galba sought to achieve a signal success, not only for the sake of Rome's strategic interests in the region, or because of the growing consternation of the Senate, but also for the advantages he would gain in future electoral contests.[15]

With these political pressures like a wind at his back, Galba rushed to the relief of beleaguered settlements in Further Spain. Our sources do not say where the Romans encountered the latest batch of Lusitanian raiders, but Appian suggests that Galba force-marched his men right into battle. The praetor's men were successful at first, as they punched through the enemy line and sent the Lusitanians running. The Romans followed the fleeing foe in hot pursuit, but they became "weak and disorderly as a result of tiredness, [and] the barbarians, seeing that they were scattered and pausing in groups for a rest, regrouped and attacked, killing some seven thousand." Galba fled.[16]

The episode is strikingly familiar. As we have seen, Mummius's army also put the Lusitanians to flight, only to have them rally and rout the Romans in turn. These parallel episodes are not just a refrain on Roman incompetence, nor are they generic set pieces meant to enliven battle narratives. Rather, these snapshots hint at the special features of Lusitanian warfare—features that made them particularly slippery foes. Classical authors repeatedly comment on the Lusitanian way of war and distinguish their fighting style from other Spanish tribes. The Greek geographer Strabo, for example, stresses the Lusitanian tendency toward mobility, writing that they are "skilled at ambush, good at scouting, swift, nimble, and adept at maneuvering." Similarly, the historian Diodorus differentiates Lusitanian and Celtiberian warfare, stating that the Lusitanians are inferior in close combat, but on the other hand, "they are agile and lightly-armed, [and] they readily flee and give chase." This last point—that they "readily flee and give chase"—is echoed elsewhere. In several accounts, Lusitanian armies feign retreat in the face of Roman advances, drawing the hapless legionaries into traps. And it is not hard to understand why these tactics were so effective: they played on the characteristic aggression of Roman armies and the legions' desire to push relentlessly forward and to pursue enemies in flight. This tendency toward aggression is doubtlessly what doomed Galba and Mummius, as it would doom many more Roman commanders before the war was over.[17]

Lusitanian capabilities were amplified by other advantages, including their masterful use of cavalry and terrain. Strabo writes that the Lusitanians trained their swift horses to traverse mountains and follow commands, and mingled cavalry with their infantry in battle. Appian also describes Lusitanian horsemen harassing the legions with rapid attacks and lightning retreats, then melting into mountains that Roman horses were incapable of crossing. A recently excavated Roman army camp in Badajoz, Spain, may attest to Roman concerns about skilled Lusitanian horsemen. The hilltop camp, which geographically and chronologically coincides with the Lusitanian War, was protected by a system of parallel stone walls built on the mid- and lower reaches of the hill. These protective walls have been interpreted as anti-cavalry defenses, perhaps reflecting Roman fears of cunning Lusitanian horsemen.[18]

Of course, we must approach ancient descriptions of "barbarian" warfare with caution, laden as they are with stereotyping and the ideological need to depict the enemy as mere bandits fighting in an unmanly fashion. The Lusitanians were doubtlessly skilled at fighting pitched battles, too. Nonetheless, it was the Lusitanians' use of ambush, tactical mobility, and rough terrain that gave them key advantages in the war with Rome. Thus Diodorus sums up the situation:

> For using as they do light arms and being altogether nimble and swift, they are a most difficult people for other men to subdue. And, speaking generally, they consider the fastnesses and crags of the mountains to be their native land and to these places, which large and heavily equipped armies find hard to traverse, they flee for refuge. (Diodorus Siculus *Library of History* 5.34.6–7; trans. Oldfather)

These tactics would have been exasperating for Roman commanders. Even if the legions managed to rout a Lusitanian army, that same force might lead them into an ambush or simply fade into the landscape. Or another raiding party would take its place. The result was not just humiliating defeat at the hands of so-called barbarians, but an inability to achieve any kind of significant victory.[19]

Galba had not adapted to the Lusitanian way of war, and so his army was heavily defeated. He and a handful of survivors fell back to Carmo (modern Carmona, northeast of Seville). There he collected other refugees from the battle and drafted local allies to the number of 20,000; if he had begun the campaign with a praetorian army of around 10,000 troops, then lost 7,000 in battle, the new recruits would have brought his total force to about 23,000 soldiers. Regardless of his renewed numbers, Galba declined to take another swing at the Lusitanians, instead retreating further west into the friendly territory of the Conii (modern south Portugal). There he went into winter quarters, and another year passed without significant Roman gains.

The Senate decided to renew Galba's command for 150. This may appear unusual to modern readers, but of course Mummius, too, had been defeated before going on to destroy several Lusitanian armies. Given a second chance, Galba might also rally his troops and achieve some success. Meanwhile, the consul Lucius Licinius Lucullus marched west from his province, Nearer Spain, in order to link up with Galba. In the next campaigning season, these two men would join forces, responding to the Lusitanians with further escalating force and changing the development of the war.[20]

In 150, much like Marcellus and Atilius two years prior, Galba and Lucullus launched a two-pronged invasion of Lusitania. Galba marched north from the territory of the Conii, while Lucullus moved out of Turdetania (the modern Guadalquivir valley) further east. As they drove their respective armies across the frontier, the commanders systematically ravaged local territory. Appian does not narrate these raids, but Tacitus and Polybius describe how Roman armies devastated landscapes in frightening detail. Describing a retaliatory invasion of Germania in the early imperial period, Tacitus says the Romans "divided their army into four columns in order to ravage more widely, and devastated an area of fifty miles with sword and fire." The legionaries destroyed the buildings they found in their path, sparing "neither sex nor age" and killing the "half-asleep, the unarmed, and stragglers." Polybius similarly describes a raid in Africa during the First Punic War in which the Roman army marched together, destroying houses, looting livestock, and capturing thousands of slaves. Whether or not Lucullus and Galba wrought such fiery destruction, their double invasion proved to be coercive enough: eventually, three Lusitanian population groups living on the southern side of the Tagus River sent envoys to Galba. These offered surrender and requested to renew the treaty they had made with Atilius.[21]

What commenced was an elaborate deception. Initially, Galba agreed to a truce and accepted the Lusitanians' *deditio in fidem*, or "surrender to the good faith" of the Romans. As discussed in chapter 4, this was a formalized ritual in which the defeated party entrusted themselves unconditionally to Roman power, often with the expectation that they would be treated fairly. And here indeed Galba acted as if he was sympathetic, proclaiming that poor soil and poverty compelled the Lusitanians "to practice banditry, to make war, and to break treaties." He then promised to resettle them all on separate tracts of rich farmland. Such resettlements were frequent during the wars in Spain, and the surrendering Lusitanians may not have been surprised at the offer; maybe they even expected it. In any case, they went along quietly, picking up their whole communities and gathering separately in three prearranged locations. The three groups may have numbered about 30,000, judging from the numbers given in our sources. Since even medium-sized *oppida* in Spain seldom had more than two or three thousand inhabitants, we must imagine the populations of several hamlets, villages, and hilltowns evacuating their homes to answer Galba's summons.[22]

When they arrived at their respective rendezvous points, Galba went to the first group and asked them "as friends" to hand over their weapons. The sources do not tell us if they protested, but certainly this was an ominous sign. Nonetheless they complied. Next, according to Appian, the Romans "surrounded them with a ditch." The comment is somewhat odd, but this probably means that Galba moved the Lusitanians into a Roman military camp; the legions built their camps according to a quadrangular, standardized plan, with palisades and ditches forming the perimeter. In fact, Livy says explicitly that the Lusitanians were in an army camp (*castra*) when Galba dealt with them. In a later period, moreover, a Roman commander named Titus Didius reportedly gained control over a Spanish tribe by herding it into a Roman camp. It is quite likely that Galba did something similar here. Shepherded inside, the Lusitanians would have found themselves penned in by the camp's outer defenses as they passed inside the timber gates. One source says that Galba separated the men of military age from the rest of the populace, and possibly he sorted out the males and collected their weapons while they shuffled into the encampment, gathering the adult men in one group and the women, children, and elderly in another. In the meantime, Galba's troops encircled the now-helpless crowd. At a signal, they entered the camp and began to butcher the unarmed males.[23]

Most of our texts just say the Lusitanians were killed, with little elaboration, so the killings may have proceeded as an organized mass execution, with restrained prisoners led out one by one and dispatched in succession. Appian, however, says that Galba "sent in men with swords," implying that the Romans simply attacked the crowd. Forming a cordon of heavily armed soldiers around the victims, Galba's troops would have been able to overpower any token resistance offered by the outnumbered and unarmed mass. Some of the Lusitanians may have frozen in abject despair, aware that the situation was hopeless, with others "vainly begging for mercy, or blindly clawing at . . . their own comrades to try to get further into the heart of the press, away from the executioners and towards an illusory way out." As the Romans pressed in, slashing and stabbing with their swords, the crowd would have transformed into a mass of panic and bodies, much like the victims of the massacre at Enna (see chapter 5). Adding to the chaos of the moment, the Roman troops likely got caught up in the disarray. A participant in the Rwandan genocide described the violent mayhem that resulted when he and his compatriots attacked an unarmed crowd. "Because of the uproar," he stated, "I remember I began to strike without seeing who it was, taking pot luck with the crowd, so to speak. Our legs were much hampered by the crush, and our elbows kept bumping." Another man told a related story, saying that he started "killing several [people] without seeing their faces. I mean, I was striking, and there was screaming, but it was on all sides, so it was a mixture of blows and cries coming in a tangle from everyone." Similarly, one must imagine the Romans striking wildly at the multitude; some legionaries getting in one another's way; the crush of Lusitanians; victims and killers tripping over the living and the dead; the ground, blades, and clothing painted red. The Lusitanians must have gradually shrunk inward, "those around the outside [of the crowd] continually killed and the others gradually pressed together." It is hard to say how long the whole affair lasted, but presumably it was not quick. And like the Carthaginians who surrounded and killed the Romans at Cannae, Galba's troops must have become "nearly exhausted, more from slaughter than fighting."[24]

At some point, the killing ceased, whereupon Galba raced his men to the second and third groups of Lusitanians to repeat the process. He acted quickly so that each group was ignorant of the others' fate. When all was said and done, several thousand Lusitanians were dead—around 8,000, according to Valerius Maximus—most of them adult males, and Galba sold the rest. All three communities were effectively destroyed.[25]

There was little spontaneous about the massacre of the Lusitanians, which necessarily required labor, organization, and coordination. The killings clearly stemmed from Galba's command-level decisions. As demonstrated in chapter 3, surrounding and eliminating a crowd of thousands would have been a difficult procedure for a Roman army, requiring significant physical effort and outlays of military resources. Galba doubtlessly involved most or all of his men in the process. If the Romans also manually disarmed the warriors, built camps, separated the military-aged males from the rest, and then massacred thousands as the sources suggest, the labor and organizational needs would have multiplied considerably. This was not the work of frenzied troops, but a carefully planned military operation orchestrated from the top down.

It is worth asking why Galba chose this laborious and difficult solution—overtly breaching Roman military norms about the treatment of surrendering enemies—when he could have renewed the Atilian treaty or sought some other, less resource-intensive penalty. Appian suggests that Galba was motivated by greed, and he claims that the operation yielded plunder (which the praetor mostly kept for himself). Yet greed is an unlikely motive. If Galba sought profit, surely he would have sold the entire population or forced them to pay an indemnity rather than orchestrating elaborate massacres. Appian is probably more interested in casting Galba as a treacherous villain than in understanding his motives, so we must look elsewhere to explain the destruction of the Lusitanians. Fortunately for us, Galba's actions ignited one of the most infamous political battles over foreign relations in the history of the Roman Republic, providing us with important indications about his motives.

Galba was not greeted with acclamation upon his return to Rome in 149, and his senatorial rivals leaped at the opportunity to censure his blatant breach of military norms. A tribune of the plebs, L. Scribonius Libo, introduced a legislative bill (*rogatio*) that deliberately undermined Galba's acts in the field. The bill proposed to find and free the Lusitanians sold into slavery, and probably also proposed a special court (*quaestio*) to prosecute Galba for his conduct. Libo's bill was enthusiastically supported by the octogenarian Cato the Elder—"a severe and bitter enemy of Galba" who was still wading into public controversies in the twilight of his life. We do not know the precise charges leveled by Cato and Libo, but presumably their accusations were similar to those lobbed against Popilius Laenas twenty-four years earlier, when he had enslaved a Ligurian tribe after its unconditional surrender (*deditio*). If so, they claimed that the massacre of the Lusitanians not only disregarded the moral and ethical norms of Roman warfare, but also "established the worst possible precedent and issued a warning that no one should ever dare [surrender] in the future." But Galba, remember, was the premier orator of his age, and he was ready to respond. He justified himself by claiming that his actions were preemptive:

> He massacred the Lusitanians who were encamped near him, because, he says, he had discovered that they had sacrificed a horse and a man according to their custom and planned to attack his army under cover of the truce. (Livy *Periochae* 49; trans. Schlesinger)

The acerbic Cato, ever clever and no slouch in the oratory department, wittily opposed Galba's argument. As Cato puts it, he himself *wanted* to have the knowledge of pontiffs and augurs, Rome's leading priests, but *wishing* for priestly knowledge did not make him a priest. By analogy, just because the Lusitanians *desired* to revolt, that did not make them *guilty* of revolt:

> But they say that [the Lusitanians] wished to revolt. At this moment I wish that I had an outstanding knowledge of pontifical law; for that reason should I now be taken as a pontifex? If I wish to have an excellent command of augury, would anyone on that account take me as an augur? (Cato fragment 105; trans. Cornell)

In the end, Galba stood before the Roman people with his children in tow, putting his fate—and the fate of his soon-to-be-orphaned children—in their hands. The assembly voted not to prosecute him.[26]

According to Cato, Galba only got off because he had pitifully paraded his children before the people. However, with due respect to Cato, we can assume that his view of Galba was not charitable. And as we have seen, although Roman norms dictated that generals should treat surrendering enemies with moderation, these conventions were relatively weak; violations of military norms might be seized upon for political purposes, as apparently happened here, but commanders were rarely punished for behavior like Galba's. As the historian Arthur Eckstein writes, generals like Galba "who perpetrated what we would call atrocities upon [surrendered enemies] gave reasons in their own defence that were accepted (at least by most Romans) as sufficient." In other words, Galba probably would have gotten off whether or not he had pulled the theatrics with his children. Indeed, he went on to hold the consulship in 144. The Roman public was not too terribly bothered by his actions.[27]

For our purposes, this political controversy is less interesting for its outcome than for the fact that Galba was forced to defend his conduct, leaving us clues about his thinking. His claim that he slaughtered the Lusitanians preemptively may sound like after-the-fact justification, and so it may have been. Yet to understand his decision, it behooves us to take him seriously. After all, Galba's campaign took place in the midst of a military crisis. Lusitanian warriors had defeated no less than four Roman armies between 155 and 151. In that space of time, they had ranged across Further Spain with virtual impunity, plundering Rome's allies and emboldening other raiders to try their luck. The Lusitanian way of war, which emphasized ambush, flight, and mobility, made them especially difficult to subdue and robbed Roman generals of the battlefield victories they craved. Of course, Mummius managed to win a few battles. Marcellus and Atilius also counterattacked, captured two towns, and wrangled terms of surrender from several tribes. Still, the Lusitanian raids resumed. On top of everything, the Romans were waging an equally disastrous war against the Celtiberians. In this crucible of failure, Roman frustration is palpable in the sources, as the impatient Senate refused to compromise with groups like the Celtiberian Arevaci, and Rome's citizenry began to resist recruitment for the wars in Spain. Galba would have been expected to achieve a *lasting* victory when he arrived in his province, not only for the benefit of Rome's strategic interests or to assuage growing senatorial impatience, but also for the sake of his own political prospects. His massive defeat in 151 would have raised the stakes further, showing him firsthand that the Lusitanians were not so easily swept aside. Under all of these weighty political and military concerns, it is entirely plausible that Galba had an "itchy trigger finger" as he dealt with the surrendered Lusitanians, nervously looking out for any sign of trouble—including, perhaps, ritual sacrifices that he associated (rightly or wrongly) with Lusitanian warfare. If he came to believe that they were planning to attack, as he later claimed, then he also feared that they would rob his victory of any permanence. Rather than run that risk, he could instead destroy the communities responsible for the raids when they were under his control, when their warriors could not scatter off to mountain hideaways, and when they could not simply renege at the earliest opportunity. Doing so would send a powerful signal to other would-be raiders across the frontier, warning them that there were dire consequences for challenging Rome.[28]

Even if Galba's claims of preemption were nothing more than empty excuses, his extreme solution to the Lusitanian raids manifestly arose out of his own political and military objectives. However ruthless his logic, massacring the recalcitrant tribes might finally restore Roman security in Further Spain and end the war to his credit, where other measures had been resoundingly unsuccessful.

THE RISE OF VIRIATHUS

For the three years following Galba's mass killing, the sources say almost nothing about the Lusitanians, and the violent campaign of 150 likely forced them into momentary submission. Events, however, would show that Rome's gains were ephemeral. Galba's extreme measures had in fact exacerbated the situation, fanning the flames of war into a fierce blaze that would burn for several more years, consuming many more lives. It did not help that Galba's successors repeated the pattern of Roman behavior that we saw between 155 and 150, escalating their application of force and violence when other strategies failed.

And so, in 147, a familiar theme played out in Further Spain when a band of about 10,000 Lusitanian warriors descended from the north and overran the countryside in Turdetania. As in previous years, the invaders set about raiding local fields until the Roman governor, a praetor named Gaius Vetilius, rushed to the defense of the province. Vetilius fell upon the enemy as they were foraging, killed many of them, and outmaneuvered the remaining force. The Lusitanians sensed that they had been backed into a corner, so they sent envoys asking for land on which they could settle and agreed to obey the Romans in the future.

Everything had unfolded largely as before. The Lusitanians invaded, the Romans responded; when the Romans were victorious, the Lusitanians made overtures for peace. Nevertheless, there were several important differences here. First, according to Appian, this new warband consisted of men who had escaped the depredations of Galba and Lucullus. Many of these raiders would have been made "rootless" (in the words of one scholar) after the Romans had destroyed their homes, which explains both their invasion and their request for land. Second, there was one individual among these desperate warriors who would shape events as much, if not more, than Galba. He was known to classical authors as Viriathus, and he was not interested in surrendering.[29]

In Appian's account, Viriathus reminded his fellow Lusitanians of Galba's treachery and the Romans' bad faith, promising that he would lead them all to safety if only they followed his lead. The whole lot was stirred to action at his words and agreed to heed his commands, which may point to the enormous resentment generated by Galba's atrocity. With the force of Lusitanian warriors united behind him, Viriathus rapidly orchestrated a complex stratagem to escape the Romans' clutches. In accordance with his plan, the Lusitanians drew up their battle line and faced Vetilius's legions. Rather than fight, they suddenly scattered in all directions, with orders to rendezvous at a town called Tribola. Viriathus and a thousand cavalrymen then stood alone, facing the much larger Roman force while most of the other Lusitanians were peeling away. Vetilius was left with the hard choice of dividing his own army to pursue the fugitives or attacking this single group of horsemen. He chose the latter, throwing his legions against Viriathus and his outnumbered companions. But the swift and skilled Lusitanian riders did not receive the Roman charge; instead they "ran away and then stopped and attacked, and so used up that day and the whole of the next running around the same battlefield." The Romans were utterly unable to come to grips with the mobile Lusitanians, and it is easy to imagine their growing exasperation (and exhaustion) as the enemy danced around them. Under the cover of night, Viriathus and his followers slipped past the Romans and made their escape toward Tribola. When Vetilius realized that he had lost his quarry, he pursued the horsemen into a thicket where his army walked right into an ambush. Attacking on both sides, the Lusitanians cut down the Romans in droves. They pushed some of the legionaries over a nearby precipice and captured others in the melee. As for Vetilius, he "was captured, and the man who took him did not recognize him, seeing only a fat old man, and killed him as someone of no importance." With their commander unceremoniously killed, 4,000 Romans fell or were captured while 6,000 more fled to a town by the sea.[30]

The architect of Vetilius's defeat would be remembered as one of Rome's greatest foes. Nonetheless, much of our information about Viriathus should be taken with a very large pinch of salt. Ancient sources serve us with a highly romanticized image of Viriathus as a tragic hero, with a healthy dollop of "noble savage" to round out their problematic portrait. He is at once physically powerful, a skilled combatant, wise, moderate in dress and food, self-possessed, just, generous, and an exemplar of the rags-to-riches hero—once a shepherd, then a bandit, then a leader of men. Or so the story goes. It is certainly possible that some or all of these things were accurate reflections of Viriathus's origins and character, but it is difficult to separate the ideals projected upon him from the historical person with any degree of confidence. What matters most here is that Viriathus was, or became, a redistributive war leader whose charisma and battlefield prowess would continually attract armed followers with whom he shared the spoils of victory. Warfare, in fact, was essential to Viriathus's leadership and prestige among the Lusitanians. There may be an element of truth in Cassius Dio's claim that "[Viriathus] carried on the war not for the sake of personal gain or power nor through anger, but for the sake of warlike deeds in themselves; hence he was accounted at once a lover of war and a master of war." And with his victory over Vetilius, coming on the heels of Galba's treachery, Viriathus's following rapidly grew.[31]

Not one to rest on his laurels, the Lusitanian commander spent the next eight years invading Rome's Spanish provinces. For long periods, he moved freely through Roman territory, plundering and extorting tribute from Iberian communities that were subject to Rome. He also garrisoned some provincial towns and collaborated with others, further stripping away the illusory veil of Roman control in the region. Viriathus's success also encouraged the Celtiberians to renew their conflict with Rome, throwing more fuel on the "fiery war." The Romans, for their part, did not sit back and idly allow these things to happen. They repeatedly sent forces against the Lusitanians just as they had before, but the results were often catastrophic: Viriathus and his warriors vanquished at least five more Roman armies, killing thousands of legionary and allied troops. In many of these encounters, Viriathus masterfully utilized the Lusitanian talent for irregular warfare. In a battle with the Roman praetor Gaius Plautius, Viriathus "pretended to flee" only to turn on his unsuspecting pursuers and cut them down. The Roman writer Frontinus describes an incident in which Viriathus "pretended to flee from Roman cavalry, and led them to a place full of very deep holes" where the Romans subsequently fell and were slain. At other times, he harassed the enemy with light troops and fast horses, attacking at night and generally making life miserable for his foes. When hard pressed, he retreated to his base at a place called the mountain of Venus (possibly located in the Sierra San Vicente range north of Toledo) and crushed the Romans who were foolish enough to follow. When necessary, he also fell back deep into Lusitania, where the long frontier and rugged terrain facilitated his escape.[32]

Viriathus was not invincible, but Roman battlefield successes did little to change the overall strategic situation. According to Cicero, the praetor Gaius Laelius had some success in Nearer Spain in 145. The consul Fabius Maximus Aemilianus was dispatched that same year, and he apparently managed a battlefield victory and probably forced Viriathus to withdraw into Lusitania. But as in the 150s, these wins were not decisive, and Viriathus eluded death or capture. So, as in the 150s, the Romans turned their swords against communities, prisoners, and noncombatants when they could not end the war by other means. This pattern is best illustrated by the consul Fabius Maximus Servilianus, who governed Further Spain between 142 and 140. He first fought two battles with Viriathus outside the town of Itucca, gaining an inconclusive victory in the first encounter and falling into an ambush in the second. Constant Lusitanian harassment eventually drove Servilianus inside the walls of Itucca, while Viriathus decided to withdraw into Lusitania. Servilianus followed him with a ruthless campaign of suppression. According to Appian, the Romans sacked five towns that had supported Viriathus in Baeturia (a region between the Baetis and Anas Rivers), captured the towns Escadia, Gemella, and Obulcula, which Viriathus had garrisoned, and alternatively sacked or forgave other unnamed settlements. Ultimately Servilianus took 10,000 prisoners, beheading five hundred and selling the rest as slaves. He also obtained the surrender of a chieftain named Connobas, whom he spared—but only after severing the hands of all his warriors. Other sources say that Servilianus cut off the hands of surrendered chieftains and deserters, too. It is worth noting that these tactics, like Galba's massacre, would have involved significant organization and planning. Mass execution, mass amputation, mass enslavement: these acts of methodical violence were choices that Servilianus made, not the random passions of soldiers run amok.[33]

Southwest of Cordoba at Cerro de la Cruz, excavators unearthed an ancient settlement that may provide silent testimony to Servilianus's campaign. Most of the settlement was destroyed by an intense fire that burned for several days and probably was not accidental; no one came to search through the rubble afterward, corpses remained unburied, and the site was not reoccupied, despite the fact that the village was relatively prosperous. Additionally, several bodies were found bearing evidence of violent deaths. In one dwelling, excavators found the bones of multiple individuals in a layer of collapsed material. Many of these remains had been exposed to extreme heat from the blaze, while others were apparently crushed in the collapse of the building. At least one of the deceased was identified as a woman. Lying in the middle of the nearby street, excavators also uncovered the remains of two young adult males. The shoulder blade of one body was cleaved in two, a perimortem wound indicating a top-down sword blow. A second wound was found on the individual's lower pelvis, again cutting from high to low. The second individual suffered a powerful blow to the left leg near the ankle, completely severing the fibula bone and partly severing the tibia. He also may have received a wound on his thigh, though a later-built medieval wall distorts that part of the body. His right radius bone was also marked by a diagonal cut, possibly a defensive wound. The limbs of both individuals were in a "forced" position, which might suggest agonized death throes as they were killed on the ground. Their wounds, taken together, indicate that they were killed by a wickedly sharp sword capable of slicing through muscle and then bone, while one of the stab wounds suggests a narrow sword point. All of these signs might point to the Roman sword, the *gladius*. According to ancient authors, the Roman *gladius* was eminently suitable for slashing and stabbing. And it left ghastly wounds, "with arms hewed off from the shoulder, with heads cut off from the body with the necks completely severed, with entrails exposed." Finally, the site provided no evidence of battle, such as arrowheads or other projectiles, and none of the deceased were found with metallic objects like brooches or belts, so they may have been stripped of belongings or even clothing. The bodies in the street were left where they lay, without the burial or the cremation rites so well attested to in Iron Age Spain. For the site investigators, all these signs might evince a massacre. It is also possible that a foreign enemy tried to expel the population. In which case, those who tried to resist were unceremoniously killed, while those who hid were burned or crushed in their dwellings as the village was torched.[34]

The violence uncovered at Cerro de la Cruz is spatially and chronologically consistent with Servilianus's tenure in Spain, and Fernando Quesada Sanz and other scholars argue that this was probably the handiwork of his campaign. If so, it gives vivid color to the intensity of the Roman effort and to Servilianus's larger objectives. Significantly, he campaigned not only in Lusitania, but also within the province of Further Spain, where Viriathus drew support from several communities that were theoretically subject to Rome. Servilianus's use of destruction, enslavement, and massacre were probably a direct response to these wavering loyalties. His ruthlessness demonstrated that there was a price for betrayal, and sooner or later the Romans would collect. Furthermore, conspicuous displays of violence acted as deterrents, sending a piercingly clear message to any community—allied or enemy—that might continue fighting for Viriathus or join him in the future. The mutilation of deserters and captives did much the same, only with the message frighteningly inscribed on their bodies. All of this acted to terrorize unbeaten foes. As one Roman author put it, Servilianus wanted "to crush and cripple the spirit" of the Lusitanians, and his actions indicate that this was indeed his aim. There were more immediate benefits, too. The amputation of hands, for example, meant the loss of enemy fighters able to grasp weapons. Much as Galba's massacre had determinedly taken Lusitanian warriors out of play, mutilations prevented captives from challenging Rome in any future encounters. Destroying towns and populations, both inside and outside of the province, similarly kept them out of Viriathus's control and meant that no further challenges would come from those communities. There is a certain finality to all of it that bespeaks Servilianus's and the Romans' overriding aim to subdue a troublesome enemy, prevent defections, and reduce Viriathus's network of support by whatever means necessary.[35]

Yet the escalating pressure and violence was not yet sufficient to end the war, and Viriathus returned. Servilianus was blockading a town called Erisane when the Lusitanians slipped through his siege lines at night, attacking the legions and putting them to flight. In the rout, the disordered Romans were forced back to some cliffs and surrounded. In that moment, Viriathus decided not to press the attack, and seeing an opportunity to end the war on favorable terms, he opened negotiations. For his part, Servilianus must have seen no way out, and, realistically, he had few options other than diplomacy. Thus he became one of those very rare Roman commanders to make a treaty from a position of weakness. In another oddity, the Roman people actually ratified the agreement—a possible sign of their war weariness. The final treaty acknowledged Viriathus as a friend of Rome and allowed him and his followers to keep the lands they already controlled.[36]

"THE YEAR WAS PEACEFUL WITH THE DEFEAT OF VIRIATHUS"

Servilianus's settlement lasted all of one year. In 139, the consul Servilius Caepio arrived to take over the governorship of Further Spain. Once there, he pestered the Senate with constant missives, urgently requesting permission to resume the war with Viriathus and disparaging Servilianus's dishonorable treaty. Eventually the senators relented. They may have concluded that the treaty was strategically untenable, or perhaps the shameful reality of it—an agreement made with an avowed enemy from a position of weakness—was intolerable after all. In any case, Caepio launched a fresh assault into Carpetania, a region in central Spain where Viriathus controlled some territory. At this point, Viriathus was outnumbered. The sources imply that he was experiencing a shortage of manpower in the latter years of the war, and he may have concluded that victory over the Romans, with their greater resources, was impossible. Certainly the Romans' tenacity and willingness to accept casualties were legendary. Pyrrhus of Epirus, who defeated Rome in several battles, reportedly said that fighting the Romans was like fighting a hydra: like the heads of that legendary beast, a new Roman army would spring up as soon as one was destroyed. It is possible that Viriathus was getting desperate in the face of this unrelenting foe, so he continually retreated in the face of Caepio's onslaught and sent multiple envoys in a futile effort to reopen talks. Ultimately, he would never get another chance to negotiate. Some of his trusted friends were bought off by Caepio, and they killed Viriathus in his tent. It was a disgraceful way to end the war, and deeply uncharacteristic of Rome's ideal martial virtues (which regarded guile as dishonorable). Yet the assassination of Viriathus was part and parcel with the Romans' use of extraordinary violence against the Lusitanians, whom they had been unable to vanquish through set-piece battles or coercive diplomacy.[37]

The Lusitanian War was burning itself out just as the fires of Viriathus's pyre settled into embers. Viriathus's followers briefly rallied around another leader, but they soon surrendered to Caepio and were peacefully resettled in Roman territory. Meanwhile, other communities in western Spain were inspired by Viriathus's example and continued to raid the provinces. Caepio's successor, the consul Decimus Junius Brutus, led his army deep into Lusitania to stop these copycat raids. He also drove northwest into the territory of the Callaeci, a neighboring tribe that had reportedly supported the Lusitanians. Throughout these campaigns, which took place between 138 and 136, Brutus used a strategic logic that was deeply rooted in Rome's experience of the Lusitanian War. As discussed earlier, most of Brutus's predecessors had marched straightaway against Lusitanian raiders in order to confront them in pitched battles. This response was typical of Roman armies during the Middle Republic, stemming from both the martial culture of Rome's soldiers and commanders and the senatorial aristocracy's desire for decisive battlefield victories. While this approach had proven successful in many previous conflicts, their impatient aggression had been ineffective against the Lusitanians. Highly mobile Lusitanian warriors lured the legions into ambushes or just scattered in the face of Roman forces.

Brutus took a different approach. He pointedly refused to chase the raiders throughout the countryside, "thinking it difficult to catch up with men who moved rapidly from place to place, as bandits do, disgraceful to fail to catch them and to conquer them a matter of no great glory." Instead, he adopted a strategy reminiscent of modern counter-guerrillas and counter-insurgents: bypassing the warriors entirely and assaulting their communities directly, he hoped to draw the raiders back to their homes where they could not so easily melt away or spring a trap. To quote Appian, "he turned against their towns, in the expectation that he would be able to inflict punishment on them, that there would be much profit in it for the army and that the bandits would scatter each to his own homeland when it was in danger.

With these notions in mind, he plundered everything he came across, the women fighting and dying alongside the men and not uttering a cry, even in the midst of the slaughter."[38]

Galba and Servilianus had also targeted enemy populations—not just capturing settlements, but killing or enslaving their inhabitants—but only after failures on the battlefield. But Brutus, drawing from the accumulated knowledge of his predecessors, attacked enemy settlements as a tactic of first resort. His severe approach encouraged several Lusitanian populations to surrender, whom he generally spared and occasionally resettled within Roman territory. He gave no such quarter to recalcitrant communities, and destroyed many. In the end, his approach worked, surely in part because Viriathus was out of the picture, and the Romans' fortunes had improved. Major Lusitanian raiding appears to have stopped, or at least tapered off enough that it attracted fewer comments from ancient authors. Brutus returned to Rome with a war-torn, exhausted, and pacified landscape behind him. He would be remembered as "the man who subjugated Lusitania." The poet Ovid went further, writing that Brutus "soaked Spanish soil with blood."[39]

CONCLUSION

The mid-second-century Lusitanian Wars did not showcase Rome's finest hour. Throughout the conflict, several Roman commanders targeted populations, prisoners, and noncombatants. Most notably, Galba systematically eliminated several Lusitanian populations in 150. He had been unable to defeat the Lusitanians in battle, and they had supposedly broken a truce with his predecessor. In addition, he may have feared an imminent attack as he later claimed, and he certainly sought a more definitive resolution to the war, since other approaches had failed. To this end, he orchestrated a highly organized massacre, destroying the communities he viewed as responsible for the raids and warning others against challenging Rome. Servilianus was similarly unable to defeat the Lusitanians in battle and turned to mass violence in order to "crush and cripple the spirit" of the enemy. His enslavements, killings, and mutilations punished those who had supported Viriathus, deterred defections, and removed hostile warriors from the war. Finally, Brutus refused to fight the Lusitanians on their own terms, and instead assaulted settlement after settlement to nip the raids in the bud.

These commanders attacked populations and noncombatants as a deliberate strategy in order to resolve their immediate military problems. Their choices were also clearly shaped by the larger circumstances of the war. Between 155 and 138, enemy warbands repeatedly invaded Rome's Spanish provinces to plunder both town and country. Under skilled leaders like Viriathus, these attackers used ambushes and mobility to crush legionary armies or to escape overwhelming forces. Roman commanders such as Mummius and Laelius fought a few successful battles, and Atilius forced a treaty after capturing a major settlement. But no battlefield and diplomatic successes brought an end to the war or to Lusitanian incursions, and frustration deepened. Meanwhile, the enemy's success invited more raids, and at times the Romans practically lost control of Further Spain. In this turbulent environment, both the Roman Senate and public were growing weary of the seemingly endless fighting in Spain. The Roman generals sent to Spain would have been keenly aware of senatorial impatience, an anxious public, and their own electoral interests. This deadly cocktail of military and political pressures incentivized extraordinary measures. And so, when they could not master the Lusitanians by means of battle or negotiations, Roman commanders used massive violence to achieve their objectives.

NOTES

1. For chronology/sequence of commanders: Richardson 1986, 184–91. Agricultural devastation in ancient (Greek) warfare: Hanson 1998. "Pillaging" in Roman warfare: Roth 1999, 148–54. Personal, sexual violence in ancient raids/warfare: Gaca 2011, 73–88. Note that this chapter draws heavily from Appian, whose reputation is spotty. Still, recent research has demonstrated that he could follow his sources closely and faithfully, at least for the main outline of events. See esp. Rich 2015; cf. Richardson 1986, 194–98; Gómez Espelosín 1997; Bucher 2000.

2. Pre-Roman Lusitania: Sánchez Moreno 2006, 57. Spanish geography/peoples: Richardson 1996, 10–12. Roman reputation, reprisals: Goldsworthy 2007, 93. Pacing the Forum: Cic. Orat. 3.133. Manilius's praetorship: Broughton 1951, 448, 451 n.1. Aristocratic military experience: Rosenstein 2011b, 132–40. Size of praetorian armies in Spain: Brunt 1971, 427–32, 663–64. Manilius and Piso defeated, casualties: App. Hisp. 56.

3. App. Hisp. 56–57; Diod. 31.42.1.

4. Roman failures: Livy Per. 47; Obs. 17. Mummius's second campaign: App. Hisp. 57; Eutrop. 4.9.1. Triumphs marking decisive victories: Clark 2014, esp. 96–97.

5. Early "administrative" developments: Richardson 1986, 73–94; Brennan 2000, 154–73. Irregular taxes/tribute: Richardson 1976, 139–52. Provinces: Gargola 2010, 157. Ongoing fighting in Spain: Dyson 1985, 185–98. Spain as area for military activity/glory: Richardson 1996, 58.

6. Roman involvement in central/western Spain: Richardson 1986, 95–98. Iberian settlements: Hoyos 2003, 55–56; Edmonson 1997, 16–18; Richardson 1986, 9–16. Roman/Greek difficulty identifying tribes: e.g., Polyb. 3.37.11; Strabo 3.3.5; 3.4.19. Artificiality of tribal names: Edmonson 1997, 26–27.

7. Activities of Gracchus/Albinus and results: Richardson 1996, 58–60, 70–76; App. Hisp. 43–44; Livy Per. 41; Flor. 1.33.9.

8. Militarization, centralization in western Spain: Sánchez Moreno 2005, 107–25, 2006, 57–69; Álvarez Sanchís 2005, 255–77; González García 2011, 184–94. Habitual "brigandage" in western Iberia: Strabo 3.3.5, 3.4.15; Diod. 5.34.6–7; Varro RR 1.16.2.

9. Raids and prestige: Sánchez Moreno 2006, 57–69; 2005, 107–25; Dyson 2014, 207. Lusitanians as mercenaries: e.g., Livy 21.43.8; 21.57.5. Effects of Roman interference: González García 2011, 185–86; Sánchez Moreno 2006, 57–69; Edmonson 1997, 26–27.

10. Exhaustion encouraging peace: Dyson 1985, 198.
11. Retaliation/deterrence as important features of Roman war/strategy: Goldsworthy 1996, 95–100; Gilliver 1996, 226–30; Reisdoerfer 2007, 75–77; Ñaco del Hoyo et al. 2009, 44–45; Mattern 1999, esp. 211–22.
12. Nertobriga: Polyb. 35.2.2. Destruction near Nertobriga: Berrocal-Rangel et al. 2014, 274–79; Marco Simón 2016, 239. Atilius's campaign: App. *Hisp.* 58. Possible terms of Atilian treaty: García Riaza 2002, 102.
13. Concurrent Celtiberian Wars: App. *Hisp.* 44–55. "Spread like fire": Walbank 1979, 641. Roman confusion in Iberian wars: Clark 2014, 151. Recruitment problems: Polyb. 35.4.1–7; Livy *Per.* 55. Attitudes toward conscription: Rich 1983, 316–18.
14. Galba and Paullus: Livy 45.35–39; Plut. *Aem.* 30.4–31.10. "Fruit of gratitude": Livy 45.35.9. "Unscarred skin": Livy 45.39.19. Galba's career: Broughton 1951, 470. Best orator: Cic. *Brut.* 82, 86–94; *Orat.* 1.58, 240; *Rep.* 3.42; Suet. *Galba* 3.2. Galba's approximate birthdate: Evans and Kleijwegt 1992, 193.
15. Electoral prospects of victorious praetors: Waller 2011, 29; Harris 1979, 32–33. Praetorian triumphs in Spain, 200–167: Rich 1993, 51–52.
16. Galba's arrival/defeat: App. *Hisp.* 58; Oros. 4.21.3; Livy *Per.* 48. "Weak and disorderly": App. *Hisp.* 58.
17. "The greatest": Strabo 3.3.3. "Skilled at ambush": Strabo 3.3.6. "Agile and lightly-armed": Diod. 5.34.5. Lusitanians tactics: e.g., App. *Hisp.* 64; Front. *Strat.* 2.5.7; Dio 37.52.3. Roman aggression: Lendon 2005, 172–211. Roman pursuit of fleeing foes: Sabin 2000, 5–6, 15.
18. Lusitanian horses: Strabo 3.4.15; App. *Hisp.* 62; Just. 44.3. Anti-cavalry defenses: Morillo Cerdán et al. 2011, 59–78.
19. Stereotyping Iberian warfare: Quesada Sanz 2002, 2011. Roman difficulty winning significant victories in Spain: Clark 2014, 151.
20. Galba regroups: App. *Hisp.* 58–59. Proroguing defeated commanders: Clark 2014.
21. Galba, Lucullus's campaign: App. *Hisp.* 59. "Divided their army": Tac. *Ann.* 1.51. Raid in Africa: Polyb. 1.29.5–7. Three communities (*civitates*): Val. Max. 9.6.2. South of the Tagus: Oros. 4.21.10.
22. Surrender: App. *Hisp.* 59; Cic. *Brut.* 23.89; Val. Max. 8.1abs.2, 9.6.2; Oros. 4.21.10; Livy *Per.* 49. *Deditio* in Roman warfare: Polyb. 20.10; Livy 1.38.1–2; Baker 2013, 111–16, with sources. "To practice banditry": App. *Hisp.* 59. Promise of resettlement: App. *Hisp.* 59; Val. Max. 9.6.2; Oros. 4.21.10. Resettlements in Spain: Richardson 1996, 75–76. Thousands of Lusitanians: Suet. *Galba* 3.2 (30,000 killed); Val. Max. 9.6.2 (8,000 Lusitanian men killed); App. *Hisp.* 61 (about 10,000 survivors). It may be that Galba killed 8,000 men out of a total population of 30,000, and Suetonius garbled the figures. Population sizes: Quesada Sanz et al. 2014, 265. See also García Moreno 1988, 377; García Riaza 2002, 105.
23. Lusitanians disarmed: App. *Hisp.* 60; Val. Max. 9.6.2; Oros. 4.21.10. The ditches: App. *Hisp.* 60. Lusitanians in *castra*: Livy *Per.* 49. Titus Didius: App. *Hisp.* 100. In narrating Didius's massacre, Appian was perhaps deliberately recalling Galba's more famous massacre; the same similarity was notice by García Riaza 2002, 106 n.343. Roman camps: Keppie 1998, 37–38, 46–51. Military-aged males: Val. Max. 9.6.2. Encirclement: Oros. 4.21.10. Killing: App. *Hisp.* 60; Cic. *Brut.* 23.89; Val. Max. 9.6.2; Suet. *Galba* 3.2; Livy *Per.* 49; Oros. 4.21.10.
24. "Sent in men with swords": App. *Hisp.* 60. "Vainly begging": Sabin 1996, 77, describing the post-battle massacre of a surrounded army. Paralysis in the face of

atrocity: Collins 2008, 103–4. Rwanda quotations: Hatzfeld 2005, 21. "Around the outside": Polyb. 3.116.11, describing the massacre after Cannae. "Nearly exhausted": Livy 22.48.6.

25. Second/third groups: App. *Hisp.* 60. Males killed, selling survivors: Val. Max. 9.6.2; Livy *Per.* 49. For the reliability of the source tradition, which likely stemmed from senatorial debates in Rome, see Barrandon 2018, ch. 2.

26. "Severe and bitter": Cic. *De Orat* 1.227. "Worst precedent": Livy 42.8.5. Reaction to Galba's massacre: Livy 39.40.12, *Per.* 49; Val. Max. 8.1.2; Gell. *NA* 1.12.17; Ps.-Ascon. 203 (St.) on Cic. *div. in Caec.* 20.66; Cic. *Brut.* 23.89–90; *Orat.* 1.227–8; *pro Mur.* 59; Quint. 2.15.8. See also Burton 2011, 323–26; Nörr 1989, 129–35; Dmitriev 2011, 259; Hölkeskamp 2000, 246; Eckstein 2009, 266–67; 1994, 83–84; Muñiz Coello 2004, 109–13; Ferrary 2007.

27. Cato's position: e.g., Cic. *Brut.* 90; Fronto *ad. M. Caes.* 3.20. Note that Cato exercised an outsized influence on this episode by recording it in his history of Rome (*Origines*), which was frequently referenced by later authors; while Cato's text is mostly lost, surely it was not favorable to Galba. Consul in 144: Broughton 1951, 470. "Commanders who perpetrated": Eckstein 2009, 266–67.

28. Lusitanians' ritual sacrifice: García Riaza 2002, 106–13. Lusitanian rituals: e.g., Strabo 3.3.6; Polyb. 12.4b.2–3.

29. Temporary suppression of Lusitanians: Richardson 1996, 64; 2000, 150. Warband of 147: App. *Hisp.* 61. "Rootless": Dyson 1985, 208.

30. Vetilius's campaign: App. *Hisp.* 61–62; Diod. 33.1.3; Livy *Per.* 49; Oros. 5.4. Viriathus's rise and Galba's massacre: Val. Max. 9.6.2; Suet. *Galba* 3.2; Oros. 4.21.10. "Ran away and stopped": App. *Hisp.* 62 (Richardson). "Fat old man": App. *Hisp.* 63 (Richardson).

31. Viriathus's attributes: Dio. fr. 73; Diod. 33.1.1–5, 33.7.1–7; App. *Hisp.* 75; Cic. *Off.* 2.40; Livy *Per.* 49. Viriathus and Greco-Roman ideals: Grünewald 2004, 37–47; Yarrow 2006, 334–36; Lens Tuero 1986, 253–72. Viriathus as redistributive war leader: Sánchez Moreno 2006, 65–69; 2005, 111–13. "Lover of war": Dio fr. 73.

32. Wavering allegiance of Roman subjects: Harris 1989, 136–37. Viriathus's military activity: App. *Hisp.* 64–70; Diod. 33.1.1–5, 33.2.1, 33.19–21; Livy *Per.* 52, 54; Oros. 5.4; Dio fr. 73, 75, 78; Flor. 1.33.15–17; Aur. Vict. *Vir. Ill.*71; *FrGH* 103 fr. 27. "Pretended to flee": App. *Hisp.* 64. "Deep holes": Front. *Strat.* 2.5.7. Long frontier, rugged terrain: Dyson 1985, 204.

33. Laelius's victory: Cic. *Off.* 2.11; *Brut.* 21.84. Aemilianus's victory: App. *Hisp.* 65. Roman adoption of extreme methods: Clark 2014, 158. Servilianus's campaigns: App. *Hisp.* 67–69; Livy *Per.* 53–54. Baeturia's location: Keay 1988, 35; Richardson 2000, 160. Chieftains: Oros 5.4.12. Deserters: Val. Max. 2.7.11; Front. *Strat.* 4.1.42.

34. Cerro de la Cruz: Quesada Sanz et al. 2014, 231–71; Quesada Sanz 2015, 14–18. Roman *gladius*: Polyb. 6.23.6–7. "Arms hewed": Livy 31.34.4.

35. Communities defecting: Diod. 33.7.5–6; Harris 1989, 136–37. "Crush and cripple": Val. Max. 2.7.11. Roman terrorism in Spanish wars: Marco Simón 2016, 222–47.

36. Romans tire of Spanish wars: Rubinsohn 1981, 200–201; Harris 1989, 134. Treaty with Viriathus: App. *Hisp.* 69; Livy *Per.* 54; *Ep. Ox.* 54; Diod. 33.1.3; Obs. 23; Aur. Vict. *Vir. Ill.*71; Charax Perg. fr. 27.

37. "Year was peaceful": Obs. 23. Caepio petitions Senate: App. *Hisp.* 70. Strategic problems with treaty, Roman manpower vs. Viriathus: Rubinsohn 1981, 200–202. Roman manpower advantages: Fronda 2010, 37–39, with sources. Romans like hydra: Plut. *Pyrrh.* 19.5; App. *Sam.* 10.3; Dio. 9.40.28. Assassination of Viriathus: Diod. 33.21;

33.1.4; App. *Hisp.* 74; Livy *Per.* 54; Vell. Pat. 2.1.3; Val. Max. 9.6.4; Flor. 1.33.17; Eutr. 4.16; Aur. Vict. *Vir. Ill.* 71; Oros. 5.4.14. Roman conduct in Lusitania: Clark 2014, 158–59. Roman martial values: Lendon 2005, 193–211.

38. Viriathus's funeral, successor: Diod. 33.1.4, 33.21–21a; App. *Hisp.* 75; Livy *Per.* 54. Brutus's campaigns, App. *Hisp.* 71–73; Val. Max. 6.4.ext.1; Livy *Per.* 55–56; Strabo 3.3.5; Diod. 33.26.1–2; Oros. 5.5.12; Vell. Pat. 2.5.1; cf. Curchin 1991, 38, with sources. Callaeci support Lusitanians: Oros. 5.5.12. Impatient aggression of Romans: Lendon 2005, 206–8. "Thinking it difficult," "against their towns": App. *Hisp.* 71 (Richardson). Modern counter-guerillas targeting civilian populations: Valentino, Huth, and Balch-Lindsay 2004, 375–407.

39. "The man who subjugated Lusitania": Livy *Per.* 59; see also Plut. *Ti. Gracch.* 21.2; Eutrop. 4.19; Strabo 3.3.2; Val. Max. 8.14.2. "Soaked Spanish soil": Ov. *Fast.* 6.461.

8

Conclusion

The destruction of Carthage and Corinth in 146 BCE* shocked contemporaries. The armies of the Roman Republic massacred or enslaved much of the population in both cities, and burned to ash, devastated, or dismantled parts of their urban fabrics. As the previous chapters have shown, many other communities met similar ends. In 211, the Romans scourged and beheaded Capua's leaders, enslaved or forcibly removed its people, and, while they did not physically destroy its buildings, they symbolically ended its independent civic existence. In 167, the Romans sacked seventy towns and villages in Epirus, carried off many thousands of their inhabitants for sale as slaves, and demolished their walls. In 150, the Romans massacred the adult males of several Lusitanian communities and sold the rest. Additional examples in between could be enumerated, but the point is clear enough: Rome's treatment of Carthage and Corinth was not extraordinary as such. What struck contemporaries and later ancient authors—and what gave these postwar atrocities their outsized place in ancient and modern imaginations—was the fame and antiquity of the two cities, their opulence and size, and the dreadful coincidence that they were destroyed in the same year.[1]

For the most part, the forms of mass violence that Roman armies employed in Carthage, Corinth, Capua, Epirus, Lusitania, and so many other places required organization, labor, time, and resources. In mass executions, they would detain and restrain captives, often taking them to special killing sites and keeping them under guard. Executioners would be assigned to the ghastly work of killing, which could be difficult to perform. Roman soldiers

---

* Henceforth, all dates in this chapter are BCE.

might massacre crowds of unarmed people in a more haphazard fashion, but even this could involve significant organization and planning; one thinks of Galba's massacre of the Lusitanians and Pinarius's massacre of the Ennaeans. To enslave large groups of people en masse, Roman soldiers forcibly herded the victims together, kept them under guard and under control, and in some cases inspected, sorted, and documented the collected captives before selling them off. Urban destruction could also involve considerable exertion. Burning cities might have been relatively easy, since Roman troops could simply throw torches or firebrands onto buildings. Yet mass arson would necessitate organization if the Romans wanted to create larger blazes or spread damage more widely. To destroy stone fortification walls or other solidly built structures, Roman soldiers would be tasked with laborious (and potentially dangerous) demolition work. Other violent behaviors, like forcible relocation and mass mutilation, would not be easy or quick, either. With the partial exception of city-sacking, the methods of mass violence available to Roman armies normally demanded command-level organization. These behaviors arose from decisions made by Roman military and political leaders in war, who marshaled their legions to carry out destructive and brutal acts.

To show why Roman commanders decided to use these methods in war, the case studies (chapters 5–7) examined mass violence from a "strategic perspective." For the purposes of this book, the strategic perspective is a theoretical framework adapted from work by political scientists. It situates mass violence firmly in the context of warfare to show how military circumstances shaped violent outcomes, and highlights the perspectives, goals, and decision-making of the leaders who ordered or directed violence. Each of the case studies examined mass violence in unique and dynamic situations, but they nonetheless uncovered some general patterns that will guide the remaining discussion.

MASS VIOLENCE AS INSTRUMENTAL STRATEGY

The clearest finding in the case studies is that Roman commanders saw massacres, executions, enslavements, and urban destruction as flexible tools that they could use for diverse purposes. Indeed, Roman leaders employed massive violence to achieve political aims, reward soldiers, punish rebellion, deter further defections, enforce obedience, quash military threats, remove hostile populations, or eliminate enemy strongholds. Several of these objectives could overlap and simultaneously shape Roman conduct. More often than not, then, the case studies indicate that the Romans used mass violence instrumentally.

This finding is consistent with the conclusions of ancient authors, as we saw in chapter 4: ancient Latin and Greek historians usually explain mass violence by pointing to the motives and goals of Roman leaders. However, many ancient authors offer simplistic or monocausal explanations for violent behavior. In contrast, the case studies indicate that Roman decision-making was complex. Military objectives, aristocratic politics, personal goals, emotion, and the threats and challenges encountered in the field could all shape the decisions of Roman commanders—often at the same time. Consider Aemilius Paullus's devastation of Epirus, when his soldiers sacked settlements, carried off tens of thousands into slavery, and demolished urban fortifications. On the one hand, this was surely punitive, since the Epirotes had defected to Perseus in the early years of the Third Macedonian War. Paullus's actions both punished the defectors and sent a sharp message to other communities who might betray Rome. On the other hand, Paullus's soldiers were dissatisfied over a lack of loot, and he used this opportunity to reward them and possibly to deflect political blowback at home. Finally, fortified Epirote communities had blocked important marching and communications routes during the war, and the Romans learned firsthand that Epirote fortresses were difficult to capture. The demolition of their walls neutralized these strongholds, ensuring that the Romans would not face this challenge again.

It is worth reiterating that "deliberate" and "instrumental" violence is not necessarily "rational," as the case studies show. Frustration certainly shaped Roman tactics in the Lusitanian War, when the Lusitanians repeatedly ambushed Roman forces or evaded them altogether. Such frustration is palpable in the sources, reflected especially in the uncompromising position of Roman senators vis-à-vis their foes in Spain. When Galba, Servilianus, and Brutus elected to destroy Lusitanian populations and prisoners outright, anger and frustration surely influenced their choices. Similarly, when Carthage broke the armistice in late 203, Polybius is explicit that Scipio began to attack Punic cities out of anger. Such emotion does not preclude his desire to put strategic pressure on Carthage or his political concern to reap victory for himself.

Aside from frustration or anger, the violence of city-sacking could stem from the greed, rage, vengefulness, or sadism of individual Roman legionaries more than the deliberate plans of their leaders. Having said that, Roman commanders repeatedly engineered city-sacking opportunities: Paullus gave the loot over to his men in Epirus, Scipio "gave" his troops two African cities to loot, Lucretius and Hortensius let their men run loose in Chalcis. While the soldiers themselves likely acted from personal motivations, their leaders were concerned with morale, discipline, and the possible political repercussions of unhappy troops. (Note the Roman legate in Illyria who let his soldiers sack Ceremia so that "his soldiers had not been exhausted from two sieges for nothing.") And as the Epirus episode shows, commanders could use the chaotic violence of city-sacking to punish recalcitrant enemies.[2]

Furthermore, military culture and ideology must have informed Roman decision-making. The historian John Lynn argues that the discourse of war (i.e., all of its representations in a given culture) influences the reality of practice. In the military culture of the Roman Republic, devastated cities and enemy casualties (the latter enumerated without distinction between combatants and noncombatants) all appear as metrics of victory and, in turn, of aristocratic excellence. This is not to say that Roman commanders employed destruction and mass killing simply to glorify their records and win triumphs. However, it is quite possible that the connection between violence and achievement in aristocratic culture influenced the decision-making calculus of Roman generals, including their understandings of such "rational" matters as strategy, tactics, and objectives. As the historian Harry Sidebottom writes, structural factors at a level "below . . . conscious perceptions and actions" could dictate Roman behavior: "[t]he elite . . . needed glory and gain to maintain their status as the elite, and these factors defined, *consciously or not*, the parameters of individuals' actions" (emphasis added). As John North, another historian, puts it, the need for glory and "the pressure of aristocratic career expectations" would have "predisposed the Romans to certain forms of action" whether they were aware of it or not. These scholars both refer to Roman decisions to wage war, but their comments equally apply to decisions to kill, destroy, or enslave during wars. Though the nature of our evidence prevents firmer conclusions, Rome's military culture and ideology of victory may have incentivized mass violence.[3]

THE EFFECTIVENESS OF MASS VIOLENCE

Judging from the case studies, mass violence was sometimes effective—that is, it could accomplish Roman goals, subdue enemies, and contribute to Rome's ultimate success in a given war. Mass violence appears to have been most effective when Rome's military fortunes were improving, or Roman victory was imminent. For example, when the Romans retook Agrigentum in the Second Punic War, beheading the leaders and enslaving the rest, forty Sicilian communities flipped back to the Roman side. Agrigentum was the last major Carthaginian stronghold in Sicily, so when it fell, there was no significant force to counterbalance the Romans on the island, and Roman victory looked all but certain. In those circumstances, the violent treatment of Agrigentum precipitously cowed other cities into submission. By the same token, after the capture and punishment of Capua and the retreat of Hannibal, the other Campanian rebels quickly surrendered. Last, Brutus's brutal campaign subdued the Lusitanians after Viriathus, the main threat to Rome, was out of the picture. In all of these cases, resistance must have appeared increasingly futile for Rome's foes, especially when the terror wrought by the legions gave urgency to surrender. Roman forces also used mass violence effectively to subdue individual cities, especially when a besieged city had little expectation of outside relief, was starving from a long blockade, or was no longer capable of offering worthwhile resistance. To give a few examples: in Spain, Scipio's breach of the walls and indiscriminate massacre at New Carthage precipitated the surrender of the citadel; and in Italy, an ongoing siege coupled with the Romans' mass mutilation of captives compelled the Capuan population to open its gates. In both instances, the defenders had little hope of successful resistance—either because their main fighters had been overcome or they were starving—and Roman victory must have seemed imminent. These findings track with some political science research on mass violence. As Downes and Cochran assert, "civilian victimization" has sometimes worked historically, especially when "the target's level of military vulnerability was also high. It may therefore be the case that civilian victimization works only in situations when the targeted state would have lost anyway or when it was combined with strategies that target military assets." They further argue that violence against noncombatants is "more effective against smaller targets," such as cities under siege.[4]

On the other hand, it is difficult to confirm that Roman violence *caused* Roman victory in any of the case studies. It may be that mass violence did not precipitate victory, so much as military success made it easier for Roman armies to inflict violence. In the Third Macedonian War, most notably, the Romans devastated several Greek and Macedonian communities *after* Perseus's defeat, when there was no further resistance. These depredations contributed very little to their ultimate victory; the war was already won. And while it might reap immediate benefits, mass violence could also stoke resentment, encourage defection, or inspire new foes to take up arms. In the early years of the Second Punic War, Rome's brutal tactics sparked more defections to the Carthaginians. Early in the Third Macedonian War, the harsh treatment of Greek cities inflamed anti-Roman sentiment, pushed sympathies toward Perseus, and led to diplomatic tension between the Romans and their Greek allies. And in the Lusitanian War, Galba's massacre stirred up deep resentment that may have cleared the way for Viriathus's rise. As Valentino writes, rather than coerce, terrorize, or subdue targeted people, mass violence can "stiffen their resolve or convince them they have nothing to lose by joining the opposing side." The evidence from Roman warfare is too incomplete to make firm conclusions about the effectiveness of mass violence, but the verdict of the case studies is decidedly mixed.[5]

MASS VIOLENCE TO EXPEDITE VICTORY, TO COUNTER THREATS, AND AS A MEANS OF CONTROL

As noted above, Roman decisions to use mass violence were dynamic, shaped by multiple pressures and aimed at various goals. Nonetheless, the case studies point to a few broad patterns. First, the Romans used mass violence to apply pressure and expedite victory when they were struggling to defeat an enemy's military forces. When Roman armies could not beat Hannibal, Perseus, and Viriathus and other Lusitanians on the battlefield, they could still attack (and destroy) the towns and populations that supported the enemy. This approach was essential to the Roman strategy in the

Second Punic War after Cannae: they sought to avoid battle with Hannibal while targeting those Italian and Sicilian towns that had changed sides. Likewise, Galba and Servilianus resorted to attacks on enemy populations in the Lusitanian War when they could not win decisive victories in the field. In the first two years of the Third Macedonian War, too, Roman commanders like Licinius turned against Greek towns when they were unable to best Perseus in battle. (Though other Roman commanders targeted pro-Macedonian towns, such as Haliartus, from the beginning.) There was also a wider explosion of atrocity against the Greeks in 171–170 as Perseus was enjoying a string of military successes. As Downes and Valentino argue, desperation and failure on the battlefield can encourage military forces to target noncombatants and populations. Such strategies are designed to coerce surrender, undermine the enemy's ability to fight, and expedite victory when more conventional military approaches are not working. A similar logic is suggested here by the Roman cases.[6]

Second, the case studies indicate that Roman generals often used mass violence to neutralize threats. On a tactical level, mass violence could eliminate immediate threats, as when Roman soldiers burst into enemy cities like New Carthage and massacred the population. In the chaos after the breach, it would have been difficult for the legions to distinguish between armed and unarmed defenders, and virtually any urban inhabitant could participate in the defense of their homes. Massacres suppressed the remaining defenders and cowed others into quicker surrender. Mass violence could also eliminate larger strategic challenges as well as future threats. During the Second Punic War, killing and enslaving defectors (or suspected waverers) kept Italian and Sicilian communities out of enemy hands and preempted subsequent defection. Similarly in the Third Macedonian War, the destruction of urban fortifications and the removal of pro-Macedonian leaders in Greece denied Perseus useful strongholds in Boeotia and the Malian Gulf. The postwar destruction of fortifications in Epirus preempted future threats from these centers and denied their use to other potential enemies. The massacres and mass amputations seen in the Lusitanian War took tough warriors out of play, preventing them from challenging Rome in later encounters. Destroying towns and populations inside and outside of Further Spain snatched them away from Viriathus and ensured that no further trouble would arise from those communities.

As a third point, the case studies repeatedly show the Romans using mass violence to punish rebellion and disobedience, and to deter such behavior in the future. These tactics loom large in the Second Punic War, when the Romans frequently beheaded rebel leaders and occasionally enslaved rebel populations. These actions took rebel towns away from the Carthaginians, but additionally, such terrifying displays of violence punished defectors and deterred other revolts. Roman generals used that same punitive violence in Greece during the Third Macedonian War. Early in the war, for example, a Roman praetor assaulted the allied city Abdera, beheaded its leaders, and enslaved its inhabitants when they would not obey his orders. After the defeat of Perseus, Aemilius Paullus and the senatorial commissioners inflicted numerous punishments on Greeks who allegedly sympathized with Macedonia or openly defected to the enemy. As mentioned above, the Romans might have destroyed Epirote communities to remove future threats from that quarter—but additionally, the devastation punished perceived betrayal and sent a sharp message to the rest of Greece. One also thinks of the Lusitanian War. When the Romans recaptured towns that had joined Viriathus, they executed some of the rebels and enslaved others. Galba's massacre of the Lusitanians was perhaps preemptive. Yet Appian also speaks of Galba getting revenge on the Lusitanians, and possibly the massacre punished Lusitanian treaty violations and raids. The killings moreover sent a frightening signal to other potential raiders.[7]

The use of mass violence to remove threats and to punish, deter, and terrorize served a related function: asserting control over communities and territories. According to several political scientists, including Downes, Valentino, and Sullivan, mass violence often occurs when a state wishes to control territory. In such cases, suppressing or eliminating a hostile (or potentially hostile) population is designed to make control easier, or to remove the threat of future rebellion. In Downes's words, "when war breaks out between intermingled antagonists, or when a belligerent seeks to conquer territory but fears that the population will rebel and pose a permanent threat to its control over the area, a strategy . . . designed to eradicate that group is the likely outcome." The Romans' use of selective executions, mass enslavement, urban destruction, and massacre of military-aged males did not always rise to the point of "eradicating groups." Still, these methods surely aimed at asserting control over territories—at least for the duration of armed conflicts—and preventing subsequent rebellion and disobedience.[8]

This brings us to the relationship between mass violence and that ever-tricky concept of Roman "imperialism." Roman expansion in the Middle Republic is fraught with scholarly controversy, and it is not a topic that I wish to wade into too deeply beyond the following brief observations. Until the late second and first centuries, Roman control over the wider Mediterranean was largely indirect. Nonetheless, contemporary observers like Polybius believed that the whole Mediterranean world had fallen under Roman sway, if not direct administrative control, by the end of the Third Macedonian War. For their part, the Romans had a word for the areas within their ambit: *imperium*. Eckstein writes that "*imperium* was the legal power to command obedience that Roman public officials possessed by virtue of election to office. . . . By extension and metaphor, *imperium* came to denote the power of the polity of the Roman people to command obedience as well, obedience to its orders internationally." In the Middle Republic, this power came primarily in the form of Roman influence and authority.[9]

Surely the Romans saw Italy as beholden to their authority. As we saw in chapter 2, the Republic employed a range of positive incentives and the latent threat of force to support its dominion there. From perhaps the fourth century and certainly by the early third, the Romans would retaliate harshly against Italian communities that dared to defy them, often beheading leaders, enslaving populations, and destroying fortifications. Thus, when the town of Volsinii rebelled in 264, the Romans executed the ringleaders of the rebellion, razed the city—which was well fortified on a defensible site—and moved the remaining population to another location. Similarly, when Falerii rebelled a couple of decades later, the Romans demolished the city and relocated the community to a place that was "easier to access." Some Samnite rebels were executed or enslaved in 269, while hundreds of disobedient Campanian troops were scourged and beheaded in the Roman Forum just a year prior. As we have seen, these same tactics were employed on a very large scale during the Second Punic War, when the Romans recaptured rebel cities and preemptively struck potential defectors to reassert their control over Italy.[10]

As the Romans fought across the Mediterranean in the third and early second centuries, they began to regard other regions as subject to their *imperium*, too. In the western Mediterranean, provinces were created first in Sicily, Sardinia, and Corsica, and then in Nearer and Further Spain. All were slow to develop administrative structures, but magistrates watched over them by the late third and the early second centuries, bringing a measure of direct Roman oversight. As in Italy, the Romans would use mass violence to reassert control over rebellious communities in their overseas provinces. In Sicily during the Second Punic War, the Romans beheaded leaders and massacred and enslaved populations that rebelled or went over to Carthage. In Further Spain, when the Romans recaptured towns that had joined or been garrisoned by Viriathus, they likewise executed some of the rebels and enslaved others. Earlier in the second century, Cato the Elder destroyed the fortifications of rebellious towns all across Nearer Spain. As with the Italian cities Volsinii and Falerii, his goal was to make these places less defensible and discourage future defiance. From the Roman perspective, the provinces were subject to their authority and beholden to their orders. Just like in Italy, mass violence was one of their primary means of answering defiance and projecting control.

The Romans did not create provinces in the rest of the Mediterranean until much later. In the eastern Mediterranean, Roman interventions did not lead to the creation of new provinces until after 146. Northern and western Spain remained outside of Rome's control until the late second century and later. And so forth. Therefore it is interesting that the Romans exported the practices of mass violence that they had used in Italy and the provinces into regions that were technically independent of their control. This pattern is most noticeable in the Third Macedonian War. After the defeat of Perseus, they executed leaders for perceived disloyalty (real or suspected), demolished fortifications to make places less defensible, and relocated (or enslaved) people and populations. Much as in Italy and the provinces, displays of violence punished disloyalty and deterred future betrayal. Galba's massacre of the Lusitanians, who apparently lived beyond Rome's provincial frontiers, also may have penalized Lusitanian raids and treaty violations and served as a warning to others. We should consider Carthage and Corinth here as well, since both cities were destroyed in 146 after their capture by Roman armies. Their destruction might have resulted, at least in part, from the fact that both were defiant in the face of Roman orders. Punishment and deterrence.[11]

The parallels between the Romans' use of mass violence in Italy, in their provinces, and abroad have clear implications for Rome's *imperium* in the Middle Republic. With their lack of a complex administrative apparatus in Italy and the provinces, the Romans had few means at their disposal to bring subject peoples back into line aside from naked force. Thus, "calculated acts of brutality" (in Rosenstein's words) were an attractive way to project power and enforce obedience. It is significant that they used these tactics further afield, indicating that they believed their *imperium* extended, or should extend, to places beyond Italy and the provinces. Displays of massive violence, often at the end of armed conflicts, made manifest Roman power, enforced obedience, dissuaded defiance, and eliminated current and future threats. In places they had annexed and those they had not, mass violence allowed the Romans to leave a stark impression of strength even after the last legionary went home, cementing the Republic's authority and ensuring its security. Diodorus, in a famous comment about Roman imperialism, states that the Romans "confirmed their power by terrorism and the destruction of the most eminent cities." He grants too much importance to the destructions of Carthage and Corinth, but in a general sense he was not far off the mark.[12]

MASS VIOLENCE IN A MEDITERRANEAN PERSPECTIVE

We have seen throughout this study that Roman generals often coupled massive displays of force with ostentatious mercy, and many were lenient to foes who surrendered voluntarily. The Romans were not uniformly violent, and usually they were not purposelessly cruel. Yet, in the case studies, we have also seen that massacres, enslavements, and destructions routinely punctuated Roman campaigns, and the Romans reportedly used mass violence in almost every war that they fought in the Middle Republic (see appendix 1). Just like the agricultural raid, the set-piece battle, and the siege, mass violence was an essential tool on the Roman commander's tool belt, allowing him to pursue his objectives and put pressure on his opponents. Additionally, Rome's demanding political culture may have encouraged Roman generals to use these methods: body counts and captured cities were translated into political rewards, and greedy troops (who could become angry voters) could be appeased with plunder and opportunities to sack cities.

Despite the prominence of mass violence in Roman warfare, it is difficult to say whether the Romans were *exceptionally* prone to these behaviors relative to other ancient peoples. Undoubtedly, some contemporaries believed that mass violence was characteristic of Roman warfare. We have seen how

Polybius describes the Romans "customary" practice of "killing all they encountered" when they assaulted cities, as well as their "customary" beheadings of prisoners. The First Book of Maccabees pointedly claims that the Romans destroyed or enslaved all their enemies. Other observers thought that the Romans used mass violence excessively and beyond the bounds of normative behavior. The Third Book of Sibylline Oracles asserts that Rome plundered more wealth and enslaved more people than any previous great power. Leading Greeks complained that the Romans were "the most ruthless barbarians," bringing to Greece the "basest violence and lawlessness" by burning towns and hauling people away into hopeless slavery. Importantly, though, these surviving reactions to Roman warfare arose from the eastern Mediterranean, where the cultural norms of Hellenistic warfare held sway. As we saw in chapter 6, Hellenistic military conventions did not prevent brutality and violence in war, but they did encourage moderation. Prisoners and captured cities were often spared, and according to Polybius, no city in mainland Greece was destroyed for a century after Alexander's death. In contrast, the Roman legions enslaved, sacked, or destroyed several dozen Greek cities in the forty-seven years between the First and Third Macedonian Wars. This would have made the Romans seem comparatively extreme in the eyes of Greeks like Polybius.

Still, the Romans' military conduct must be viewed in perspective: they were not the only ancient people to use mass violence in war, by any stretch of the imagination. Greeks, Macedonians, Persians, Carthaginians, Thracians, and many others reportedly destroyed and sacked cities, enslaved prisoners, and massacred unarmed people in their military campaigns. Warrior kings like Alexander the Great and Agathocles of Syracuse arguably spilled more blood than any one Roman general in the Middle Republic (see appendix 2). Hannibal, Perseus, and Viriathus also used mass violence in their wars against Rome. Hannibal's sack of Saguntum set off the Second Punic War, and in Italy he sacked and destroyed cities, killed off or forcibly removed populations, sold prisoners, and executed pro-Roman leaders. During the Third Macedonian War, Perseus sacked the Greek city Mylae and sold the survivors into slavery. He also visited mass killing and enslavement upon towns in Illyria and Dardania. Although we know much less about Viriathus's campaigns due to fragmentary sources, he apparently slaughtered 5,000 of Rome's Iberian allies, the Belli and Tithi, and widely ravaged rural communities in Roman Spain—ravages that probably included the merciless seizure of people and property alongside the destruction of crops and villages. All ancient states existed in an unforgiving political-military environment, and many or most fought ruthlessly to survive, defeat rivals, and increase their own power. As Polybius writes, war "compels us to seize and destroy the enemy's forts, harbors, cities, men, ships, crops, and other such things, thus weakening the enemy and strengthening our own affairs and plans." His logic would have been familiar to military commanders across the ancient Mediterranean world, not just to the Romans.[13]

If the Romans appeared exceptionally violent to some observers, this likely points to the scope of their military efforts and their overwhelming military success more than to a peculiarly atrocious way of war. By the late third century, they were able to wage war more widely and with greater resources than all their Mediterranean rivals, and in the long run, they did so more successfully than their main competitors. Relative to other ancient states, this meant that more prisoners fell into Roman hands, more cities fell to Roman armies, and more territory fell under Rome's direct or indirect control. In turn, the legions gained—and often exploited—numerous opportunities to destroy, kill, and enslave in pursuit of their own objectives. The scale of the Romans' military effort was unprecedented. Mass violence, however, was grimly commonplace in ancient warfare.[14]

NOTES

1. Destructions of Carthage and Corinth: e.g., Davies 2014; Purcell 1995; Clark 2014, 138–47.

2. For the political science on mass violence/genocide, see chapter 1. "Two sieges": Livy 43.1.3.

3. Discourse of war: Lynn 2008, 359–70. "Conscious perceptions": Sidebottom 2005, 316, summarizing the arguments of several other scholars. "Aristocratic career expectations": North 1981, 6, 8; cf. Rich 1993, 58; Harris 1979, 9–53. Influence of cultural values on ancient military practices: Lendon 2005. For military culture encouraging mass violence in a modern context, see Hull 2004, 2008.

4. "Military vulnerability," "smaller targets": Downes and Cochran 2010, 55. For the possible effectiveness of mass violence, see also Downes 2007, 2008; Lyall 2009; Kalyvas 2006.

5. "Stiffen their resolve": Valentino 2014, 98–99, summarizing the literature.

6. Desperation to win/attritional wars: Downes 2008, esp. 29–33, 83–155; Valentino 2004, esp. 84–88. Valentino, Huth, and Balch-Lindsay 2004 argue that insurgencies and guerilla wars incentivize attacks on populations that support the enemy, often when frustrated military forces cannot win decisive battles. See Valentino 2004, 81–83, 196–233; Downes 2008, 156–77; Downes 2007.

7. Galba "avenged faithlessness": App. *Hisp.* 60.

8. "When war breaks out": Downes 2008, 178–79. Conquest, control of territory, prevention of rebellion, and mass violence (conventional and civil wars): Downes 2008; Valentino 2004; Sullivan 2012; Kalyvas 2006; Balcells 2010; Kim 2010.

9. Roman *imperium* in the Middle Republic: e.g., Eckstein 2008, esp. 342–81 (referring to the Greek world); Rosenstein 2012, 73–82 (referring to Italy). "Legal power": Eckstein 2007, 568.

10. Hegemony in Italy: Fronda 2010, 24. Volsinii: Zon 8.7; Aur. Vict. *Vir. Ill.* 36.2. Falerii: Zon. 8.18; cf. Livy *Per.* 20; Polyb. 1.65.2. Samnite rebels: Dion. Hal. 20.17.2. Campanian troops: Polyb. 1.7.9–12; Dion. Hal. *Ant. Rom.* 20.5.5, 20.16.1–2; App. *Sam.* 21; Livy 28.28.3, *Per.* 15; Oros. 4.3.5; Front. *Strat.* 4.1.38; Val. Max. 2.7.15.

11. Corinth punished: Cic. *Man.* 5; Livy *Per.* 52; Strabo 8.6.23. Corinth exemplary: Just. 34.2.6. Carthage defiant: Polyb. 36.9.13–17; cf. App. *Pun.* 70–93; Livy *Per.* 48–50.

Mummius's decisions at Corinth also may have been shaped by his desire to compete politically with Scipio Aemilianus, the recent destroyer of Carthage: Kendall 2009.

12. For the argument that Rome had few means of imperial control in the Middle Republic aside from force and "calculated acts of brutality," see Rosenstein 2012, 225, and 211–39 *passim*. "Confirmed their power": Diod. 32.4.5 (trans. Oldfather). The arguments of Mattern 1999 (regarding Roman imperialism) and Van Wees 2013 (regarding ancient "genocide") overlap somewhat with my claims here, though they stress slightly different motives.

13. Hannibal: e.g., Polyb. 3.86.10–11, 3.100.4, 9.26.7–9; Livy 23.15.6, 23.17.7; App. *Pun.* 134; Plut. *Flam.* 13.4–6. Perseus: Livy 42.54.6, 43.18.11, 43.19.11–12; Diod. 30.4.1. Viriathus: App. *Hisp.* 63–66. Anarchic Mediterranean: see esp. Eckstein 2006. "Compels us": Polyb. 5.11.3.

14. Unprecedented scale of Roman resources/efforts: Polyb. 1.1.5; Eckstein 2006, 245–57, 2010, 567; Erdkamp 2011, 278–79, 284–91; Rosenstein 2012, 73–118.

Appendix 1

124 Cases of Mass Violence in Roman Warfare, c. 400–100 BCE

Appendix 1

Year BCE	Roman Commanders[a]	Vanquished People, Group, or City	Mass Violence Described[b]	Source(s)
396	M. Furius Camillus*	Veii (Etruscan city)	City destroyed; some inhabitants killed; Veian survivors enslaved	Diod. 14.93.2, 14.115.2; Livy 5.21.2–14; Plut. Cam. 6.1; Flor. 1.6.10
388	M. Furius Camillus	Sutrium (Etruscan city)	Etruscan captives enslaved	Livy 6.4.2; cf. Plut. Cam. 35.1–36.1
388	Unnamed	Cortuosa (Etruscan city)	City burned	Livy 6.4.9
353	Unnamed	Tarquinians (Etruscan people)	Prisoners killed; 358 (or 260) leading men scourged, beheaded in the Forum	Livy 7.19.2–3; Diod. 16.45.8
346	M. Valerius Corvus	Satricum (Volscian city)	City burned, destroyed; 4,000 survivors sold	Livy 7.27.8
338	Unnamed	Vellitrae (Volscian city)	Walls demolished	Livy 8.14.5
329	L. Aemilius Mamercinus*	Privernum (Volscian city)	Walls demolished; some leaders executed	Livy 8.20.7–10
319	L. Papirius Cursor*	Satricum (Volscian city)	Captives responsible for defection scourged, beheaded	Livy 9.16.9–10; cf. Oros. 3.15.9–10
314	M. Poetelius; C. Sulpicius	Sora (Volscian town)	Some killed during assault; 225 defectors scourged, beheaded in the Forum	Livy 9.24.12–15
314	M. Poetelius; C. Sulpicius	Ausona, Menturnae, Vescia (Ausonian cities)	Ausonian people destroyed (or "exterminated," *deleta*)	Livy 9.25.9
314	Unnamed	Luceria (Apulian city)	Inhabitants killed during assault on city	Livy 9.26.2–3
313	Q. Fabius* or C. Poetelius*[c]	Fregellae (Latin city)	200 leading men scourged, beheaded in the Forum	Diod. 19.101.3; cf. Livy 9.28.3

a. Commanders' names are starred (*) when the sources explicitly grant them responsibility for mass violence.
b. This chart primarily includes references to mass killing, mass execution, mass enslavement, and destruction of the built environment (e.g., cities, villages, buildings).
c. Diodorus has Q. Fabius capturing Fregellae, but Livy attributes the deed to C. Poetelius.

313	C. Poetelius* or C. Iunius Bubulcus*[a]	Nola (Campanian city)	"All the buildings outside the walls" burned	Livy 9.28.5–6
311	C. Iunius Bubulcus*	Cluviae (Samnite city)	Adult males killed during or after assault on city	Livy 9.31.1–4
310	C. Marcius Rutulus	Unnamed Samnite forts and villages	Forts and villages destroyed	Livy 9.38.1
308	P. Decius*	Unnamed Volsinian forts	Fortresses destroyed	Livy 9.41.6
308	Q. Fabius	Allifae (Samnite city)	7,000 inhabitants enslaved	Livy 9.42.7–8
304	P. Sulpicius Saverrio; P. Sempronius Sophus	Aequi towns	"Most" of 31 towns destroyed, burned	Livy 9.45.17
299	M. Valerius*	Unnamed Etruscan villages	Villages burned	Livy 10.11.6
296	P. Decius	Romulea (Samnite city)	2,300 inhabitants/defenders killed during sack of city	Livy 10.17.8
293	L. Papirius Cursor*	Aquilonia (Samnite city)	City burned	Livy 10.44.2
293	Sp. Carvilius Maximus*	Cominium (Samnite city)	City burned	Livy 10.44.2
293	Sp. Carvilius Maximus	Velia, Palumbinum, Herculaneum (Samnite cities)	<5,000 killed during capture of cities	Livy 10.45.11
293	L. Papirius Cursor	Saepinum (Samnite city)	7,400 inhabitants killed during assault on city	Livy 10.45.14
283	P. Cornelius Dolabella*	Unnamed towns of the Gallic Senones	Adult males killed; women and children enslaved	App. *Sam.* 13, *Gall.* 13; Dion. Hal. *Ant. Rom.* 19.13.1; Flor. 1.8.21; cf. Strabo 5.1.6; Polyb. 2.19.10–11

(continued)

a. Livy mentions two different traditions, with two different commanders capturing Nola.

Appendix 1

Year BCE	Roman Commanders	Vanquished People, Group, or City	Mass Violence Described	Source(s)
282	Decius*[a]	Rhegium (Italiote-Greek city)	Adult males killed or expelled	Polyb. 1.7.6–8; Diod. 22.1.2; Dion. Hal. 20.4.6–8; App. *Sam.* 19; Strabo 6.1.6; Dio Fr. 40.10; Livy 28.28.2, *Per.* 12; Oros. 4.3.4
272	L. Papirius Cursor	Tarentum (Italiote-Greek city)	Walls demolished	Zon. 8.6
270	C. Genucius Clepsina*	Rhegium (Italiote-Greek city)	"More than 300" survivors taken to the Roman Forum, scourged, beheaded[b]	Polyb. 1.7.9–12; Dion. Hal. *Ant. Rom.* 20.5.5, 20.16.1–2; App. *Sam.* 21; Livy 28.28.3, *Per.* 15; Oros. 4.3.5; Front. *Strat.* 4.1.38; Val. Max. 2.7.15
269	Q. Ogulnius Gallus* C. Fabius Pictor*[c]	Samnite rebels in unnamed town	Prisoners scourged with rods and killed; others enslaved	Dion. Hal. *Ant. Rom.* 20.17.2
264	M. Fulvius Flaccus*[d]	Volsinii (Etruscan city)	Rebel leaders scourged to death; city destroyed and survivors forcibly relocated to new site	Zon. 8.7; Aur. Vict. *Vir. Ill.* 36.2; cf. Pliny *NH* 34.34; Festus 228L; *Fasti Triumph.* 264

a. The sources suggest that the Roman garrison at Rhegium, which was composed primarily of allied Campanian troops, "went rogue" under its commander Decius. In which case, the Romans' later capture of Rhegium and execution of captives would have been punishment for military mutiny, but the details are not clear. See Rosenstein 2012, 56–57.
b. Polybius says "more than 300" were brought for execution, but Valerius Maximus says "500 a day" were killed for an indeterminate number of days, Dionysius of Halicarnassus says 4,500, and Livy and Frontinus say 4,000 were executed.
c. Dionysius says "the consuls" were responsible for suppressing this revolt; the consuls for 269 BCE are listed here, but they may not have been in command for this specific event.
d. Zonaras says a "consul" had the captives executed; the Fasti and Festus indicate that M. Fulvius Flaccus was the consul who ultimately captured Volsinii.

124 Cases of Mass Violence in Roman Warfare

Year	Commander	Location	Outcome	Sources
262	L. Postumius Megellus; Q. Mamilius Vitulus	Agrigentum (Siciliote-Greek city)	25,000 inhabitants enslaved	Diod 23.7.1, 23.9.1; Zon. 8.10; Polyb. 1.19.14–15
259	L. Cornelius Scipio	Olbia, Aleria (Sardinian cities)	Cities destroyed	Flor. 1.18.16; cf. Zon. 8.11; Val. Max. 5.1.2; CIL 1².9
258	A. Atilius Caiatinus*	Myttistratus (Siciliote-Greek city)	Inhabitants killed indiscriminately during assault on city; survivors enslaved; city destroyed or burned	Diod. 23.9.4; Zon. 8.11; cf. Polyb. 1.24.11
258	Unnamed	Mazarin (Carthaginian town in Sicily)	Inhabitants enslaved	Diod. 23.9.4
258	P. Claudius*	Camarina (Siciliote-Greek city)	Inhabitants enslaved	Diod. 23.9.5, Val. Max. 6.5.1
256	L. Manlius Vulso*	Rural settlement around Aspis (North Africa)	Many homes destroyed; >20,000 inhabitants enslaved	Polyb. 1.29.7; Oros. 4.8.9
254	Cn. Cornelius Scipio Asina; A. Atilius Caiatinus	Panormus (Carthaginian town in Sicily)	Many inhabitants killed during assault on "outer" city; 13,000 survivors enslaved	Diod. 23.18.4–5; cf. Polyb. 1.38.7–10
252	C. Aurelius*	Lipara (Siciliote-Greek city)	"All" the inhabitants killed	Zon. 8.14; cf. Polyb. 1.39.13
241	Unnamed	Falerii (Italian city)	City destroyed; surviving inhabitants forcibly relocated to a new site	Zon. 8.18; cf. Livy Per. 20; Polyb. 1.65.2; Eutrop. 2.28.1
229	Cn. Fulvius Centumalus	Unnamed Illyrian tribes	Illyrian chieftains beheaded	Flor. 1.21.4
219	L. Aemilius Paullus*	Pharos (Illyrian city)	City destroyed	Polyb. 3.19.12; App. Illyr. 8
218	Ti. Sempronius Longus*	Malta (Carthaginian island)	Carthaginian soldiers enslaved	Livy. 21.51.2
217	Cn. Cornelius Scipio	Onusa (Iberian city)	Buildings burned	Livy 22.20.5
217	Cn. Cornelius Scipio	Unnamed villages on island of Ebusus	Villages burned	Livy 22.20.9

(continued)

Year BCE	Roman Commanders	Vanquished People, Group, or City	Mass Violence Described	Source(s)
215	M. Claudius Marcellus*	Nola (Campanian city)	"Over 70" conspirators beheaded	Livy 23.17.2
215	Unnamed	Casilinum (Campanian city)	Inhabitants killed	Livy 23.17.10
215	M. Valerius Laevinus	Vercellium, Vescellium, Sicilinum (Samnite towns)	Two leaders beheaded; >5,000 prisoners sold	Livy 23.37.12–13
214	M. Claudius Marcellus	Casilinum (Campanian city)	Inhabitants killed indiscriminately during assault	Livy 24.19.8–9
214	M. Claudius Marcellus; Ap. Claudius Pulcher	Leontini (Siciliote-Greek city)	Inhabitants killed indiscriminately (?);[a] 2,000 deserters scourged and beheaded	Livy 24.30.4–7; Plut. Marc. 14.1
214	Q. Fabius Maximus	Samnites	Some rural inhabitants enslaved	Livy 24.20.4
214	Q. Fabius Maximus	Compulteria, Telesia, Compsa, Melae, Fugifulae, Orbitanium, Blandae, Aecae (Samnite, Apulian, Lucanian towns)	25,000 inhabitants/defenders killed or captured during assault or siege; 300 deserters beaten and thrown from the Tarpeian Rock	Livy 24.20.5–6
213	M. Claudius Marcellus*	Megara (Siciliote-Greek city)	City destroyed	Livy 24.35.2
213	L. Pinarius*	Enna (Siciliote-Greek city)	Inhabitants killed	Livy 24.39.1–6; Front. *Strat.* 4.7.22; Polyaen. 8.21.1; cf. *CIL* 1².608
213	Q. Fabius Maximus*	Arpi (Apulian city)	Carthaginians in the city killed	App. *Hann.* 31; cf. Livy 24.47.7–11
212	Cn. Cornelius Scipio*	Unnamed city of Turdetani (Iberian city)	City destroyed; inhabitants enslaved	Livy 24.42.11, 28.39.12; Zon. 9.3
212	Unnamed	Escaped hostages from Tarentum and Thurii (Italiote-Greek cities)	Hostages scourged in the Roman Comitium and thrown from the Tarpeian Rock	Livy 25.7.14

124 Cases of Mass Violence in Roman Warfare

Year	Commander	Location	Description	Sources
212	M. Claudius Marcellus	Syracuse (Siciliote-Greek city)	Some inhabitants killed (?);[a] some inhabitants enslaved (?);[b] parts of the city burned	Livy 25.25.5–10, 25.31.8–11, 26.32.4; Diod. 26.20.1–2; Zon. 9.5
211	M. Fulvius Flaccus	70 Numidian prisoners	Prisoners beaten, mutilated	Livy 26.12.18
211	M. Fulvius Flaccus*	Capua (Campanian city)	Leading men scourged, beheaded; (some?) Campanians enslaved	Livy 26.15.8–9, 16.6; App. *Hann.* 43; Zon. 9.6; Oros. 4.17.12
211	M. Valerius Laevinus	Anticyra (Greek city)	Inhabitants enslaved	Polyb. 9.39.2–3; cf. Livy 26.26.3
210	P. Sulpicius Galba*	Aegina (Greek city)	Inhabitants enslaved	Polyb. 9.42.5–8, 22.8.9–10
210	M. Valerius Laevinus*	Agrigentum (Siciliote-Greek city)	Leading men scourged, beheaded; inhabitants enslaved	Livy 26.40.13–14
209	P. Cornelius Scipio*	New Carthage (Carthaginian city in Spain)	Inhabitants killed indiscriminately during assault; some survivors enslaved	Livy 26.46.7–10, 26.47.1–3 Polyb. 10.15.4–9, 10.17.9–14
209	Q. Fabius Maximus*	Tarentum (Italiote-Greek city)	Inhabitants killed indiscriminately during assault; 30,000 enslaved; walls separating city from its citadel torn down	Livy 27.16.5–8; Plut. *Fab.* 22.4
208	P. Cornelius Scipio*	Carthaginian soldiers	Captives sold	Livy 27.19.2; cf. Polyb. 10.40.9
207	L. Cornelius Scipio	Orongis (Iberian city)	Surrendering enemy deserters killed; 2,000 armed inhabitants killed during assault	Livy 28.3.10–16

(continued)

---

a. The sources describe Marcellus restraining his troops during the siege and assault operations, but he apparently allowed them free rein to sack the city after capturing Achradina. It was at this stage that Archimedes was cut down, suggesting that some killing took place. Moreover, Zonaras asserts that many Syracusans were slaughtered during the Roman assault on the city.

b. According to Diodorus, some Syracusans voluntarily submitted to enslavement as a survival strategy because they could not obtain enough food.

Appendix 1

Year BCE	Roman Commanders	Vanquished People, Group, or City	Mass Violence Described	Source(s)
207	P. Sulpicius Galba, with Attalus I of Pergamum	Oreum (Greek city)	Some inhabitants surrounded, killed, or captured during assault	Livy 28.6.5
207	M. Livius Salinator; C. Claudius Nero	Carthaginian and Gallic soldiers	Captives enslaved or executed	Polyb. 11.3.1–3
206	P. Cornelius Scipio	Iliturgi/Ilurgia (Iberian city)	Inhabitants killed indiscriminately during assault; city and buildings burned, destroyed	Livy 28.20.6–7; App. *Hisp.* 32; Zon. 9.10
205	P. Cornelius Scipio*	Locri (Italiote-Greek city)	Leaders responsible for defection executed	Livy 29.8.2
203	P. Cornelius Scipio	Locha (Carthaginian city)	Inhabitants killed indiscriminately during assault on city	App. *Pun.* 15
203	P. Cornelius Scipio*	Unnamed Carthaginian towns	Inhabitants enslaved	Polyb. 15.4.1–2
200	C. Claudius Centho	Chalcis (Greek city)	Adult males killed indiscriminately during assault on city; city "half-ruined"; fires set around the agora; granaries and arsenal burned	Livy 31.23.4–10, 31.24.3
200	L. Apustius*	Antipatrea (Macedonian city)	Adult males killed; walls demolished; city burned	Livy 31.27.4
200	L. Apustius (with Attalus I of Pergamum)	Oreum (Greek city)	Inhabitants enslaved	Livy 31.46.16
198	P. Villius Tappulus*	Hestiaea/Oreum, Anticyra (Greek cities)	Cities destroyed	Paus. 7.7.9; cf. Livy 31.46.14–16, 32.18.6
198	T. Quinctius Flamininus	Phaloria (Greek city)	City burned	Livy 32.15.3
198	Unnamed	Dyme (Greek city)	Inhabitants enslaved	Livy 32.22.10–11
197	Q. Minucius Rufus	Clastidium (Ligurian city)	City burned	Livy 32.31.4

124 Cases of Mass Violence in Roman Warfare

Year	Commander	Target	Outcome	Sources
197	T. Quinctius Flamininus	Macedonian soldiers	Some soldiers killed while attempting to surrender; some taken captive and sold	Livy 33.10.3–7, 33.11.2; Polyb. 18.26.9–12; 18.33.8
195	M. Helvius	Iliturgi	Adult males killed	Livy 34.10.2
195	M. Porcius Cato	7 Bergistani strongholds (Iberian forts/towns)	Inhabitants sold	Livy 34.16.10
195	M. Porcius Cato*	Unnamed Iberian towns in Nearer Spain	Walls demolished	Livy 34.17.11; Front. Strat. 1.1.1; Polyb. 19.1.1 (= Plut. Cat. Mai. 10.3); App. Hisp. 41; Polyaen. 8.17.1; Aur. Vict. Vir. Ill. 47.2; Zon. 9.17
195	M. Porcius Cato*	Bergium (Iberian fort/town)	Some inhabitants sold; some "bandits" executed	Livy 34.21.5–6
189	M. Fulvius Nobilior	Same (Greek city)	Inhabitants enslaved	Livy 38.29.11
189	C. Manlius Vulso	Galatian camp on Mysian Olympus	40,000 (?) Galatians killed indiscriminately or sold[a]	Livy 38.23.5–10; App. Syr. 42
185	Ap. Claudius Pulcher*	Six unnamed Ligurian towns	43 leaders beheaded	Livy 39.32.3–4
184	A. Terentius*	Corbio (Iberian town)	Inhabitants enslaved	Livy 39.42.1
181	L. Aemilius Paullus*	Unnamed Ligurian towns	Walls demolished	Plut. Aem. 6.3
178	M. Iunius Brutus; A. Manlius Vulso	Mutila, Faveria, Nessatium (Istrian towns)	Some inhabitants killed during assault on Nessatium; 5,632 captives sold; leaders scourged and beheaded; cities destroyed	Livy 41.11.7–9

a. Livy says that there were conflicting numbers in his sources, and he does not explicitly state that the captured Galatians were sold. However, he says that the consul (Manlius Vulso) sold some of the booty and distributed some to his troops, and both he and other authors mention a Galatian noblewoman (Chiomara) who became the slave of one of Vulso's centurions. This suggests that the captives were numbered along with the rest of the plunder. See Livy 38.24.1–3; Polyb. 21.38 (= Plut. Mor. 258 E–F); Val. Max. 6.1e.2; Flor. 1.27.6.

(continued)

Appendix 1

Year BCE	Roman Commanders	Vanquished People, Group, or City	Mass Violence Described	Source(s)
177	C. Claudius Pulcher	Mutina (Roman colony taken by Ligurians)	8,000 killed (during assault on city?)	Livy 41.16.8
177	Ti. Sempronius Gracchus	Sardinia (island in rebellion)	80,000 killed or enslaved	Livy 41.28.4; Aur. Vict. Vir. Ill 57.2
173	M. Popilius Laenas*	Carystus (Ligurian town)	Inhabitants enslaved; city destroyed	Livy 42.8.1–3
171	C. Lucretius Gallus	Haliartus (Greek city)	Inhabitants ("old men" and "boys") killed indiscriminately in assault; city destroyed	Livy 42.63.10–12; Strabo 9.2.30; Polyb. 30.20.1–8
171	C. Licinius Crassus*	Pteleum (Greek city)	City destroyed	Livy 42.67.9
170	C. Licinius Crassus*	Unnamed Greek cities	Cities destroyed; inhabitants enslaved	Zon. 9.22; Livy Per. 43
170	L. Hortensius*	Abdera (Greek city)	Leading men beheaded; inhabitants enslaved	Livy 43.4.9–13
169	Q. Marcius Philippus	Dium (Macedonian town)	Walls (partly?) destroyed (?)[a]	Livy 44.8.5
167	Q. Fabius Labeo*	Antissa (Greek city)	City destroyed; inhabitants forcibly removed	Livy 45.31.13–14
167	L. Aemilius Paullus*	Demetrias (Macedonian/Greek city)	Walls destroyed	Diod. 31.8.6
167	L. Aemilius Paullus*	70 Epirote towns	Cities destroyed; walls demolished; 150,000 inhabitants enslaved or "carried off"	Strabo 7.7.3; Livy 45.34.1–6; Trog. pr. 33; Plut Aem. 29; Pliny NH 4.39

a. Livy says that when the Macedonian king Perseus recaptured Dium from the Romans, he refortified the city by repairing whatever the Romans had "thrown down or devastated" (disiecta ac vastata) and replacing the "battlements of the walls which had been cast down" (pinnas moenium decussas). The language, and Perseus's military needs in the moment, suggest that the Romans had partly destroyed the fortifications when they abandoned the city.

156	C. Marcius Figulus*	Delminium (Dalmatian town)	City burned	App. *Ill.* 11; Flor. 2.25.11
155	P. Cornelius Scipio Nasica*	Delminium, unnamed Dalmatian towns	Inhabitants enslaved	Zon. 9.25; Strab. 7.5.5
154	Q. Opimius*	Aegitna (Ligurian town)	Inhabitants enslaved	Polyb. 33.10.3
151	M. Licinius Lucullus*	Cauca (Celtiberian city)	Adult males killed after city surrendered	App. *Hisp.* 52
150	Ser. Sulpicius Galba*	Lusitanians	Lusitanians surrounded with a ditch, disarmed, killed with swords; survivors sold	App. *Hisp.* 59–60; Cic. *Brut.* 89; Livy *Per.* 49; Val. Max. 8.1.2, 9.6.2; Oros. 4.21.10; Suet. *Gal.* 3.2
147	L. Calpurnius Piso*	Neapolis (Carthaginian city)	City destroyed	Diod. 32.18.1; Zon. 9.29
146	P. Cornelius Scipio Aemilianus*	Carthage	City burned or destroyed; walls demolished; inhabitants killed during assault; survivors forced to depart, imprisoned, or enslaved	Diod. 32.4.5; App. *Pun.* 1, 75, 81, 83, 133–136, *BC* 24; Strabo 6.4.2, 14.5.2, 17.3.13; Oros. 4.23.6, 5.3.1; Zon. 9.26, 30; Obs. 20; Vell. Pat. 1.12.4–5, 2.4.2–3; Cic. *Agr.* 2.51, 87, *Catil.* 4. 21, *Mur.* 58, *Man.* 60, *Har.* 6, *Rep.* 6.11, *Off.* 1.35; [Cic.] *Herr.* 4.27.37;

(*continued*)

Appendix 1

Year BCE	Roman Commanders	Vanquished People, Group, or City	Mass Violence Described	Source(s)
146	P. Cornelius Scipio Aemilianus*	Carthage	City burned or destroyed; walls demolished; inhabitants killed during assault; survivors forced to depart, imprisoned, or enslaved	Sall. *Iug.* 41; Livy 44.44.2, *Per.* 51; Suet. *De Poet.* 11.1; Vell. Pat. 1.12.5, 1.13.1, 2.4.1–3; Just. 38.6.5; Plut. *Aem.* 22.4; Curt. 4.3.23
146	P. Cornelius Scipio Aemilianus	Unnamed Carthaginian towns	Cities destroyed	App. *Pun.* 135
146	L. Mummius*	Corinth (Greek city)	City destroyed or burned; walls demolished; theater and other unnamed buildings destroyed; male inhabitants killed; women and children sold	Diod. 32.4.5; Livy 45.28.2, *Per.* 52; Paus. 7.16.7–10; Zon. 9.31; Just. 34.2.5–6; Flor. 1.32.5–6; Vitr. 5.5.8; Strabo 8.6.23; Pliny *NH* 34.6–7; Cic. *Agr.* 2.87, *Fam.* 4.5.4, *Off.* 1.35; [Cic.] *Herr.* 4.27.37; Vell. Pat. 1.13.1; Oros. 5.3.6; *CIL* 1².626; Trog. 34 pr.
146	L. Mummius*	Unnamed Greek (Achaean?) cities	Walls destroyed	Paus 7.16.9
144	Fabius Maximus Aemilianus*	Unnamed Lusitanian town	City burned	App. *Hisp.* 65

124 Cases of Mass Violence in Roman Warfare

Year	Commander	Location/Target	Action	Sources
142	Fabius Maximus Servilianus*	Escadia, Gemella, Obolcola, and several unnamed towns (Iberian or Lusitanian towns)	500 prisoners beheaded; 9,500 prisoners sold	App. *Hisp.* 68
142	Fabius Maximus Servilianus*	Lusitanian prisoners and deserters	Captives' hands severed	Oros 5.4.12; Val. Max. 2.7.11; Front. *Strat.* 4.1.42.
141	Q. Pompeius*	Lagni/Malia (Celtiberian town)	"All nobles" killed; city destroyed	Diod. 33.17.3; cf. App. *Hisp.* 77
139	D. Iunius Brutus	Unnamed Lusitanian communities	Inhabitants killed during attack on towns	App. *Hisp.* 71
134	P. Cornelius Scipio Aemilianus*	Loutia (Celtiberian town)	400 captives' hands severed	App. *Hisp.* 94
133	L. Calpurnius Piso*	Rebel slaves in Sicily	Captives crucified	Oros. 5.9.6
133	P. Cornelius Scipio Aemilianus*	Numantia (Celtiberian town)	City burned or destroyed; inhabitants sold	App. *Hisp.* 98; Livy *Per.* 59; Diod. 31.26.3, 32.4.5; Val. Max. 2.7.1; Amm. Marc. 23.5.20; Vell. Pat. 2.4.2; Flor. 1.34.15; Oros. 5.7.16; Eutrop. 4.17.2; Cic. *Cat.* 4.21, *Mur.* 58, *Man.* 60, *Rep.* 6.11, *Off.* 1.35; [Cic.] *Herr.* 4.27.37; Sall. *Iug.* 8; Plut. *Aem.* 22.4
132	P. Rupilius*	Tauromenion (Siciliote-Greek town held by rebel slaves)	Captives scourged and thrown off a cliff	Diod. 34.2.20–21

(continued)

Year BCE	Roman Commanders	Vanquished People, Group, or City	Mass Violence Described	Source(s)
125	L. Opimius*	Fregellae (Italian town)	City destroyed	Livy *Per.* 60; Obs. 30; Vell. Pat. 2.6.4; [Cic.] *Rhet. Herr.* 4.27.37
118	Q. Marcius Rex	Styni (Alpine tribe)	Surviving captives enslaved	Oros. 5.14.5–6; cf. Livy *Per.* 62
108	Q. Caecilius Metellus*	Unnamed Numidian towns and fortresses	Towns and fortresses burned; adult males killed	Sall. *Iug.* 54.6
107	C. Marius*	Capsa (Numidian town)	Adult males killed; city burned; inhabitants enslaved	Sall. *Iug.* 91.6

Appendix 2

181 Cases of Mass Violence in Ancient Mediterranean Warfare (excluding Rome), c. 500–100 BCE

Appendix 2

Year BCE	Commanders and/or People[a]	Vanquished People, Group, or City	Mass Violence Described[b]	Source(s)
498	Ionian Greeks	Sardis (Persian-Lydian city)	City, temples, houses burned	Hdt. 5.101–102, 5.105; 6.101; 7.11
494	Persian Empire	Miletus (Greek city)	"Most" men killed; women and children enslaved; temple burned	Hdt. 6.19
493	Unnamed Persian generals,* Persian Empire	Unnamed Ionian Greek cities	Cities, temples burned; women and children enslaved	Hdt. 6.32
493	Persian Empire	Several Ionian Greek cities	Cities burned	Hdt. 6.33
490	Persian Empire	Naxos (Greek island/city)	City, temples burned; inhabitants enslaved	Hdt. 6.96
490	Persian Empire	Eretria (Greek city)	Temples burned; inhabitants enslaved	Hdt. 6.101
484	Gelon,* Syracuse	Camarina (Siciliote-Greek city)	City destroyed	Hdt. 7.156
484	Gelon,* Syracuse	Megara, Euboea (Siciliote-Greek cities)	Inhabitants enslaved	Hdt. 7.156
480	Persian Empire	Many Greek cities in Phocis	Cities, temples burned	Hdt. 8.32–33, 8.35
480	Persian Empire	Plataea, Thespiae (Greek cities)	Cities burned	Hdt. 8.50; Diod. 11.14.5
480	Xerxes,* Persian Empire	Athens (Greek city)	Inhabitants killed indiscriminately during assault; acropolis and temples destroyed, burned	Hdt. 8.53; Diod. 11.14.5, 11.15.2
479	Artabazus,* Persian Empire	Olynthus (Greek city)	Inhabitants executed	Hdt. 8.127
479	Mardonius,* Persian Empire	Athens	City burned; walls, houses, and remaining temples destroyed	Hdt. 9.13 Diod. 11.28.6; Just. 6.5.10; Arr. Anab. 3.18

a. Commanders' names are marked (*) when the sources explicitly grant them responsibility for mass violence.
b. This chart primarily includes references to mass killing, mass execution, mass enslavement, and destruction of the built environment (e.g., cities, villages, buildings).

181 Cases of Mass Violence in Ancient Mediterranean Warfare

468	Argos (Greek city)	Mycenae (Greek city)	City destroyed; inhabitants enslaved	Diod. 11.65.5
463	Athens	Thasos (Greek city/island)	Walls demolished (on Athenian orders)	Thuc. 1.101.3
440	Athens	Samos (Greek city/island)	Walls demolished (on Athenian orders)	Thuc. 1.117.3; Diod 12.28.4
440	Syracuse (Siciliote-Greek city)	Trinaciē (Siculian town)	Inhabitants sold; city destroyed	Diod. 12.29.4
433	Athens	Potideia (Greek city)	Walls demolished (on Athenian orders)	Thuc. 1.56.2
431	Plataea (Greek city)	Thebes (Greek city)	180 Thebans (who had tried and failed to capture Plataea) executed	Thuc. 2.5.7; Polyaen. 6.19
430	Sparta (Greek city)	Sailors, merchants of Athens and Athenian allies	Captives executed, thrown into pits	Thuc. 2.67.4
427	Paches, Athens	Notium (Greek city)	Arcadian and "barbarian" mercenaries massacred during assault on city	Thuc. 3.34.3
427	Paches, Athens	Mytilene (Greek city)	Walls destroyed; 1,000 "most culpable" for revolt taken to Athens and executed	Thuc. 3.50.1
427	Sparta, Thebes	Plataea (Greek city)	Men executed; women enslaved; city destroyed	Thuc. 3.68.2–3; Diod. 12.56.6
424	Nicias,* Athens	Thyrea (Greek city)	Inhabitants enslaved; city destroyed; Aeginetan prisoners taken to Athens and executed	Thuc. 4.57.3; Diod. 12.65.9
423	Thebes	Thespiae (Greek city)	Walls destroyed	Thuc. 4.133.1
421	Athens	Skione (Greek city)	Adult males killed; women and children enslaved	Thuc. 5.32.1; Diod. 12.76.3
417/6	Sparta	Hysiae (Greek city)	Male inhabitants killed; city destroyed	Diod. 12.81.1; Thuc. 5.83.2
416	Argos	Orneae (Greek city)	City destroyed	Thuc. 6.7.2

(continued)

Appendix 2

Year BCE	Commanders and/or People	Vanquished People, Group, or City	Mass Violence Described	Source(s)
416	Byzantium, Chalcedon (Greek cities); Thracians	Unnamed Bithynian settlements	All captives executed	Diod. 12.82.2
416/5	Athens	Melos (Greek island/city)	Adult males killed; women and children enslaved	Thuc. 5.116; Diod. 12.80.5
413	Thracians	Mycalessus (Greek city)	Inhabitants killed indiscriminately	Thuc. 7.29.4–5
413	Syracusans	Athens	Captive Athenian generals and Athenian allies executed; other Athenians enslaved in Syracusan quarries	Diod. 13.33.1
409	Hannibal,* Carthage	Selinous (Siciliote-Greek city)	Inhabitants killed indiscriminately; some houses burned; walls demolished; city destroyed; women and children enslaved	Diod. 13.57–58, 13.59.4, 13.80.1
409	Hannibal,* Carthage	Himera (Siciliote-Greek city)	Inhabitants killed indiscriminately; temples burned; 3,000 male survivors executed; women and children taken; city destroyed	Diod. 13.62.3–4, 13.80.1; Cic. Ver. 2.2.86
405	Lysander,* Sparta	Iasus (Greek city)	Male inhabitants killed; women and children sold; city destroyed	Diod. 13.104.7
405	Lysander,* Sparta	Athenian captives	Captives executed/throats cut	Xen. Hell. 2.1.32; Plut. Lys. 13.1–2
405	Himilco,* Carthage	Acragas (Siciliote-Greek city)	City destroyed; temples "mutilated" (periékopsen)	Diod. 13.108.2
404	Sparta	Athens	Long walls destroyed	Xen. Hell. 2.2.20–23; Plat. Menex. 244c; Polyb. 38.2.6; Strabo 9.1.15; Plut. Lys. 15.4

Date	Aggressor	Victim	Action	Source
403	Dionysius,* Syracuse	Naxos, Catania (Siciliote-Greek cities)	Inhabitants enslaved; walls and buildings destroyed	Diod. 14.15.2–3, 14.68.2
400	Seuthes,* Thracians	Unnamed villages of the Thracian Thynians	Captured inhabitants killed	Xen. *Anab.* 7.4.6
397	Dionysius,* Syracuse	Motya (Carthaginian city)	Inhabitants killed indiscriminately during assault; some captives executed; survivors sold	Diod. 14.53.1–4
397	Himilco,* Carthage	Messene (Siciliote-Greek city)	Walls, buildings destroyed	Diod. 14.58.3, 14.68.5
390	Gauls	Rome	City, houses burned; some inhabitants killed	Livy 5.41.8–5.42.4; Diod. 14.116.8
389	Dionysius I,* Syracuse	Caulonia (Italiote-Greek city)	City destroyed; population forcibly removed	Diod. 14.106.3
388	Dionysius I,* Syracuse	Hipponium (Italiote-Greek city)	City destroyed; population forcibly removed	Diod. 14.107.2
387	Dionysius I,* Syracuse	Rhegium (Italiote-Greek city)	City destroyed	Strabo 6.1.6
385	Sparta	Mantineia (Greek city)	Walls destroyed	Xen. *Hell.* 5.2.7; Diod. 15.12.2
377	Latins	Satricum (Volscian city)	City, temples burned	Livy 6.33.4
372	Thebes	Plataea	City destroyed	Diod. 15.46.6; Paus 9.1.8
369	Lycomedes, Arcadians	Pellana (Greek city)	Spartan garrison killed; inhabitants enslaved	Diod. 15.67.2
368	Sparta	Caryae (Greek city)	All captives killed	Xen. *Hell.* 7.1.28
367	Alexander (tyrant of Pherae),* Pheraeans	Scotussa (Greek city)	Adult males surrounded in assembly, massacred, and dumped in a ditch outside the city; women and children sold	Diod. 15.75.1; Paus. 6.5.2; Plut. *Pelop.* 29.4

(continued)

Appendix 2

Year BCE	Commanders and/or People	Vanquished People, Group, or City	Mass Violence Described	Source(s)
365	Eleans	Elean Pylos (Greek city)	Elean exiles killed; other captives sold	Xen. *Hell.* 7.4.26
364	Thebans	Orchomenus (Greek city)	Male inhabitants killed; women and children enslaved; city destroyed	Diod. 15.79.5–6
358/7	Philip II,* Macedonia	Potideia (Greek city)	Inhabitants enslaved	Diod. 16.8.5
354	Boeotians	Phocian captives	Captives executed	Diod. 16.31.1
354	Philomelus,* Phocians	Boeotian captives	Captives executed	Diod. 16.31.2
354	Philip II,* Macedonia	Methone (Greek city)	City destroyed; inhabitants forced to depart	Diod. 16.31.6, 16.34.5
353	Chares,* Athens	Sestus (Greek city)	Adult male inhabitants killed; survivors sold	Diod. 16.34.3
353/2	Philip II,* Macedonia	Phocians	3,000 captives executed	Diod. 16.35.6
352/1	Phaÿllus,* Phocians	Naryx (Greek city)	City destroyed	Diod. 16.38.5
348	Philip II,* Macedonia	Olynthus (Greek city)	City destroyed	Diod. 32.4.2; Trog. pr.8
346	Philip II,* Macedonia	Greek cities in Phocis	Cities, houses, walls destroyed	Diod. 16.60.2; Dem. 18.36, 18.39, 19.65, 19.325
346/5	Elea (Greek city); Arcadians	Elean exiles	Some prisoners executed; some prisoners sold	Diod. 16.63.5
337	Sidicini	Suessa (Auruncian city)	Walls destroyed; city destroyed	Livy 8.15.4
335	Alexander III ("the Great"),* Macedonia	Unnamed city of the Getae	City destroyed	Arr. *Anab.* 1.4.5

181 Cases of Mass Violence in Ancient Mediterranean Warfare

Year	Perpetrator	Location	Action	Sources
335	Alexander III,* Macedonia; Boeotians	Thebes	Inhabitants massacred indiscriminately during assault; surviving males executed; women and children sold; city destroyed	Arr. *Anab.* 1.8.8–1.9.10; Diod. 17.13.1–17.14.4; Plut. *Alex.* 11.10–12; Polyb. 38.2.13, 5.10.6; Just. 11.3–4
332	Alexander III,* Macedonia	Tyre (Persian-Phoenician city)	Inhabitants killed indiscriminately during assault; surviving men crucified after capture of city	Arr. *Anab.* 2.24.3; Curt. 4.4.17; Diod. 17.46.4; Just. 18.3.18
332	Alexander III,* Macedonia	Gaza (Persian city)	Adult males killed; women and children sold	Arr. *Anab.* 2.27.7
331	Alexander III,* Macedonia	Halicarnassus (Greek city)	City destroyed	Diod. 17.27.6; Arr. *Anab.* 1.23.6
330	Alexander III,* Macedonia	Persepolis (Persian city)	(Male?) inhabitants killed indiscriminately during assault; (some?) women enslaved; palace, city burned	Curt. 5.6.19–7.11; Arr. *Anab.* 3.18.11–12; Diod. 17.72.5–6; Plut. *Alex.* 38
329	Alexander III, Macedonia	Unnamed town of Branchidae	Inhabitants killed indiscriminately; walls destroyed	Curt. 7.5.33
329	Alexander III, Macedonia	Unnamed Bactrian town	Adult males killed; city destroyed	Curt. 7.6.16
328	Alexander III,* Macedonia	Gabae/Gaza and another unnamed Sogdian city	"All" men killed; women and children enslaved	Arr. *Anab.* 4.2.4
328	Alexander III,* Macedonia	Cyropolis (Persian city)	Inhabitants killed; city destroyed	Curt. 7.6.21; Arr. *Anab.* 4.3.4
328	Alexander III,* Macedonia	Unnamed city of the Memaceni	City destroyed	Curt. 7.16.23

(continued)

Appendix 2

Year BCE	Commanders and/or People	Vanquished People, Group, or City	Mass Violence Described	Source(s)
327	Alexander III,* Macedonia	Unnamed Aspasian city	Inhabitants killed indiscriminately; city destroyed	Arr. *Anab.* 4.23.5
326	Alexander III,* Macedonia	Sangala (Indian city)	City burned, destroyed	Arr. *Anab.* 5.24.8; Diod. 17.91.4
326	Alexander III,* Macedonia	Unnamed towns of the Agalasseis	Inhabitants sold	Diod. 17.96.3
326	Alexander III,* Macedonia	Unnamed town of the Agalasseis	Adult males killed; survivors sold; city burned	Diod. 17.96.4–5; Curt. 9.4.1–8
326	Alexander III,* Macedonia	Unnamed town of the Mallians ("city of the Brahmans")	2,000 inhabitants who took refuge in the citadel killed; fugitives hunted down and killed	Arr. *Anab.* 6.7.5–6, 6.8.1–2
326	Alexander III, Macedonians	Unnamed city of the Mallians	Inhabitants killed indiscriminately	Arr. *Anab.* 6.11.1; Curt. 9.5.19–20; Diod. 17.99.4
326/5	Alexander III,* Macedonia	Unnamed cities of King Porticanus	Cities destroyed	Diod. 17.102.5
326/5	Alexander III,* Macedonia	Unnamed cities of King Sambus	Inhabitants killed or enslaved; cities destroyed	Diod. 17.102.6
325	Peithon,* Alexander III,* Macedonia	Unnamed cities of King Musicanus	Inhabitants enslaved; cities destroyed; some leaders executed	Arr. *Anab.* 6.17.1–2
323	Peithon,* Macedonians	Rebel Greek mercenaries	Several thousand surrendering Greeks dispersed among the Macedonians, killed with javelins	Diod. 18.7.8–9
323	Perdiccas,* Macedonians	Laranda (Lycaonian city)	Adult males killed; other inhabitants enslaved; city destroyed	Diod. 18.22.2
314	Aristodemus, mercenaries	Aegium (Greek city)	Many inhabitants killed; some buildings destroyed	Diod. 19.66.3

181 Cases of Mass Violence in Ancient Mediterranean Warfare   235

314	Aristodemus, mercenaries; Dyme (Greek city)	Garrison of Alexander (son of Polyperchon) in Dyme	"Most" of the garrison killed; Dymaeans loyal to Alexander killed	Diod. 19.66.6
314	Aetolians	Agrinium (Greek city)	Inhabitants forced to depart; "all but a few" pursued and killed	Diod. 19.68.1
313	Ptolemy I,* Macedonians	Marion (Cypriote city)	City destroyed; inhabitants forcibly relocated	Diod. 19.79.4
313	Ptolemy I,* Macedonians	Malus (Cypriote city)	Inhabitants sold	Diod. 19.79.6
312	Antigonus I,* Macedonians	Elis (Greek city)	Citadel destroyed	Diod. 19.87.3
312	Lyciscus,* Macedonians	Eurymenae (Epirote city)	City destroyed	Diod. 19.88.6
312	Ptolemy I,* Macedonians	Ake, Syria, Ioppe, Samaria, Gaza (Syrian cities)	Cities destroyed	Diod. 19.93.7
312	Agathocles,* Syracuse	Tauromenium, Messene (Siciliote-Greek cities)	600 inhabitants executed	Diod. 19.102.6
311	Agathocles,* Syracuse	Gela (Siciliote-Greek city)	4,000 inhabitants executed; bodies dumped in ditch outside the walls	Diod. 19.107.4–5
311	Samnites	Cluviae (Italian city)	Roman garrison scourged, executed	Livy 9.31.1–4
310	Agathocles,* Syracuse	Megalopolis, "White" Tunis (Carthaginian cities)	Cities destroyed	Diod. 20.8.7
307	Demetrius I,* Macedonians	Athens	Fortress Munychia destroyed	Diod. 20.46.1; Plut. *Demtr.* 10.1
307	Agathocles,* Syracuse	Utica (Carthaginian city)	Inhabitants killed during assault; captives executed	Diod. 20.55.1–2
307	Agathocles,* Syracuse	Apollonia (Siciliote-Greek city)	"Most" inhabitants killed	Diod. 20.56.4

(continued)

Appendix 2

Year BCE	Commanders and/or People	Vanquished People, Group, or City	Mass Violence Described	Source(s)
307	Eumachus,* Syracuse	Acris (Libyan city)	Inhabitants enslaved	Diod. 20.57.6
306	Agathocles,* Syracuse	Segesta (Siciliote-Greek city)	Inhabitants tortured; males killed; women and children enslaved	Diod. 20.71.1–5
304	Agathocles,* Syracuse	Leontini (Siciliote-Greek city)	10,000 prisoners surrounded, killed	Polyaen. 5.3.2
303	Demetrius I,* Macedonians	Orchomenus (Greek city)	80 leaders executed	Diod. 20.103.6
302	Lysimachus,* Macedonians	Autariatae/Illyrian troops	5,000 soldiers suspected of disloyalty killed	Polyaen. 4.12.1
301	Agathocles, Syracuse	Ligurian and Etruscan troops	2,000 mutinous soldiers killed	Diod. 21.3.1
295	Agathocles, Syracuse	Croton (Italiote-Greek city)	Male inhabitants killed	Diod. 21.4.1
291	Demetrius I,* Macedonians	Thebes (Greek city)	10–14 leaders executed[a]	Diod. 21.14.1–2; Plut. Demetr. 40.6
288	Mamertines/Campanian mercenaries	Messena (Siciliote-Greek city)	Male inhabitants killed; women and children taken[b]	Polyb. 1.7.3–4; Diod. 21.18.3; Dio fr. 40.6
281	Lysimachus, Macedonians	Astacus (Greek city)	City destroyed	Strabo 12.4.2
281	Seiles,* Seleucid Empire	Persians	3,000 rebels surrounded, killed	Polyaen. 7.39.1
279	Orestorius and Combutis, Gauls	Callium (Greek city)	Male inhabitants killed; mass rape of women	Paus. 10.22.3–4

a. Diodorus mentions the execution of ten individuals (along with fourteen others in unnamed Boeotian cities), while Plutarch notes the execution of thirteen individuals.
b. Polybius writes that the Mamertines took possession of the families of the dead Messenian men, implying a form of enslavement.

181 Cases of Mass Violence in Ancient Mediterranean Warfare

Year	Aggressor	Victim	Event	Source
274	Heiro,* Syracuse	Ameselum (Siciliote-Greek city)	City destroyed	Diod. 22.13.1
272		Rhegium (Italiote-Greek city)	City destroyed	Zon. 8.6
259	Hamilcar,* Carthage	Eryx (Sicilian city)	City destroyed; inhabitants relocated	Diod. 23.9.4; Zon. 8.11
254	Carthalo,* Carthage	Acragas (Siciliote-Greek city)	City burned; walls demolished	Diod. 23.18.2
250	Carthage	Selinous (Siciliote-Greek city)	City destroyed; inhabitants forcibly relocated	Diod. 24.1.1
249	Hamilcar Barca,* Carthage	Eryx (Sicilian city)	Roman garrison (?) killed; inhabitants relocated	Diod. 24.8.1
243	Aratus,* Achaeans	Lechaeum (Greek port)	400 prisoners sold	Plut. *Arat.* 24.1
240	Rebel mercenaries	Carthaginians in Sardinia	"All" Carthaginians killed	Polyb. 1.79.4
239	Carthaginians	Rebel mercenaries	Captives executed, some trampled by elephants	Polyb. 1.82.2
239	Rebel mercenaries	Carthaginians	700 captives mutilated, thrown in a trench, left to die	Polyb. 1.80.13
239	Utica, Hippo (Rebel Carthaginian cities)	Carthaginian garrisons	500 soldiers killed and thrown from the walls	Polyb. 1.82.10; Diod. 25.3.2
237	Carthage	Micitani/Numidians	Captured men, women, and children executed	Diod. 26.23.1
226	Cleomenes, Sparta; Mantineia (Greek city)	Mantineia	Achaean garrison killed	Polyb. 2.58.4
223	Antigonus Doson, Macedonia; Achaeans	Mantineia (Greek city)	Some inhabitants killed; survivors sold	Polyb. 2.56.7, 2.58.12; Plut. *Arat.* 45.4

(continued)

Appendix 2

Year BCE	Commanders and/or People	Vanquished People, Group, or City	Mass Violence Described	Source(s)
223	Cleomenes,* Sparta	Megalopolis (Greek city)	Some inhabitants killed; city destroyed	Polyb. 2.55.7; Livy 38.34.7; Plut. Cleom. 25.1, Phil. 5.1; Paus. 4.29.8, 8.27.15–16
222	Antigonus Doson, Macedonia; Achaeans	Sellasia (Greek city)	Inhabitants enslaved	Paus. 2.9.2, 3.10.7
221	Aetolians	Macedonian prisoners	Prisoners enslaved	Polyb. 4.6.1
220	Aetolians	Cynaetha (Greek city)	Inhabitants tortured, killed, or enslaved	Polyb. 4.18.7–8, 9.38.8
220	Aetolians	Unnamed villages near Sparta	Inhabitants enslaved	Polyb. 4.34.9
220	Cnossos (Cretan-Greek city)	Lyttos (Cretan-Greek city)	Women and children taken; city burned and demolished	Polyb. 4.54.1–2
219	Scopas,* Aetolians	Dium (Macedonian city/sanctuary)	City/sanctuary burned; destroyed	Polyb. 4.62.2–3
219	Philip V,* Macedonia	Ithoria (Greek city)	City destroyed	Polyb. 4.64.10
219	Dorimachus,* Aetolians	Dodona (Epirote city/sanctuary)	City/sanctuary burned; demolished	Polyb. 4.67.3–4
219	Hannibal,* Carthage	Saguntum (Iberian city)	Inhabitants killed indiscriminately during assault	Livy 21.14.4, 21.15.1, 28.39.12
218	Antiochus III,* Seleucid Empire	Trieres, Calamus (Ptolemaic cities)	Cities burned	Polyb. 5.68.8
218	Philip V, Macedonia	Thermum (Aetolian city/sanctuary)	City/sanctuary burned, demolished	Polyb. 5.9.2–3, 5.18.5
218	Philip V,* Macedonia	Metapa (Greek city)	City destroyed	Polyb. 5.13.8
218	Hannibal,* Carthage	Unnamed Italian city	City destroyed; Roman prisoners killed	Zon. 8.24

181 Cases of Mass Violence in Ancient Mediterranean Warfare

Year	Perpetrator	Victim/Location	Event	Source
217	Hannibal,* Carthage	Rural Italians	All adult captives killed	Polyb. 3.86.11
217	Hannibal,* Carthage	Geronium (Italian city)	Inhabitants killed	Polyb. 3.100.4
217	Philip V,* Macedonia	Phthiotic Thebes (Greek city)	Inhabitants sold	Polyb. 5.100.8
216	Prusias I,* Bithynia	Galatians	Women and children in Galatian camp killed	Polyb. 5.111.6
216	Hannibal, Carthage	Nuceria (Italian city)	City burned, destroyed	Livy 23.15.6, 27.3.6
216	Hannibal,* Carthage	Acerrae (Italian city)	City burned	Livy 23.17.7, 27.3.6
214	Antiochus III, Seleucid Empire	Sardis (Rebel Seleucid city)	(Some?) inhabitants killed indiscriminately during assault on city; houses burned	Polyb. 7.18.9
212	Hannibal,* Carthage	Tarentum (Italiote-Greek city)	Romans in the city killed	Polyb. 8.30.4
205	Mago Barca,* Carthage	Genua (Italian city)	City destroyed	Livy 30.1.10
204	Philip V, Macedonia	Dardanians	10,000 prisoners killed?[a]	Diod. 28.2.1
203[b]	Hannibal, Carthage	Roman prisoners	Prisoners sold	Plut. *Flam.* 13.4–6
202	Philip V,* Macedonia	Cius (Greek city)	Inhabitants enslaved; city destroyed	Polyb. 15.23.7–9, 18.3.12, 18.44.5; Livy 32.33.16; Strabo 12.4.3
202	Philip V,* Macedonia	Thasus (Greek island/city)	Inhabitants enslaved	Polyb. 15.24.1

(continued)

a. Diodorus writes that Philip killed 10,000 Dardanians after defeating them in battle, suggesting that he killed them after their defeat and surrender, or killed captives.
b. Plutarch writes that Hannibal sold the Roman prisoners taken in Italy, so these were presumably captives taken throughout his Italian campaigns from 218 to 203 BCE.

Appendix 2

Year BCE	Commanders and/or People	Vanquished People, Group, or City	Mass Violence Described	Source(s)
200	Philip V,* Macedonia	Sciathus, Peparethus (Greek cities/islands)	Cities destroyed	Livy 31.28.6
200	Gauls	Placentia (Latin colony)	City burned, destroyed	Livy 31.10.3; Zon. 9.15
200	Aetolians; Amynander, Athamania	Cercinium (Greek city)	Inhabitants killed or enslaved; city burned	Livy 31.41.3
200	Aetolians	Cyretiae (Greek city)	City destroyed	Livy 31.41.5
197	Thracians	Lysimacheia (Greek city)	City burned, destroyed; inhabitants enslaved	Livy 33.38.11; 33.40.6; App. *Syr.* 1
192	Achaeans	Aetolian prisoners	Prisoners enslaved	Livy 35.36.10
188	Achaeans	Sparta	63, 80, or 350 Spartans executed; walls destroyed	Polyb. 22.3.1, Diod. 29.17.1; Livy 38.33.11–38.34.1; Plut. *Phil.* 16.3–4, *Comp. Phil. Flam.* 1.3; Paus. 7.8.5
185/4	Thracians[a]	Maronea (Greek city)	Many inhabitants killed	Polyb. 22.13.5–6; Livy 39.34.2–3
182	Achaeans	Messenia (Greek city)	Some leaders tortured, executed	Plut. *Phil.* 21.2; cf. Polyb. 23.16.12
176	Ligurians	Mutina (Roman colony)	Captives tortured, killed	Livy 41.18.3
171	Perseus,* Macedonia	Mylae (Greek city)	Some inhabitants killed during assault; survivors enslaved; city destroyed	Livy 42.54.6
170	Perseus,* Macedonia	Chalestrum (Dardanian town)	Inhabitants killed	Diod. 30.4.1

a. According to Polybius and Livy, the Thracians raided the city at the instigation of Philip V of Macedonia.

181 Cases of Mass Violence in Ancient Mediterranean Warfare

Year	Perpetrator	Location	Description	Source
169	Cydonia (Cretan-Greek city)	Apollonia (Cretan-Greek city)	Male inhabitants killed; women and children taken	Polyb. 28.14.1; Diod. 30.13.1
169	Perseus,* Macedonia	Uscana (Illyrian city)	Inhabitants enslaved	Livy 43.19.2
169	Perseus, Macedonia	Oaeneum (Illyrian city)	Adult males killed; women and children taken	Livy 43.19.12
168	Antiochus IV,* Seleucid Empire	Jerusalem	Inhabitants killed indiscriminately; survivors enslaved; city, buildings burned; walls destroyed	Jos. AJ 12.251–252; Jos. BJ 1.32 Suda A.2693; 2 Macc. 5.12–14
163	Judas Maccabeus,* Judaeans	Jazer (Ammonite city)	City burned; women and children taken	Jos. AJ 12.329
163	Judas Maccabeus,* Judaeans	Hebron (Idumaean city)	Walls destroyed	Jos. AJ 12.353
147	Hadrubal,* Carthage	Roman soldiers	Prisoners mutilated, thrown from Carthage's walls	App. Pun. 118; Zon. 9.29; Suda A.4134
145	Demetrius II,* Seleucid Empire	Antioch	City burned; some inhabitants killed indiscriminately	Diod. 33.4.2; Jos. AJ 13.139
144	Diēgylis,* Thracians	Lysimacheia (Greek city)	City burned; some inhabitants tortured to death	Diod. 33.14.2
135	Eunus,* rebel slaves in Sicily	Enna (Siciliote-Greek city)	Inhabitants killed indiscriminately during assault; most survivors executed	Diod. 34.2.11–15, 34.2.24b
132	Antiochus VII,* Seleucid Empire	Jerusalem	Walls destroyed	Jos. AJ 13.248; Diod. 34.1.5
128	Euhimerus,* Parthian Empire	Babylon	Marketplace, temples burned; "finest" parts" of city destroyed; inhabitants enslaved	Diod. 34.21.1
112	Jugurtha,* Numidia	Cirta (Numidian city)	Adult males killed indiscriminately; Italian traders killed	Sall. Jug. 1.26; Diod. 34.31.1
108	Aristobulus and Antigonus,* Judaeans	Samaria	City destroyed; inhabitants enslaved	Jos. BJ 1.64

Appendix 3

The Government and Army in the Middle Republic

THE ROMAN GOVERNMENT

The Roman people were theoretically sovereign in the Roman Republic, and they primarily exercised their political rights in Rome's citizen assemblies. In these assemblies, Roman citizens* elected their magistrates, voted on laws, declared war, and made peace, among other responsibilities. However, despite the rhetoric of popular sovereignty that prevailed in Republican Rome, the power of the citizenry was curtailed in several ways. First, the citizen assemblies had no scope for independent decision-making. They could only vote "yes" or "no" on proposals brought forward by magistrates, could not debate or offer amendments to proposals, and only met when summoned by a magistrate. Second, wealthier citizens' votes tended to carry more weight in the assemblies, both in practice and by design. The imbalance in voter power was especially manifest in the Centuriate Assembly (*comitia centuriata*), which elected senior magistrates, conducted capital trials, and was responsible for declaring war and making peace. The Centuriate Assembly divided citizens into voting groups called "centuries." Citizens first voted

---

* Only adult male Romans could actively participate in politics by, for example, voting or running for office. In contrast, women in the Roman Republic were largely excluded from political life and often fell under the legal authority of a man, usually a husband or father. However, Roman women exercised several rights that were often unavailable to their counterparts in other ancient (and modern) states. Mary Beard states, "A woman did not take her husband's name or fall entirely under his legal authority. After the death of her father, an adult woman could own property in her own right, buy and sell, inherit or make a will and free slaves—many of the rights that women in Britain did not gain till the 1870s" (Beard 2015, 308). Women also had a prominent role in Rome's public rituals and religious festivals and could wield considerable political influence. For women in the Roman Republic, see Culham 2004 and Rawson 2006.

within their individual century until they had reached a majority decision; then each century cast a single vote based on the decision of its members. Significantly, the centuries were not distributed proportionately among the population: membership in a century was determined by property qualification, that is, wealth, and their distribution heavily favored those Romans who owned more property. Out of 193 centuries, Rome's wealthiest citizens were divided into seventy "First Class" centuries and eighteen "Cavalry" centuries (with some fluctuation in these numbers depending on the period). Well-off Romans thus controlled about 45 percent of the centuries/votes, even though they certainly made up much less than 45 percent of the population. Although most Roman citizens felt free to voice their opinion and to criticize their public officials, the structures of citizen participation meant that their actual political power was limited.

The Roman Senate was far more powerful and far more important in Rome's government. Unlike the modern United States Senate, which was inspired by and named after its ancient predecessor, Roman senators were not elected. Instead, the (roughly) three hundred senators were enrolled by officials called "censors." Senators themselves were usually current or former magistrates and primarily belonged to Rome's leading aristocratic families. Although it had little formal power, by the early third century BCE, the Senate had acquired essential public functions. It received and dispatched ambassadors, oversaw finances, and allocated military resources. In times of war, the Senate often determined broad Roman strategy and policy. Indeed, the senators debated declarations of war before taking such momentous decisions to the citizen voters (who rarely rejected the senators' recommendations); they determined where and against whom Roman armies would march; they often sent directives to generals in the field or outlined basic military objectives; and they ratified the acts of Roman commanders returning from campaign. Finally, the senators managed Rome's relationship with deities, ensuring that the "peace of the gods" (*pax deorum*) was maintained through the proper rituals, festivals, and sacrifices.

It can seem strange to modern students that the Senate had so much authority, given its lack of formal powers. But in this period, Roman citizens were deeply deferential to the collective authority of the Senate—and it is worth stressing that the Senate consisted of Rome's most prestigious, well-connected, and powerful individuals, all of whom had repeatedly and publicly proved their mettle by the time they became senators. Furthermore, Rome's elected officials had brief terms of office and served alongside one or more coequal colleagues, which limited their potential influence. In contrast, the Senate's permanence made it central to the Republican system of government.

While the Senate was undoubtedly supreme, elected magistrates were also essential to Roman government. As noted in chapter 2, the most important officials were the two annually elected consuls, each of whom possessed *imperium*, or the power of command. The consuls could propose laws to the assemblies, convene and chair meetings of the Senate, and lead Roman armies in war. Every year, the Senate would assign a province (*provincia*) to each consul; in this period, "province" meant something like "area of responsibility" or "area of operations" rather than a defined administrative territory. A province designated where the consuls would wage war and against whom. Since the consuls could be found marching their armies to their provinces and battling against Rome's foes nearly every year, the consulship was the best, most coveted avenue to prestige and military glory. The praetors—magistrates ranked just below the consuls—also possessed a lesser form of *imperium*, and they too could lead armies when they were not serving in administrative or judicial roles. In the early third century BCE, there was only one praetor elected each year, but by the early second century BCE, there were six. Praetors increasingly acted as governors of overseas territories when Rome began to conquer the Mediterranean, and their numbers were sometimes increased to keep pace with military expansion.

Both consuls and praetors enjoyed enormous latitude when leading Rome's armies, in terms of both decision-making and strategy. While the Senate determined broad foreign and military policies, primitive communications and a lack of complex bureaucracy limited its ability to micromanage war. Consequently, Roman generals often acted on their own initiative on campaign, albeit within wide senatorial guidelines, and the Senate usually deferred to commanders' judgments and decisions in the field.

Aside from consuls and praetors, other important elected magistrates included the two censors, who conducted the census and drew up the rolls of the Senate; four aediles, who administered Rome's public buildings and festivals; the quaestors, whose numbers grew from four to ten in this period, and who had financial and administrative responsibilities; and the ten tribunes of the plebs, who could propose legislation in the Plebeian Assembly and veto the decisions of other magistrates, and were charged with protecting the interests of the mass of citizens.[1]

THE ROMAN LEGIONS

The Romans fought their battles with the legions, which were the basic building blocks of the Roman army for centuries. In the Middle Roman Republic, a single legion numbered about 5,000 heavy infantrymen and was subdivided into thirty smaller units called "maniples" (*manipuli*, a Latin word meaning "handfuls"). On the battlefield, these maniples were arranged in three separate lines, one behind the other, with gaps between each unit. This gave Roman armies a sort of checkerboard appearance when they were arrayed for action (see figure A3.1). The legion's odd structure—with many modular units arranged into multiple battle lines—probably allowed the maniples to move into and out of combat more easily. Roman commanders could send fresh maniples from the second and third lines to reinforce the front line, or withdraw first-line maniples that were flagging by pulling them back to the rear. This gave the legions enormous staying power in combat, since the fighting line was constantly refreshed with new Roman troops. In contrast, most enemy armies deployed their infantry in a single line of battle, which was more exposed to exhaustion, wounds, and casualties as the fight wore on. Time and again the manipular legions proved their superiority over other armies, from the dense phalanx formation that was preferred by Hellenistic kings to the warbands of Spain and Gaul.

The primary weapon for most individual legionaries was a short sword, the famed *gladius*. This sword had a sturdy blade, was effective at cutting and thrusting, and exemplified the Roman preference for close combat. Legionaries also carried heavy javelins and a curved oval shield, and were equipped with a metal helmet and some kind of body armor (a small bronze chest plate or mail, depending on what the individual soldier could afford). Several sizeable cavalry contingents also accompanied the legions, usually deploying on the wings, and light infantry skirmishers screened the front of the heavy infantry lines. In the Middle Republic, a consul normally commanded a force of two legions and two similarly sized divisions of Italian allies, altogether numbering about 20,000 to 25,000 troops (give or take a few thousand).[2]

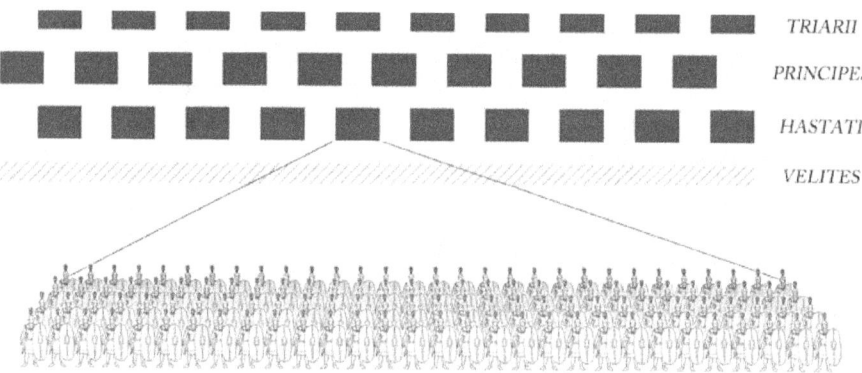

Figure A3.1. The manipular legion in battle array, top-down view with close-up of a single maniple. In the Middle Republic, the legion's heavy infantry was divided into three lines of ten maniples each. Maniples in the first two lines (manned by soldiers called the *hastati* and *principes*) numbered about 120 soldiers per maniple, while soldiers in the rear line (the *triarii*) were grouped about 60 per maniple. In battle, about 1,200 light-armed troops (*velites*) formed an irregular screen in front of the heavy infantry lines, and cavalry (*equites*, not pictured) usually posted on the wings. A consular army consisted of two legions and two equivalent-sized divisions of Italian allies. (Illustration by author.)

NOTES

1. Republican government/institutions: e.g., Nicolet 1980; Lintott 1999; North 2010; Flower 2010; Brennan 2000, 2014. Debates about popular power in the Republic: e.g., Mouritsen 2017, esp. 15–31, 54–104; Morstein-Marx 2004; North 1990, 3–21; Yakobson 2010, 1–21; Millar 1998. Consuls' *imperium*: Beck 2011, 77–96. Senate and generals at war: Eckstein 1987.

2. Manipular legion: Keppie 1984, 19–56; Sabin 2000; Goldsworthy 2000, 48–60; Lendon 2005, 163–92; Taylor 2014.

Bibliography

Adam, J.-P. 2003. *Roman Building: Materials and Techniques*. Routledge.
Alapont Martín, L., M. Calvo Gálvez, and A. Ribera i Lacomba. 2009. *La destrucción de Valentia Por Pompeyo (75 a.C.)*. Ajuntament de València.
Almagro-Gorbea, M., and A. J. Lorrio. 2002. "War and Society in the Celtiberian World." *E-Keltoi: The Journal of Interdisciplinary Celtic Studies* 6: 73–112.
Álvarez Sanchís, J. R. 2005. "Oppida and Celtic Society in Western Spain." *E-Keltoi: The Journal of Interdisciplinary Celtic Studies* 6: 255–85.
Arafat, K. W. 1996. *Pausanias' Greece: Ancient Artists and Roman Rulers*. Cambridge University Press.
Armstrong, J. 2016. *War and Society in Early Rome*. Cambridge University Press.
Assenmaker, P. 2013. "Poids symbolique de la destruction et enjeux idéologiques de ses récits: Réflexion sur les sacs d'Athènes et d'Ilion durant la première guerre mithridatique." In *Destruction: Archaeological, Philological and Historical Perspectives*, edited by J. Driessen, 391–414. Universitaires de Louvain.
Astin, A. E. 1964. "The Roman Commander in Hispania Ulterior in 142 B.C." *Historia: Zeitschrift für Alte Geschichte* 13, no. 2: 245–54.
Austin, M. M. 1986. "Hellenistic Kings, War, and the Economy." *The Classical Quarterly* 36, no. 2: 450–66.
Austin, R. P. 1931–1932. "Excavations at Haliartos, 1931." *The Annual of the British School at Athens* 32: 180–212.
Awasthi, D., and U. Mahurkar. 1992. "Orchestrated Onslaught." *India Today*, May 31, 1992.
Azam, J.-P., and A. Hoeffler. 2002. "Violence against Civilians in Civil Wars: Looting or Terror?" *Journal of Peace Research* 39, no. 4: 461–85.
Baatz, D. 1982. "Hellenistische Katapulte aus Ephyra (Epirus)." *Mitteilungen des Deutschen Archäologischen Instituts, Athenische Abteilung* 97: 211–33.
Badian, E. 1958. *Foreign Clientelae, 264–70 B.C.* Clarendon.

———. 1968. *Roman Imperialism in the Late Republic*. Blackwell.
Baker, G. 2013. "Sallust, Marius, and the Alleged Violation of the *ius belli*." *Ancient History Bulletin* 27, no. 3–4: 108–29.
Baker, H. D. 2015. "'I Burnt, Razed (and) Destroyed Those Cities': The Assyrian Accounts of Deliberate Architectural Destruction." In *Architecture and Armed Conflict*, edited by J. M. Mancini and K. Bresnahan, 45–57. Routledge.
Balcells, L. 2010. "Rivalry and Revenge: Violence against Civilians in Conventional Civil Wars." *International Studies Quarterly* 54: 291–313.
Balmaceda, C. 2017. *Virtus Romana: Politics and Morality in the Roman Historians*. University of North Carolina Press.
Bankoff, G., U. Lübken, and J. Sand, eds. 2012. *Flammable Cities: Urban Conflagration and the Making of the Modern World*. University of Wisconsin Press.
Barnett, R. D. 1976. *Sculptures from the North Palace of Ashurbanipal at Nineveh (668–627 B.C.)*. Trustees of the British Museum.
Barrandon, N. 2016. "Les gouvernants de la République romaine et le massacre: de la tactique militaire aux vices." In *La pathologie du pouvoir: vices, délits et crimes des gouvernants (Antiquité, Moyen Âge, époque modern)*, edited by P. Gilli, 13–41. Brill.
———. 2018. *Les massacres de la République romaine*. Fayard. Kindle edition.
Barry, W. D. 1996. "Roof Tiles and Urban Violence in the Ancient World." *Greek, Roman, and Byzantine Studies* 37, no. 1: 55–74.
Bauer, Y. 1984. "The Place of the Holocaust in Contemporary History." In *Studies in Contemporary Jewry*, edited by J. Frankel, 201–24. Indiana University Press.
Beard, M. 2007. *The Roman Triumph*. Harvard University Press.
———. 2015. *SPQR: A History of Ancient Rome*. Liveright.
Beck, H. 2011a. "Consular Power and the Roman Constitution: The Case of *Imperium* Revisited." In *Consuls and Res Publica: Holding High Office in the Roman Republic*, edited by H. Beck, A. Duplá, M. Jehne, and F. Pina Polo, 77–96. Cambridge University Press.
———. 2011b. "The Reasons for the War." In *A Companion to the Punic Wars*, edited by D. Hoyos, 225–41. Wiley-Blackwell.
Becker, J. 2007. "The Building Blocks of Empire: Civic Architecture, Central Italy, and the Roman Middle Republic." PhD dissertation. University of North Carolina, Chapel Hill.
Beckmann, M. 2011. *The Column of Marcus Aurelius: The Genesis and Meaning of a Roman Imperial Monument*. University of North Carolina Press.
Bell, M. 1981. *Morgantina Studies*. Vol. I: *The Terracottas*. Princeton University Press.
Bellamore, J. 2012. "The Roman Concept of Massacre: Julius Caesar in Gaul." In *Theatres of Violence: Massacre, Mass Killing, and Atrocity throughout History*, edited by P. D. Dwyer and L. Ryan, 38–49. Berghahn Books.
Berrocal-Rangel, L., J. L. de la Barrera Antón, R. Caso Amador, and G. C. Cabanillas de la Torre. 2014. "Nertobriga Concordia Iulia. La conquête de la Béturie." In *La Guerre et ses traces: Conflits et sociétés en Hispanie à l'époque de la conquête romaine (IIIe-Ier s. a.C.)*, edited by F. Cadiou and M. Navarro Caballero, 273–96. Ausonius.
Beste, H.-J. 2016. "The Castle Euryalos of Syracuse." In *Focus on Fortifications: New Research on Fortifications in the Ancient Mediterranean and Near East*, edited by R. Frederiksen, S. Müth, et al., 193–206. Oxbow Books.
Billows, R. 2003. "Cities." In *A Companion to the Hellenistic World*, edited by Andrew Erskine, 196–215. Blackwell.

———. 2007. "International Relations." In *The Cambridge History of Greek and Roman Warfare*. Vol. 1: *Greece, the Hellenistic World, and Rome*, edited by P. Sabin, H. Van Wees, and M. Whitby, 303–24. Cambridge University Press.
Bintliff, J., and A. Snodgrass. 1988. "Mediterranean Survey and the City." *Antiquity* 62: 62–64.
Bispham, E. 2006. "Coloniam Deducere: How Roman Was Roman Colonization During the Middle Republic?" In *Greek and Roman Colonization: Origins, Ideologies and Interactions*, edited by G. Bradley and J. P. Wilson, 73–160. Classical Press of Wales.
———. 2007. *From Asculum to Actium: The Municipalization of Italy from the Social War to Augustus*. Oxford University Press.
———. 2010. "Literary Sources." In *A Companion to the Roman Republic*, edited by N. Rosenstein and R. Morstein-Marx, 29–50. Wiley-Blackwell.
Blackie, J. S. 1885. *The Scottish Highlanders and the Land Laws: An Historico-economical Enquiry*. Chapman and Hall.
Blackman, D. 1998–1999. "Archaeology in Greece 1998–99." *Archaeological Reports* 45: 1–124.
———. 1999–2000. "Archaeology in Greece 1999–2000." *Archaeological Reports* 46: 3–151.
———. 2000–2001. "Archaeology in Greece 2000–2001." *Archaeological Reports* 47: 1–144.
———. 2001–2002. "Archaeology in Greece 2001–2002." *Archaeological Reports* 48: 1–115.
Blackman, D., J. Baker, and N. Hardwick. 1997–1998. "Archaeology in Greece, 1997–98." *Archaeological Reports* 44: 1–136.
Blánquez Pérez, J., and L. Roldán Gómez. 2009. "La Muralla de Casernas de la Ciudad Púnica de *Carteia* (San Roque, Càdiz)." *Almoraima* 39: 93–104.
Bonfante, L., and J. Swaddling. 2006. *Etruscan Myths*. University of Texas Press.
Bosworth, B. 1988. *Conquest and Empire: The Reign of Alexander the Great*. Cambridge University Press.
———. 1996. *The Legacy of Alexander. Politics, Warfare, and Propaganda under the Successors*. Oxford University Press.
———. 2003. "Plus ça change . . . Ancient Historians and Their Sources." *Classical Antiquity* 22: 167–98.
———. 2012. "Massacre in the Peloponnesian War." In *Theatres of Violence: Massacre, Mass Killing, and Atrocity throughout History*, edited by P. D. Dwyer and L. Ryan, 17–26. Berghahn Books.
Bradley, K. 2004. "On Captives under the Principate." *Phoenix* 58, nos. 3/4: 298–318.
Braund, D. 2011. "The Slave Supply in Classical Greece." In *The Cambridge World History of Slavery*. Vol. 1: *The Ancient Mediterranean World*, edited by K. Bradley and P. Cartledge, 112–33. Cambridge University Press.
Brennan, C. T. 1995. "Notes on the Praetors in Spain in the Mid-Second Century BC." *Emerita* 63: 47–76.
———. 2000. *The Praetorship in the Roman Republic*, vol. 1. Oxford University Press.
———. 2014. "Power and Process under the Republican 'Constitution.'" In *The Cambridge Companion to the Roman Republic*, 2nd ed., edited by H. Flower, 19–53. Cambridge University Press.
Brice, L. 2014. *Warfare in the Roman Republic: From the Etruscan Wars to the Battle of Actium*. ABC-Clio.
Briscoe, J. 1973–2012. *A Commentary on Livy*, 4 vols. Clarendon Press.

———. 1989. "The Second Punic War." In *The Cambridge Ancient History* 8.2, edited by A. E. Astin, F. W. Walbank, et al., 44–80. Cambridge University Press.
———. 2010. "Livy's Sources and Methods of Composition in Books 31–33." In *Livy*, edited by J. D. Chaplin and C. S. Kraus, 461–75. Oxford University Press.
Broughton, T. R. S. 1951. *The Magistrates of the Roman Republic*, vol. 1. American Philological Association.
Browning, C. 1992. *Ordinary Men: Reserve Police Battalion 101 and the Final Solution in Poland*. Penguin Books.
Brunt, P. A. 1971. *Italian Manpower, 225 B.C.–A.D. 14*. Oxford University Press.
Buchanan, J. 2009. *Up in Flames: Humanitarian Law Violations and Civilian Victims in the Conflict over South Ossetia*. Human Rights Watch.
Bucher, G. S. 2000. "The Origin, Program, and Composition of Appian's *Roman History*." *Transactions of the American Philological Association* 130: 411–58.
Buck, R. J. 1979. *A History of Boeotia*. University of Alberta Press.
Buckler, J. 2006. "Boeotia." In *Encyclopedia of Ancient Greece*, edited by N. Wilson, 129–30. Routledge.
Buitenwerf, R. 2003. *Book Three of the Sibylline Oracles and Its Social Setting*. Brill.
Burton, P. J. 2007. "Genre and Fact in the Preface to Cicero's *De Amicitia*." *Antichton* 41: 13–32.
———. 2009. "Ancient International Law, the Aetolian League, and the Ritual of Surrender during the Roman Republic: A Constructivist View." *The International History Review* 31, no. 2: 237–52.
———. 2011. *Friendship and Empire: Roman Diplomacy and Imperialism in the Middle Republic (353–146 BC)*. Cambridge University Press.
———. 2017. *Rome and the Third Macedonian War*. Cambridge University Press.
Cabanes, P. 1976. *L'Épire de la mort de Pyrrhos à la conquête romaine (272–167 av. J.C.)*. Les Belles Lettres.
Cabanes, P., and J. Andréou. 1985. "Le règlement frontalier entre les cités d'Ambracie et de Charadros." *Bulletin de correspondance hellénique* 109, no. 1: 499–544.
Camp, J. M. 2000. "Walls and the Polis." In *Polis and Politics: Studies in Ancient Greek History Presented to Mogens Herman Hansen on His Sixtieth Birthday*, edited by M. H. Hansen, P. Flensted-Jensen, T. H. Nielsen, and L. Rubinstein, 41–58. Museum Tusculanum Press.
Campbell, D. B. 2002. "Aspects of Roman Siegecraft." PhD dissertation, University of Glasgow.
Canter, H. V. 1932. "Conflagrations in Ancient Rome." *The Classical Journal* 27, no. 4: 270–88.
Caprino, C., A. M. Colini, G. Gatti, M. Pallottino, and P. Romanelli. 1955. *La Colonna di Marco Aurelio*. Erma di Bretschneider.
Carpenter, R., A. Bon, and A. W. Parsons. 1936. "The Defenses of Acrocorinth and the Lower Town." *Corinth* 3, no. 2: 303–15.
Casson, L. 1991. *The Ancient Mariners: Seafarers and Sea Fighters of the Mediterranean in Ancient Times*. Princeton University Press.
Catling, H. W. 1984/1985. "Archaeology in Greece, 1984-85." *Archaeological Reports* 31: 3–69.
———. 1985/1986. "Archaeology in Greece, 1985–1986." *Archaeological Reports* 2: 3–101.
Cerchiai, L., L. Jannelli, F. Longo, and M. E. Smith. 2007. *The Greek Cities of Magna Graecia and Sicily*. Arsenale Editrice.

Champion, C. 1996. "Polybius, Aetolia, and the Gallic Attack on Delphi (279 B.C.)." *Historia: Zeitschrift für Alte Geschichte* 45, no. 3: 321–25.

———. 1997. "The Nature of Authoritative Evidence in Polybius and Agelaus' Speech at Naupactus." *Transactions of the American Philological Association (1974–)* 127: 111–28.

———. 2000. "Romans as Barbaroi: Three Polybian Speeches and the Politics of Cultural Indeterminacy." *Classical Philology* 95, no. 4: 425–44.

———. 2004. *Cultural Politics in Polybius's Histories*. University of California Press.

Chaniotis, A. 2005. *War in the Hellenistic World*. Blackwell Publishing.

Chaplin, J. D. 2000. *Livy's Exemplary History*. Oxford University Press.

Chaplin, J. D., and C. S. Kraus, eds. 2010. *Livy*. Oxford University Press.

Churchill, J. B. 1999. "Ex qua quod vellent facerent: Roman Magistrates' Authority over Praeda and Manubiae." *Transactions of the American Philological Association* 129: 85–116.

Cichorius, C. 1896–1900. *Die Reliefs der Traianssaule*. G. Reimer.

Citter, C., G. M. Amato, V. Di Natale, and A. Patacchini. 2017. "A Stratified Route Network in a Stratified Landscape. The Region of Enna (Central Sicily) from the Bronze Age to the 19th c. AD." *Open Archaeology* 3: 305–12.

Clark, E. D. 1992. "A Historical Commentary on Plutarch's *Marcellus*." PhD dissertation, University of British Columbia.

Clark, J. H. 2014. *Triumph in Defeat: Military Loss and the Roman Republic*. Oxford University Press.

Collins, J. J. 1974. "The Provenance of the Third Sibylline Oracles." *Bulletin of the Institute of Jewish Studies* 2: 1–18.

Collins, R. 2008. *Violence: A Micro-Sociological Theory*. University of Michigan.

Combs, N. A. 2007. *Guilty Pleas in International Criminal Law: Constructing a Restorative Justice Approach*. Stanford University Press.

Conner, W. R. 1985. "The Razing of the House in Greek Society." *Transactions of the American Philological Association* 115: 79–102.

Conwell, D. 2008. *Connecting the City to the Sea: The History of Athenian Walls*. Brill.

Cooper, C. 2006. *A Village in Sussex: The History of Kingston-near Lewes*. I.B. Tauris.

Cornell, T. J. 1989. "The Conquest of Italy." In *The Cambridge Ancient History* 7.2, edited by A. E. Astin, F. W. Walbank, M. W. Frederiksen, and R. M. Ogilvie, 351–419. Cambridge University Press.

———. 1995. *The Beginnings of Rome: Italy and Rome from the Bronze Age to the Punic Wars (c. 1000–264 BC)*. Routledge.

Coster, W. 1999. "Massacre and Codes of Conduct in the English Civil War." In *The Massacre in History*, edited by M. Levene and P. Roberts, 89–106. Berghahn Books.

Crawford, M., and L. Keppie. 1984. "Excavations at Fregellae, 1978–1984: An Interim Report on the Work of the British Team." *Papers of the British School at Rome* 52: 21–35.

Croker, J. W. 1857. *Essays on the Early Period of the French Revolution*. J. Murray.

Culham, P. 2004. "Women in the Roman Republic." In *The Cambridge Companion to the Roman Republic*, edited by H. Flower, 127–48. Cambridge University Press.

Curchin, L. 1991. *Roman Spain: Conquest and Assimilation*. Routledge.

Dakaris, S. 1962. "The Dark Palace of Hades." *Archaeology* 15, no. 2: 85–93.

———. 1971a. *Archaeological Guide to Dodona*. Translated by E. Kirk-Deftereou. Cultural Society/The Ancient Dodona.

———. 1971b. *Cassopaia and the Elean Colonies*. Athens Center for Ekistics.

Daly, E. 2002. "Between Punitive and Reconstructive Justice: The Gacaca Courts in Rwanda." *NYU Journal of International Law and Politics* 34: 355–96.

Daly, G. 2002. *Cannae: The Experience of Battle in the Second Punic War.* Routledge.

Dausse, M.-P. 2007. "Les villes molosses: bilan et hypotheses sur les quatre centres mentionnés par Tite-Live." In Épire, *Illyrie, Macédoine: Mélanges offerts au Professeur Pierre Cabane*, edited by D. Berranger-Auserve, 197–234. Presses Universitaires Blaise Pascal.

Davies, G. 2006. *Roman Siege Works.* Tempus.

Davies, S. H. 2014. "Beginnings and Endings: 146 BCE as an Imperial Moment, from Polybius to Sallust." *Epekeina* 4, no. 2: 177–218.

Davis, R. H. 1914. "Horrors of Louvain Told by Eyewitness; Circled Burning City." *New York Tribune*, August 31.

Dechambre, A., et al. 1884. *Dictionnaire encyclopédique des sciences médicales.* G. Masson.

Degrassi, A. 1947. *Inscriptiones Italiae.* Vol. 13.1: *Fasti et Elogia: Fasti consulares et triumphales.* Libreria della Stato.

Derow, P. S. 1989. "Rome, the Fall of Macedon, and the Sack of Corinth." In *The Cambridge Ancient History* 8.2, edited by A. E. Astin, F. W. Walbank, et al., 290–323. Cambridge University Press.

———. 2003. "The Arrival of Rome: From the Illyrian Wars to the Fall of Macedon." In *A Companion to the Hellenistic World*, edited by A. Erskine, 51–70. Oxford.

De Souza, P. 2008. "*Parta victoriis pax*: Roman Emperors as Peacemakers." In *War and Peace in Ancient and Medieval History*, edited by P. de Souza and J. France, 76–106. Cambridge University Press.

D'Huys, V. 1987. "How to Describe Violence in Historical Narrative. Reflections of the Ancient Greek Historians and Their Ancient Critics." *Ancient Society* 18: 209–50.

Dillon, S. 2006. "Women on the Columns of Trajan and Marcus and the Visual Language of Roman Victory." In *Representations of War in Ancient Rome*, edited by S. Dillon and K. Welch, 244–71. Cambridge University Press.

Dixon, M. D. 2014. *Late Classical and Early Hellenistic Corinth: 338–196 BC.* Routledge.

Dmitriev, S. 2011. *The Greek Slogan of Freedom and Early Roman Politics in Greece.* Oxford University Press.

Dobkin, M. H. 1972. *Smyrna 1922: The Destruction of a City.* Faber.

Docter, R., et al. 2003. "Carthage Bir Massouda: Preliminary Report on the First Bilateral Excavations of Ghent University and the Institut National du Patrimoine (2002–2003)." *BABesch. Annual Papers on Mediterranean Archaeology* 78: 43–70.

———. 2006. "Carthage Bir Massouda. Second Preliminary Report on the Bilateral Excavations of Ghent University and the Institut National du Patrimoine (2003–2004)." *BABesch. Annual Papers on Mediterranean Archaeology* 81: 37–89.

Douglas, R. M. 2012. *Orderly and Humane: The Expulsion of the Germans after the Second World War.* Yale University Press.

Dowling, M. B. 2006. *Clemency & Cruelty in the Roman World.* University of Michigan Press.

Downes, A. B. 2007. "Draining the Sea by Filling the Graves: Investigating the Effectiveness of Indiscriminate Violence as a Counterinsurgency Strategy." *Civil Wars* 9, no. 4: 420–44.

———. 2008. *Targeting Civilians in War.* Cornell University Press.

Downes, A. B., and K. M. Cochran. 2010. "Targeting Civilians to Win? Assessing the Military Effectiveness of Civilian Victimization in Interstate War." In *Rethinking Violence: States and Non-State Actors in Conflict*, edited by E. Chenoweth and A. Lawrence, 23–56. MIT Press.

Ducrey, P. 1968. *Le Traitement Des Prisonniers De Guerre Dans La Grèce Antique: Des Origines A La Conquête Romaine*. Ecole Française D'Athènes.
Dyson, S. L. 1985. *The Creation of the Roman Frontier*. Princeton University Press.
Eckstein, A. M. 1976. "T. Quinctius Flamininus and the Campaign against Philip in 198 B.C." *Phoenix* 30, no. 2: 119–42.
———. 1987. *Senate and General: Individual Decision Making and Roman Foreign Relations, 264–194 B.C.* University of California Press.
———. 1994. Review of *Aspekte des romischen Volkerrechts: Die Bronzetafel von Alcantara* and *Die Fides im romischen Volkerrecht* by Dieter Nörr, *Classical Philology* 89, no. 1: 82–87.
———. 1995a. "Glabrio and the Aetolians: A Note on *Deditio*." *Transactions of the American Philological Association* 125: 271–89.
———. 1995b. *Moral Vision in* The Histories *of Polybius*. University of California Press.
———. 2006. *Mediterranean Anarchy, Interstate War, and the Rise of Rome*. University of California Press.
———. 2007. "Conceptualizing Roman Imperial Expansion under the Republic." In *A Companion to the Roman Republic*, edited by N. Rosenstein and R. Morstein-Marx, 567–89. Wiley-Blackwell.
———. 2008. *Rome Enters the Greek East: Anarchy to Hierarchy in the Hellenistic Mediterranean, 230–170 BC*. Wiley-Blackwell.
———. 2009. "Ancient 'International Law,' the Aetolian League, and the Ritual of Unconditional Surrender to Rome: A Realist View." *International History Review* 31, no. 2: 253–67.
———. 2013. "Hegemony and Annexation Beyond the Adriatic." In *A Companion to Roman Imperialism*, edited by D. Hoyos, 79–98. Brill.
Editorial Staff. 2013. "Anthropological and Forensic Research of the Skeletons from the Mass Burial Site at the Southeastern Necropolis of Scupi." *Paleopatologia*. Accessed September 27, 2014. http://www.paleopatologia.it/articoli/aticolo.php?recordID=186.
Edlund-Berry, I. E. M. 1994. "Ritual Destruction of Cities and Sanctuaries: The 'Unfounding' of the Archaic Monumental Building at Poggio Civitate (Murlo)." In *Murlo and the Etruscans*, edited by R. D. De Puma and J. P. Small, 16–28. University of Wisconsin Press.
Edmonson, J. C. 1996. "Roman Power and Provincial Administration in Lusitania." In *Pouvoir et imperium (IIIe s. av. J.-C. - Ier s. ap. J.-C.)*, edited by E. Hermon, 163–211. Jovine.
———. 1997. "Creating a Provincial Landscape: Roman Imperialism and Rural Change in Lusitania." In *Les Campagnes de Lusitanie Romaine*, edited by J.-G. Gorges and M. Salinas de Frias, 13–30. Casa de Velázquez.
Eilers, C. 2002. *Roman Patrons of Greek Cities*. Oxford University Press.
Enos, R. L. 2005. "*Rhetorica ad Herennium* (ca. 86–82 BCE)." In *Classical Rhetorics and Rhetoricians: Critical Studies and Sources*, edited by M. Ballif and M. G. Moran, 331–38. Praeger Publishers.
Epstein, S. R. 1992. *An Island for Itself: Economic Development and Social Change in Late Medieval Sicily*. Cambridge University Press.
Erdkamp, P. 2006. "Late-Annalistic Battle Scenes in Livy." *Mnemosyne* 59: 525–63.
———. 2011. "Army and Society." In *A Companion to the Roman Republic*, edited by N. Rosenstein, 147–66. Wiley-Blackwell.
Errington, M. A. 2008. *A History of the Hellenistic World, 323–30 B.C.* Blackwell Publishing.

Erskine, A. 2010. *Roman Imperialism.* Edinburgh University Press.
Evans, J. K. 1988. "Resistance at Home: The Evasion of Military Service in Italy during the Second Century B.C." In *Forms of Control and Subordination in Antiquity,* edited by T. Yuge and M. Doi, 121–40. Brill.
Evans, R. J. 1989. "Was M. Caecilius Metellus a Renegade? A Note on Livy 22.53.5." *Acta Classica* 32: 117–21.
Evans, R. J., and M. Kleijwegt. 1992. "Did the Romans Like Young Men? A Study of the Lex Villia Annalis: Causes and Effects." *Zeitschrift für Papyrologie und Epigraphik* 92: 181–95.
Ferrary, J.-L. 1988. *Philhellénisme et impérialisme: aspects idéologiques de la conquête romaine du monde hellénistique, de la seconde guerre de Macédoine à la guerre contre Mithridate.* Ecole Française de Rome.
———. 2007. "Rogatio Scribonia proposant le rachat de Lusitaniens vendus comme esclaves et la création d'une *quaestio* pour punir le(s) responsable(s) de cette atteinte à la *fides populi Romani.*" In *Lepor: Leges Populi Romani,* edited by J.-L. Ferrary and P. Moreau. IRHT-TELMA. Accessed November 28, 2014. http://www.cn-telma.fr/lepor/notice653/.
Ferris, I. 2009. *War and Hate: The Column of Marcus Aurelius.* History Press.
Fischer. 1992. "Maccabees, Books of." *The Anchor Bible Dictionary,* vol. 4, edited by David Noel Freedman, 439–50. Yale University Press.
Flamerie de Lachapelle, G. 2007. "Le sort des villes ennemies dans l'œuvre de Tite-Live: aspects historiographiques." *Revue de philologie, de littérature et d'histoire anciennes* 81, no. 1: 79–110.
Flores Gomes, J. M., and D. Carneiro. 2005. *Subtus Montis Terroso: Património Arqueológico no Concelho da Póvoa de Varzim.* Câmara Municipal.
Flower, H. 2010. *Roman Republics.* Princeton University Press.
Forsén, B. 2011. "The Emerging Settlement Patterns of the Kokytos Valley." In *Thesprotia Expedition II: Environment and Settlement Patterns,* edited by B. Forsén and E. Tikkala, 1–37. Finish Institute at Athens.
Forsythe, G. 2005. *A Critical History of Early Rome: From Prehistory to the First Punic War.* University of California Press.
Fossey, J. M. 1979. "The Cities of the Kopais in the Roman Period." *Aufstieg und Niedergang der romischen Welt* 2, no. 7: 549–91.
Frankfort, H. 1996. *The Art and Architecture of the Ancient Orient.* Yale University Press.
Friedland, R., and R. Hecht. 1998. "The Bodies of Nations: A Comparative Study of Religious Violence in Jerusalem and Ayodhya." *History of Religions* 38, no. 2: 101–49.
Fronda, M. 2010. *Between Rome and Carthage: Southern Italy during the Second Punic War.* Cambridge University Press.
Gabba, E. 1970. "Commento a Floro, II 9, 27–28." *Studi Classici E Orientali* 19/20: 461–64.
Gabrieli, F. 2009. *Arab Historians of the Crusades.* Translated by E. J. Costello. Routledge.
Gaca, K. 2010. "The Andrapodizing of War Captives in Greek Historical Memory." *Transactions of the American Philological Association* 140: 117–61.
———. 2011. "Girls, Women, and the Significance of Sexual Violence in Ancient Warfare." In *Sexual Violence in Conflict Zones,* edited by E. Heineman, 73–88. University of Pennsylvania Press.
Gaignerot-Driessen, F. 2013. "The 'Killing' of a City: A Destruction by Enforced Abandonment." In *Destruction: Archaeological, Philological and Historical Perspectives,* edited by J. Driessen, 285–98. Presses Universitaires de Louvain.

García Moreno, L. A. 1988. "Infancia, juventud y primeras aventuras de Viriato, caudillo Lusitano." In *Actas del I Congreso Peninsular de Historia Antigua*, vol. 2, edited by G. P. Menaut, 373–82. Universidad de Santiago de Compostela.
García Riaza, E. 2003. *Celtíberos y lusitanos frente a Roma: diplomacia y derecho de guerra*. Servicio Editorial Universidad del País Vasco.
García y Bellido, A. 1945. "Bandas y guerrillas en las luchas con Roma." *Hispania* 5, no. 21: 547–604.
Gargola, D. 2010. "Mediterranean Empire." In *A Companion to the Roman Republic*, edited by N. Rosenstein, 147–66. Wiley-Blackwell.
Garlan, Y. 1988. *Slavery in Ancient Greece: Revised and Expanded Edition*. Translated by J. Lloyd. Cornell University Press.
Gates, S. 2002. "Recruitment and Allegiance: The Microfoundations of Rebellion." *Journal of Conflict Resolution* 46, no. 1: 111–30.
Gebhard, E. R., and M. W. Dickie. 2003. "The View from the Isthmus, ca. 200 to 44 B.C." In *Corinth*. Vol. 20: *The Centenary: 1896–1996*, edited by C. K. Williams and N. Bookidis, 261–78. American School of Classical Studies at Athens.
Gerding, Henrik. 2013. "Roofs and Superstructures." In *Shipsheds of the Ancient Mediterranean*, edited by D. Blackman and B. Rankov, 41–58. Cambridge University Press.
Gerlach, C. 2010. *Extremely Violent Societies: Mass Violence in the Twentieth-Century World*. Cambridge University Press.
Gilliver, C. M. 1996. "The Roman Army and Morality in War." In *Battle in Antiquity*, edited by A. B. Lloyd, 219–38. Duckworth.
———. 2005. *Caesar's Gallic Wars, 58–50 BC*. Routledge.
Gleig, G. R. 1821. *A Narrative of the Campaigns of the British Army at Washington and New Orleans*. J. Murray.
Goldsworthy, A. 1996. *The Roman Army at War, 100 BC–200 AD*. Oxford University Press.
———. 2000. *Roman Warfare*. Smithsonian Books.
———. 2006. *The Fall of Carthage*. Phoenix.
———. 2007. "War." In *The Cambridge History of Greek and Roman Warfare*. Vol. 2: *Rome from the Late Republic to the Late Empire*, edited by P. Sabin, H. Van Wees, and M. Whitby, 76–121. Cambridge University Press.
Gómez Espelosín, F. J. 1997. "Appian's 'Iberiké.' Aims and Attitudes of a Greek Historian of Rome." *Aufstieg und Niedergang der römischen Welt* 34.1.2: 403–27.
González García, F. J. 2011. "From Cultural Contact to Conquest: Rome and the Creation of a Tribal Zone in the North-Western Iberian Peninsula." *Greece & Rome* 58, no. 2: 184–94.
Goring, J. 2003. *Burn Holy Fire: Religion in Lewes Since the Reformation*. Lutterworth.
Goudsblom, J. 1992. *Fire and Civilization*. Allen Lane.
Grainger, J. D. 2016. *Great Power Diplomacy in the Hellenistic World*. Routledge.
Graninger, D. 2011. *Cult and Koinon in Hellenistic Thessaly*. Brill.
Green, P. 1993. *Alexander to Actium: The Historical Evolution of the Hellenistic Age*. University of California Press.
Gruen, E. 1984a. *The Hellenistic World and the Coming of Rome*. University of California Press.
———. 1984b. "Material Rewards and the Drive for Empire." In *Imperialism of Mid-Republican Rome*, edited by W. V. Harris, 59–82. American Academy of Rome.
———. 1998. "Jews, Greeks, and Romans in the Third Sibylline Oracle." In *Jews in a Graeco-Roman World*, edited by M. Goodman, 15–36. Oxford University Press.

Grünewald, T. 2004. *Bandits in the Roman Empire: Myth and Reality.* Translated by J. Drinkwater. Routledge.

Haeussler, R. 2013. *Becoming Roman? Diverging Identities and Experiences in Ancient Northwest Italy.* Routledge.

Hammond, N. G. L. 1953. "Hellenic Houses at Ammotopos in Epirus." *The Annual of the British School at Athens* 48: 135–40.

———. 1966. "The Opening Campaigns and the Battle of the Aoi Stena in the Second Macedonian War." *Journal of Roman Studies* 56: 39–54.

———. 1967. *Epirus: The Geography, the Ancient Remains, the History and Topography of Epirus and Adjacent Areas.* Clarendon Press.

———. 1984. "The Battle of Pydna." *Journal of Hellenic Studies* 104: 31–47.

Hammond, N. G. L., and F. W. Walbank. 1988. *A History of Macedonia: 336–167 B.C.* Oxford University Press.

Hansen, M. H. 2006a. *Polis: An Introduction to the Ancient Greek City-State.* Oxford University Press.

———. 2006b. *The Shotgun Method: The Demography of the Ancient Greek City-State Culture.* University of Missouri Press.

———. 2008. "An Update on the Shotgun Method." *Greek, Roman, and Byzantine Studies* 48: 259–86.

Hansen, M. H., and T. H. Nielsen. 2004. *An Inventory of Archaic and Classical Poleis.* Oxford University Press.

Hanson, V. D. 1992. "Cannae." In *The Experience of War*, edited by R. Cowley, 42–49. Dell Publishing.

———. 1998. *Warfare and Agriculture in Classical Greece.* Revised edition. University of California Press.

Harrington, J. F. 2013. *The Faithful Executioner: Life and Death, Honor and Shame in the Turbulent Sixteenth Century.* Farrar, Straus and Giroux.

Harris, W. V. 1979. *War and Imperialism in Republican Rome, 327–70 B.C.* Oxford University Press.

———. 1989. "Roman Expansion in the West." In *Cambridge Ancient History.* Vol. 8: *Rome and the Mediterranean*, edited by A. E. Astin, F. W. Walbank, M. W. Frederiksen, and R. M. Ogilvie, 107–62. Cambridge University Press.

Hatzfeld, J. 2005. *Machete Season: The Killers in Rwanda Speak.* Translated by L. Coverdale. Farrar, Straus and Giroux.

Hewitt, H. J. 1999. "The Organization of War." In *The War of Edward III: Sources and Interpretations*, edited by C. J. Rogers, 285–302. The Boydell Press.

Hoffman, M. 2007. *"My Brave Mechanics": The First Michigan Engineers and Their Civil War.* Wayne State University Press.

Hölkeskamp, K.-J. 2000. "Fides—deditio in fidem—dextra data et accepta: Recht, Religion und Ritual in Rom." In *The Roman Middle Republic: Politics, Religion, and Historiography c. 400–133 B.C.*, edited by C. Bruun, 223–50. Institutum Romanum Finlandiae.

Holliday, P. 1997. "Roman Triumphal Painting: Its Function, Development, and Reception." *Art Bulletin* 79: 130–47.

Holloway, R. R. 1994. *The Archaeology of Early Rome and Latium.* Routledge.

Honda, K. 1999. *The Nanjing Massacre: A Japanese Journalist Confronts Japan's National Shame.* East Gate.

Honigman, S. 2014. *Tales of High Priests and Taxes: The Books of the Maccabees and the Judean Rebellion Against Antiochus IV.* University of California Press.

Hoyos, D. 1992. "Sluice-Gates or Neptune at New Carthage, 209 B.C.?" *Historia: Zeitschrift für Alte Geschichte* 41, no. 1: 124–28.
———. 1998. *Unplanned Wars: The Origins of the First and Second Punic Wars*. Walter de Gruyter.
———. 2003. *Hannibal's Dynasty: Power and Politics in the Western Mediterranean, 247–183 B.C.* Routledge.
———. 2007. "The Age of Overseas Expansion (264–146 BC)." In *A Companion to the Roman Army*, edited by P. Erdkamp, 63–79. Oxford University Press.
———. 2010. *The Carthaginians*. Routledge.
———. 2015. *Mastering the West: Rome and Carthage at War*. Oxford University Press.
Huertas, P. 2012. "Notas para un studio de la situación del castrum de Escipión en Carthago-Nova." *Revista Oficial Federación de Tropas y Legiones Carthagineses y Romanos*: 65–69.
Hull, I. V. 2004. *Absolute Destruction: Military Culture and the Practices of War in Imperial Germany*. Cornell University Press.
———. 2008. "The Military Campaign in German Southwest Africa, 1904–1907 and the Genocide of the Herero and Nama." *Journal of Namibian Studies* 4: 7–24.
Hultman, L. 2012. "Attacks on Civilians in Civil War: Targeting the Achilles Heel of Democratic Governments." *International Interactions* 38, no. 2: 164–81.
———. 2014. "Violence Against Civilians." In *The Routledge Handbook of Civil Wars*, edited by E. Newman and K. DeRouen Jr., 289–99. Routledge.
Humphreys, M., and J. M. Weinstein. 2006. "Handling and Manhandling Civilians in Civil War." *American Political Science Review* 100, no. 3: 429–47.
Hutton, W. 2005. *Describing Greece: Landscape and Literature in the Periegesis of Pausanias*. Cambridge University Press.
Isaac, B. 1998. "Hierarchy and Command Structure in the Roman Army." In *The Near East under Roman Rule: Selected Papers*, 388–402. Brill.
———. 2004. *The Invention of Racism in Classical Antiquity*. Princeton University Press.
Isager, J. 2001. "Introduction." In *Foundation and Destruction: Nikopolis and Northwestern Greece*, edited by J. Isager, 1–16. Danish Institute of Athens.
Itgenshorst, T. 2005. *Tota Illa Pompa: der Triumph in der Römischen Republik*. Vandenhoeck & Ruprecht.
Jackson, M., and F. Marra. 2006. "Roman Stone Masonry: Volcanic Foundations of the Ancient City." *American Journal of Archaeology* 110, no. 3: 403–36.
Jacobs, D. 2014. "The Images of Space in the Third Sibylline Oracle." PhD dissertation, Humboldt-Universität zu Berlin.
James, S. A. 2014. "The Last of the Corinthians? Society and Settlement from 146 to 44 BCE." In *Corinth in Contrast: Studies in Inequality*, edited by S. J. Friesen, S. A. James, and D. N. Schowalter, 17–37. Brill.
Janes, R. 2005. *Losing Our Heads: Beheadings in Literature and Culture*. New York University Press.
Jiménez, A. 2014. "Punic after Punic Times? The Case of the So-Called 'Libyphoenician' Coins of Southern Iberia." In *The Punic Mediterranean: Identities and Identification from Phoenician Settlement to Roman Rule*, edited by J. C. Quinn and N. C. Vella, 219–42. Cambridge University Press.
Johnstone, C. L. 1996. "Greek Oratorical Settings and the Problem of the Pnyx: Rethinking the Athenian Political Process." In *Theory, Text, Context: Issues in Greek Rhetoric and Oratory*, edited by C. L. Johnstone, 97–128. State University of New York Press.

Jones, A. 2000. "Gendercide and Genocide." *Journal of Genocide Research* 2, no. 2: 185–211.

———. 2006. *Genocide: A Comprehensive Introduction*. Routledge.

Jones, I. 2012. "A Sea of Blood? Massacres during the Wars of the Three Kingdoms, 1641–1653." In *Theatres of Violence: Massacre, Mass Killing, and Atrocity Throughout History*, edited by P. G. Dwyer and L. Ryan, 63–78. Berghahn Books.

Joshel, S. R. 2010. *Slavery in the Roman World*. Cambridge University Press.

Jovanova, L., F. Veljanovska, M. Velova Graorkovska, and A. Stankov. 2017. *Scupi, A Mass Grave: A Glance into a Dark Side of the Roman History*. Museum of the City of Skopje.

Kalyvas, S. N. 1999. "Wanton and Senseless? The Logic of Massacres in Algeria." *Rationality and Society* 11, no. 3: 243–85.

———. 2006. *The Logic of Violence in Civil War*. Cambridge University Press.

———. 2008. "Promises and Pitfalls of an Emerging Research Program: The Microdynamics of Civil War." In *Order, Conflict, and Violence*, edited by S. N. Kalyvas, I. Shapiro, and T. Masoud, 397–421. Cambridge University Press.

Kalyvas, S. N., and M. A. Kocher. 2009. "The Dynamics of Violence in Vietnam: An Analysis of the Hamlet Evaluation System." *Journal of Peace Resolution* 46, no. 3: 335–55.

Kanta-Kitsou, E. 2008. *Gitana Thesprotia Archaeological Guide*. Athens Ministry of Culture.

Kanta-Kitsou, E., and V. Lambrou. 2008. *Doliani Thesprotia Archaeological Guide*. Athens Ministry of Culture.

Karatzeni, V. 2001. "Epirus in the Roman Period." In *Foundation and Destruction: Nikopolis and Northwestern Greece*, edited by J. Isager, 163–79. Danish Institute of Athens.

Kay, P. 2014. *Rome's Economic Revolution*. Oxford University Press.

Keaveney, A. 2007. *The Army in the Roman Revolution*. Routledge.

Keay, S. J. 1988. *Roman Spain*. University of California Press.

Kendall, J. 2009. "Scipio Aemilianus, Lucius Mummius, and the Politics of Plundered Art in Italy and Beyond in the 2nd Century B.C.E." *Etruscan and Italic Studies* 12, no. 1: 169–81.

Kenji, O. 2007. "Massacres Near Mufushan." In *The Nanking Atrocity 1937–38: Complicating the Picture*, edited by B. T. Wakabayashi, 70–85. Berghahn Books.

Keppie, L. 1984. *The Making of the Roman Army*. University of Oklahoma Press.

Kern, P. B. 1999. *Ancient Siege Warfare*. Indiana University Press.

Kim, D. 2010. "What Makes State Leaders Brutal? Examining Grievances and Mass Killing during Civil War." *Civil Wars* 12, no. 3: 237–60.

Klee, E., W. Dressen, and V. Riess, eds. 1999. *"The Good Old Days": The Holocaust as Seen by Its Perpetrators and Bystanders*. Konecky & Konecky.

Knapp, R. C. 1977. *Aspects of the Roman Experience in Iberia, 206–100 B.C*. Universidad de Valladolid.

Knowles, D. 1976. *Bare Ruined Choirs: The Dissolution of the English Monasteries*. Cambridge University Press.

Konstan, D. 2004. *Pity Transformed*. Bloomsbury.

Koon, S. 2011. "Phalanx and Legion: The 'Face' of Punic War Battle." In *A Companion to the Punic Wars*, edited by D. Hoyos, 77–94. Wiley-Blackwell.

Krentz, P. 1985. "Casualties in Hoplite Battles." *Greek, Roman, and Byzantine Studies* 26, no. 1: 13–20.

Kuretsky, S. D. 2012. "Jan van der Jeyden and the Origins of Modern Firefighting." In *Flammable Cities: Urban Conflagration and the Making of the Modern World*, edited by G. Bankoff, U. Lübken, and J. Sand, 23–43. University of Wisconsin Press.

"The Lament of Urim." 2003–2006. In *The Electronic Text Corpus of Sumerian Literature*, edited by Faculty of Oriental Studies. University of Oxford. Accessed January 24, 2016. http://etcsl.orinst.ox.ac.uk/.

Lancel, S., and J.-P. Morel. 1992. "La Colline de Byrsa: Les Vestiges Puniques." In *Pour Sauver Carthage: Exploration et conservation de la cite punique, romaine et byzantine*, edited by A. Ennabli, 43–68. UNESCO/INAA.

Langerbein, H. 2004. *Hitler's Death Squads: The Logic of Mass Murder*. Texas A&M University Press.

Lanni, A. 2008. "The Laws of War in Ancient Greece." *Law and History Review* 26, no. 3: 469–89.

Larsen, J. A. O. 1968. *Greek Federal States: The Institutions and History*. Clarendon Press.

Laurence, R. 1996. "Ritual, Landscape, and the Destruction of Place in Roman Imagination." In *Approaches to the Study of Ritual: Italy and the Ancient Mediterranean*, edited by J. B. Wilkins, 111–21. Accordia Research Centre, University of London.

Lawrence, A. W. 1979. *Greek Aims in Fortification*. Clarendon Press.

Lazari, K., and E. Kanta-Kitsou. 2010. "Thesprotia during the Late Classic and Hellenistic Periods: The Formation and Evolution of Cities." In *Lo spazio ionico e le comunità della Grecia nord-occidentale: territorio, società e istituzioni*, edited by C. Antonetti, 35–60. Edizioni ETS.

Lazenby, J. F. 1978. *Hannibal's War: A Military History of the Second Punic War*. University of Oklahoma Press.

———. 1996. *The First Punic War: A Military History*. Stanford University Press.

Lee, J. W. I. 2010. "Urban Warfare in the Classical Greek World." In *Makers of Ancient Strategy: From the Persian Wars to the Fall of Rome*, edited by V. D. Hanson, 138–62. Princeton University Press.

Lendon, J. E. 2005. *Soldiers and Ghosts: A History of Battle in Classical Antiquity*. Yale University Press.

———. 2007. "War and Society." In *The Cambridge History of Greek and Roman Warfare*. Vol. 1: *Greece, the Hellenistic World, and Rome*, edited by P. Sabin, H. Van Wees, and M. Whitby, 498–516. Cambridge University Press.

———. 2009. "Historians without History: Against Roman Historiography." In *The Cambridge Companion to the Roman Historians*, edited by A. Feldherr, 41–61. Cambridge University Press.

Lens Tuero, J. 1986. "Viriato, héroe y rey cínico." *Estudios de Filología griega* 2: 253–72.

Levene, M., and P. Roberts, eds. 1999. *The Massacre in History*. Berghahn Books.

Levithan, J. 2013. *Roman Siege Warfare*. University of Michigan Press.

Liddell, H. G., and R. Scott, et al. 1996. *Greek-English Lexicon, with a Revised Supplement*. Oxford University Press.

Linderski, J. 1996. "Cato Maior in Aetolia." In *Transitions to Empire. Essays in Greco-Roman History, 360–146 B.C., in Honor of E. Badian*, edited by R. W. Wallace and E. M. Harris, 376–408. University of Oklahoma Press.

Lintott, A. W. 1972. "Imperial Expansion and Moral Decline in the Roman Republic." *Historia* 21, no. 4: 626–38.

———. 1999. *The Constitution of the Roman Republic*. Oxford University Press.

Lomas, K. 1993. *Rome and the Western Greeks, 350 BC–AD 200*. Routledge.

———. 2012. "The Weakest Link: Elite Social Networks in Republican Italy." In *Processes of Integration and Identity Formation in the Roman Republic*, edited by S. T. Roselaar, 197–214. Brill.

———. 2018. *The Rise of Rome: From the Iron Age to the Punic Wars*. Belknap Press of Harvard University Press.

López Castro, J. L. 2013. "The Spains, 205–72 BC." In *A Companion to Roman Imperialism*, edited by D. Hoyos, 67–78. Brill.

Lu, S. 2004. *They Were in Nanjing: The Nanjing Massacre Witnessed by American and British Nationals*. Hong Kong University Press.

Luce, T. 1977. *Livy: The Composition of His History*. Princeton University Press.

Luján Martínez, E. R. 2006. "The Language(s) of the Callaeci." *E-Keltoi: The Journal of Interdisciplinary Celtic Studies* 6: 716–48.

Lyall, J. 2009. "Does Indiscriminate Violence Incite Insurgent Attacks? Evidence from Chechnya." *Journal of Conflict Resolution* 53, no. 3: 331–62.

Lynn, J. 2008. *Battle: A History of Combat and Culture*. Basic Books.

Ma, J. 2000. "Fighting Poleis in the Hellenistic World." In *War and Violence in Ancient Greece*, edited by H. Van Wees, 337–76. Classical Press of Wales.

Marco Simón, F. 2016. "Insurgency or State Terrorism? The Hispanic Wars in the Second Century BCE." In *Brill's Companion to Insurgency and Terrorism in the Ancient Mediterranean*, edited by T. Howe and L. Brice, 221–47. Brill.

Marincola, J. 2001. *Greek Historians*. Cambridge University Press.

Martin, T. R. 2013. "Demetrius 'the Besieger' and Hellenistic Warfare." In *The Oxford Handbook of Warfare in the Classical World*, edited by B. Campbell and L. Tritle, 671–87. Oxford University Press.

Martín Camino, M., and J. A. Belmonte Marín. 1993. "La muralla púnica de Cartagena: valoración arqueológica y análisis epigráfico de sus materiales." *Aula Orientalis* 11, no. 2: 161–71.

Marvin, L. W. 2012. "Atrocity and Massacre in the High and Late Middle Ages." In *Theatres of Violence: Massacre, Mass Killing, and Atrocity throughout History*, edited by P. D. Dwyer and L. Ryan, 50–62. Berghahn Books.

Mattern, S. 1999. *Rome and the Enemy: Imperial Strategy in the Principate*. University of California Press.

McDonnell, M. 2006. *Roman Manliness: Virtus and the Roman Republic*. Cambridge University Press.

McGing, B. 1986. *The Foreign Policy of Mithridates VI Eupator, King of Pontus*. Brill.

Merger, S. 2016. *Rape Loot Pillage: The Political Economy of Sexual Violence in Armed Conflict*. Oxford University Press.

Metcalf, A. 2009. *The Muslims of Medieval Italy*. Edinburgh University Press.

Meyer, E. A. 2013. *The Inscriptions of Dodona and a New History of Molossia*. Franz Steiner Verlag.

Midlarsky, M. I. 2005. *The Killing Trap: Genocide in the Twentieth Century*. Cambridge University Press.

Millar, F. 1998. *The Crowd in Rome in the Late Republic*. University of Michigan Press.

Mintzker, Y. 2011. "What Is Defortification? Military Functions, Police Roles, and Symbolism in the Demolition of German City Walls in the Eighteenth and Nineteenth Centuries." *Bulletin of the German Historical Institute* 48: 33–58.

———. 2012. *The Defortification of the German City, 1689–1866*. Cambridge University Press.

Mirman, L. 1917. *Their Crimes*. Translated by J. Esslemont Adams. Cassell and Company.
Moderato, M. 2015. "The Walled Towns of Thesprotia: from the Hellenistic Foundation to the Roman Destruction." In *Proceedings of the 15th Symposium on Mediterranean Archaeology*, vol. 1, edited by P. M. Militello and H. Öniz, 313–20. Archaeopress.
Morgan, D. O. 2007. *The Mongols*. Wiley-Blackwell.
Morillo Cerdán, Á., G. Rodríguez Martín, E. Martín Hernández, and R. Duran Cabello. 2011. "The Roman Republican Battlefield at Pedrosillo (Casas de Reina, Badajoz, Spain): New Research." *Conimbriga* 50: 59–78.
Morris, M. 2017. *Castles: Their History and Evolution in Medieval Britain*. Pegasus Books.
Morrison, J. S. 1988. "Dilution of Oarcrews with Prisoners of War." *The Classical Quarterly* 38, no. 1: 251–53.
Morstein-Marx, R. 1995. *Hegemony to Empire: The Development of Roman Imperium in the East from 148 to 62 B.C.* University of California Press.
———. 2004. *Mass Oratory and Political Power in the Later Roman Republic*. Cambridge University Press.
Morton, P. 2014. "The Geography of Rebellion: Strategy and Supply in the Two 'Sicilian Slave Wars.'" *Bulletin of the Institute of Classical Studies* 57, no. 1: 20–38.
Mouritsen, H. 2017. *Politics in the Roman Republic*. Cambridge University Press.
Mueller, J. E. 2000. "The Banality of Ethnic War." *International Security* 25, no. 1: 42–70.
Muñiz Coello, J. M. 2004. "El proceso de Galba, las quaestiones y la justicia ordinaria (Roma, siglos II/I a.C.)." *L'antiquité classique* 73: 109–26.
Ñaco del Hoyo, T., B. Antela-Bernárdez, I. Arrayás-Morales, and S. Busquets-Artigas. 2009. "The Impact of the Roman Intervention in Greece and Asia Minor upon Civilians (88–63 BC)." In *Transforming Historical Landscapes in the Ancient Empires*, edited by B. Antela-Bernardez and T. Ñaco del Hoyo, 33–51. British Archaeological Reports.
Nicolet, C. 1980. *The World of the Citizen in Republican Rome*. Translated by P. S. Falla. University of California Press.
Noguera Celdrán, J. M., M. J. Madrid Balanza, and V. Velasco Estrada. 2011–2012. "Novedades sobre la *arx Hasdrubalis* de *Qart Hadast* (Cartagena): nuevas evidencias arqueológicas de la muralla púnica." *Cuadernos de Prehistoria y Arqueología* 37–38: 479–504.
Nörr, D. 1989. *Aspekte des römischen Völkerrechts: Die Bronztafel von Alcantara*. Verlag der Bayerischen Akademie der Wissenschaften.
North, J. A. 1981. "The Development of Roman Imperialism." *Journal of Roman Studies* 71: 1–9.
———. 1990. "Democratic Politics in Republic Rome." *Past & Present* 126: 3–21.
———. 2010. "The Constitution of the Roman Republic." In *A Companion to the Roman Republic*, edited by N. Rosenstein and R. Morstein-Marx, 256–77. Wiley-Blackwell.
Nossov, K. 2005. *Ancient and Medieval Siege Weapons*. Lyons Press.
Oakley, S. 1985. "Single Combat in the Roman Republic." *The Classical Quarterly* 35, no. 2: 392–410.
———. 1993. "The Roman Conquest of Italy." In *War and Society in the Roman World*, edited by J. Rich and G. Shipley, 9–37. Routledge.
———. 2010. "Livy and His Sources." In *Livy*, edited by J. D. Chaplin and C. S. Kraus, 439–60. Oxford University Press.
Olásolo, H. 2008. *Unlawful Attacks in Combat Situations: From ICTY's Case Law to the Rome Statute*. Brill.

O'Malley, G. E. 2014. *Final Passages: The Intercolonial Slave Trade of British America, 1619–1807*. University of North Carolina Press.
Oost, S. I. 1954. *Roman Policy in Epirus and Acarnania in the Age of the Roman Conquest of Greece*. Southern Methodist University Press.
Osborne, R., and A. Wallace-Hadrill. 2013. "Cities of the Ancient Mediterranean." In *The Oxford Handbook of Cities in World History*, edited by P. Clark, 49–65. Oxford University Press.
Patrick, E. 2005. "Intent to Destroy: The Genocidal Impact of Forced Migration in Darfur, Sudan." *Journal of Refugee Studies* 18, no. 4: 410–29.
Paul, G. M. 1982. "'Urbs Capta': Sketch of an Ancient Literary Motif." *Phoenix* 36, no. 2: 144–55.
Petersen, Eugen, et al. 1896. *Die Marcus-Säule auf Piazza Colonna in Rom*. Bruckmann. Universitätsbibliothek Heidelberg. Accessed February 24, 2016. http://digi.ub.uni-heidelberg.de/diglit/petersen1896ga.
Pettegrew, D. K. 2016. *The Isthmus of Corinth: Crossroads of the Mediterranean World*. University of Michigan Press.
Phang, S. 2008. *Roman Military Service: Ideologies of Discipline in the Late Republic and Early Principate*. Cambridge University Press.
Pitch, A. S. 1998. *The Burning of Washington: The British Invasion of 1814*. Naval Institute Press.
Pittenger, M. R. P. 2008. *Contested Triumphs: Politics, Pageantry, and Performance in Livy's Republican Rome*. University of California Press.
Preka-Alexandri, K. 1999. "Recent Excavations in Ancient Gitana." In *L'Illyrie méridionale et l'Épire dans l'Antiquité* III, edited by Pierre Cabanes, 167–69. De Boccard.
Pritchett, W. K. 1991. *The Greek State at War*, Part V. University of California Press.
Prósper, B. M. 2007. "Lusitanian: A Non-Celtic Indo-European Language of Western Hispania." In *Celtic and Other Languages in Ancient Europe*, edited by J. L. García Alonso, 53–64. Ediciones Universidad de Salamanca.
Purcell, N. 1995. "On the Sacking of Carthage and Corinth." In *Ethics and Rhetoric: Classical Essays for Donald Russell on His Seventy-Fifth Birthday*, edited by D. Innes, H. Hine, and C. Pelling, 133–48. Clarendon Press.
Quesada Sanz, F. 2002. "La evolución de la panoplia, modos de combate y tácticas de los iberos." In *La guerra en el mundo ibérico y celtibérico, ss. VI-II a. de C.*, edited by P. Moret and F. Quesada Sanz, 35–64. Casa de Velázquez.
———. 2011. "Military Developments in the 'Late Iberian' Culture (c. 237–c. 195 BC): Mediterranean Influences in the 'Far West' via the Carthaginian Military." In *Hellenistic Warfare I. Proceedings Conference Torun (Poland), October 2003*, edited by N. Sekunda and A. Noguera, 207–57. Fundacion Libertas.
———. 2015. "Genocide and Mass Murder in Second Iron Age Europe: Methodological Issues and Case Studies in the Iberian Peninsula." In *The Routledge History of Genocide*, edited by C. Carmichael and R. C. Maguire, 9–22. Routledge.
Quesada Sanz, F., I. Muñiz Jaén, and I. López Flores. 2014. "La guerre et ses traces: destruction et massacre dans le village ibérique du Cerro de la Cruz (Cordoue) et leur contexte historique au IIe s. a.C." In *La Guerre et ses traces: Conflits et sociétés en Hispanie à l'époque de la conquête romaine (IIIe-Ier s. a.C.)*, edited by F. Cadiou and M. Navarro Caballero, 231–71. Ausonius.
Raaflaub, K. 1996. "'Born to Be Wolves? Origins of Roman Imperialism." In *Transitions to Empire. Essays in Greco-Roman History, 360–146 B.C. in Honour of E. Badian*, edited by R. W. Wallace and E. Harris, 273–314. University of Oklahoma Press.

———. 2010. "Between Myth and History: Rome's Rise from Village to Empire (the Eighth Century to 264)." In *A Companion to the Roman Republic*, edited by N. Rosenstein and R. Morstein-Marx, 125–46. Wiley-Blackwell.

Rakoczy, L. 2008. "Out of the Ashes: Destruction, Reuse, and Profiteering in the English Civil War." In *The Archaeology of Destruction*, edited by L. Rakoczy, 261–86. Cambridge Scholars Publishing.

Ramakrishnan, V. 1993. "The Wrecking Crew: In Ayodhya, Demolition and After." *Frontline*, January 1.

Ramallo Asensio, S. F. 2006. "Carthago Nova: *urbs opulentissima omnium in Hispania*." In *Early Roman Towns in* Hispania Tarraconensis, edited by L. Abad Casal, S. Keay, and S. Ramallo Asensio, 91–104. Journal of Roman Archaeology.

Rátkai, S. 2015. *Wigmore Castle, North Herefordshire: Excavations 1996 and 1998*. Society for Medieval Archaeology.

Rawlings, L. 2007. "Army and Battle during the Conquest of Italy (350–264 BC)." In *A Companion to the Roman Army*, edited by P. Erdkamp. Blackwell.

———. 2011. "The War in Italy, 218–203." In *A Companion to the Punic Wars*, edited by D. Hoyos, 299–319. Wiley-Blackwell.

Rawson, B. 2006. "Finding Roman Women." In *A Companion to the Roman Republic*, edited by N. Rosentein and R. Morstein-Marx, 324–41. Wiley-Blackwell.

Reisdoerfer, J. 2007. ". . . non aetate confectis, non mulieribus, non infantibus pepercerunt: Etude sur le massacre d'Avaricum (BG VII 28)." *Göttinger Forum für Altertumswissenschaft* 10: 59–80.

Ribera i Lacomba, A. 2006. "The Roman Foundation of Valencia and the Own in the 2nd–1st c. B.C." In *Early Roman Towns in* Hispania Tarraconensis, edited by L. Adab Casal, S. Kealy, and S. Ramallo Asensio, 75–89. Roman Journal of Archaeology.

Ribera i Lacomba, A., and M. Calvo Galvez. 1995. "La primera evidencia arqueológica de la destrucción de Valentia por Pompeyo." *Journal of Roman Archaeology* 8: 19–40.

Rich, J. 1983. "The Supposed Roman Manpower Shortage of the Later Second Century B.C." *Historia: Zeitschrift für Alte Geschichte* 32, no. 3: 287–331.

———. 1993. "Fear, Greed, and Glory: The Causes of Roman War-Making in the Middle Republic." In *War and Society in the Roman World*, edited by J. Rich and G. Shipley, 55–64. Routledge.

———. 1996. "The Origins of the Second Punic War." In *The Second Punic War: A Reappraisal*, edited by T. Cornell, B. Rankov, and P. Sabin, 1–37. Institute of Classical Studies, School of Advanced Study, University of London.

———. 2012. "Roman Attitudes to Defeat in Battle under the Republic." In *Vae Victis!: Perdedores en el Mundo Antiguo*, edited by F. Marco Simón, F. Pina Polo, and J. Remesal Rodríguez, 83–111. Universitat de Barcelona, Publicacions i edicions.

———. 2015. "Appian, Polybius and the Romans' War with Antiochus the Great: A Study in Appian's Sources and Methods." In *Appian's Roman History: Empire and Civil War*, edited by K. Welch, 65–123. Classical Press of Wales.

Richardson, J. S. 1976. "The Spanish Mines and the Development of Provincial Taxation in the Second Century B.C." *Journal of Roman Studies* 66: 139–52.

———. 1986. *Hispaniae: Spain and the Development of Roman Imperialism, 218–82 BC*. Cambridge University Press.

———. 1987. "The Purpose of the Lex Calpurnia de Repetundis." *Journal of Roman Studies* 77: 1–12.

———. 1996. *The Romans in Spain*. Blackwell.

———. 2000. *Appian: Wars of the Romans in Iberia*. Aris & Philips Ltd.
———. 2011. "*Fines Provinciae.*" In *Frontiers in the Roman World*, edited by O. Hekster and T. Kaizer, 1–12. Brill.
Ridley, R. T. 1978. "Was Scipio Africanus at Cannae?" *Latomus* 34, no. 1: 161–65.
Riggsby, A. M. 2006. *Caesar in Gaul and Rome: War in Words*. University of Texas Press.
Roller, M. B. 2010. "Demolished Houses, Monumentality, and Memory in Roman Culture." *Classical Antiquity* 29, no. 1: 117–80.
Romano, D. G. 2003. "City Planning, Centuriation, and Land Division in Roman Corinth: Colonia Laus Iulia Corinthiensis & Colonia Iulia Flavia Augusta Corinthiensis." In *Corinth, The Centenary, 1896–1996*, edited by C. K. Williams and N. Bookidis, 279–301. American School of Classical Studies at Athens.
Roncaglia, C. E. 2018. *Northern Italy in the Roman World: From the Bronze Age to Late Antiquity*. Johns Hopkins University Press.
Rosenstein, N. 1990. *Imperatores Victi: Military Defeat and Aristocratic Competition in the Middle and Late Republic*. University of California Press.
———. 1993. "Competition and Crisis in Mid-Republican Rome." *Phoenix* 47: 313–38.
———. 2004. *Rome at War*. University of North Carolina Press.
———. 2010. "Aristocratic Values." In *A Companion to the Roman Republic*, edited by N. Rosenstein and R. Morstein-Marx, 365–82. Wiley-Blackwell.
———. 2011a. "War, Wealth, and Consuls." In *Consuls and Res Publica: Holding High Office in the Roman Republic*, edited by H. Beck, A. Dupla' Ansuátegui, M. Jehne, and F. Pina Polo, 133–58. Cambridge University Press.
———. 2011b. "Military Command, Political Power, and the Republican Elite." In *A Companion to the Roman Army*, edited by P. Erdkamp, 132–47. Wiley-Blackwell.
———. 2012. *Rome and the Mediterranean, 290–146 B.C.: The Imperial Republic*. Edinburgh University Press.
Roth, J. 1999. *The Logistics of the Roman Army at War (264 B.C.–A.D. 235)*. Brill.
———. 2006. "Siege Narrative in Livy: Representation and Reality" In *Representations of War in Ancient Rome*, edited by S. Dillon and K. Welch, 49–67. Cambridge University Press.
———. 2007. "War." In *The Cambridge History of Greek and Roman Warfare*. Vol. 1: *Greece, the Hellenistic World, and Rome*, edited by P. Sabin, H. Van Wees, and M. Whitby, 368–98. Cambridge University Press.
Rubinsohn, Z. W. 1981. "The Viriatic War and Its Roman Repercussions." *Revista Storica dell'Antichità* 11: 161–204.
Ruiz Valderas, E., and M. J. Madrid Balanza. 2002. "Las Murallas de Cartagena en la Antigüedad." In *Estudio y catalogación de las defensas de Cartagena y su bahía*, edited by A. I. Sanmartín and J. A. Martínez López, 19–84. Comunidad Autónoma de la Región de Murcia.
Rüpke, J. 1995. "Wege zum Töten, Wege zum Ruhm: Krieg in der römischen Republik." In *Töten im Krieg*, edited by H. von Stietencron and J. Rüpke, 213–40. K. Alber.
———. 1999. "Kriegsgefangene in der römischen Antike." In *In der Hand des Feindes: Kriegsgefangenschaft von der Antike bis zum Zweiten Weltkrieg*, edited by R. Overmans, 85–98. Böhlau.
Russell, J. M. 1992. *Sennacherib's Palace without Rival at Nineveh*. University of Chicago Press.
Rutledge, S. H. 2007. "The Roman Destruction of Sacred Sites." *Historia: Zeitschrift für Alte Geschichte* 56, no. 2: 179–95.

Sabin, P. 1996. "The Mechanics of Battle in the Second Punic War." *Bulletin of the Institute of Classical Studies* 41, no. 67: 59–79.

———. 2000. "The Face of Roman Battle." *Journal of Roman Studies* 90: 1–17.

Sakowicz, K. 2005. *Ponary Diary, 1941–1943: A Bystander's Account of a Mass Murder*, edited by Y. Arad, translated by L. Weinbaum. Yale University Press.

Salerian, A. J., P. Tuglaci, G. Salerian, J. B. Edwards, A. Baum, and B. Mendelsohn. 2007. "Review of Mass Homicides of Intelligentsia as a Marker for Genocide." *Forensic Examiner* 16, no. 3: 34–41.

Sánchez Moreno, E. 2002. "Algunas notas sobre la guerra como estrategia de interacción social en la Hispania prerromana: Viriato, jefe redistributive (y II)." *Habis* 33: 149–69.

———. 2005. "Warfare, Redistribution, and Society in Western Iberia." In *Warfare, Violence and Slavery in Prehistory. Proceedings of a Prehistoric Society Conference at Sheffield University*, edited by M. Parker-Pearson and I. J. N. Thorpe, 107–25. Archaeopress.

———. 2006. "Ex pastore latro, ex latrone dux . . . Medioambiente, guerra y poder en el occidente de Iberia." In *Guerra y territorio en el mundo romano*, edited by T. Ñaco del Hoyo and I. Arrayás Morales, 55–79. John and Erica Hedges.

Saunders, J. J. 1971. *The History of the Mongol Conquests*. University of Pennsylvania Press.

Scheidel, W. 2011. "The Roman Slave Supply." In *The Cambridge World History of Slavery*. Vol. 1: *The Ancient Mediterranean World*, edited by K. Bradley and P. Cartledge, 287–310. Cambridge University Press.

Scopacasa, R. 2016. "Rome's Encroachment on Italy." In *A Companion to Roman Italy*, edited by A. Cooley, 35–56. Wiley-Blackwell.

Scranton, R. L. 1951. *Corinth: Monuments in the Lower Agora and North of the Archaic Temple*. American School of Classical Studies at Athens.

Scullard, H. H. 1930. *Scipio Africanus in the Second Punic War*. Cambridge University Press.

Sear, F. 2006. *Roman Theatres: An Architectural Study*. Oxford University Press.

Sekunda, N., and P. De Souza. 2007. "Military Forces." In *The Cambridge History of Greek and Roman Warfare*. Vol. 1: *Greece, the Hellenistic World, and Rome*, edited by P. Sabin, H. Van Wees, and M. Whitby, 325–67. Cambridge University Press.

Sémelin, J. 2014. *Purify and Destroy: The Political Uses of Massacre and Genocide*. Hurst and Company.

Serrati, J. 2007. "Warfare and the State." In *The Cambridge History of Greek and Roman Warfare*. Vol. 1: *Greece, the Hellenistic World, and Rome*, edited by P. Sabin, H. Van Wees, and M. Whitby, 461–97. Cambridge University Press.

———. 2013. "The Hellenistic World at War: Stagnation or Development?" In *The Oxford Handbook of Warfare in the Classical World*, edited by B. Campbell and L. A. Tritle, 179–98. Oxford University Press.

Shatzman, I. 2011. "The Roman Republic: From Monarchy to Julius Caesar." In *The Practice of Strategy: From Alexander the Great to the Present*, edited by J. A. Olsen and C. S. Gray, 36–56. Oxford University Press.

Shaw, M. 2003. *War and Genocide: Organized Killing in Modern Society*. Polity Press.

———. 2004. "New Wars of the City: Relations of 'Urbicide' and 'Genocide.'" In *Cities, War, and Terrorism*, edited by S. Graham, 141–53. Blackwell.

———. 2009. "Genocide in the Global Age." In *The Routledge Handbook of Globalisation Studies*, edited by B. S. Wilson, 312–27. Routledge.

———. 2013. *Genocide and International Relations: Changing Patterns in the Transitions of the Late Modern World*. Cambridge University Press.

Sherwin-White, A. N. 1980. "Rome the Aggressor?" Review of *War and Imperialism in Republican Rome, 327–70 B.C.*, by W. V. Harris. *Journal of Roman Studies* 70: 177–81.

Shipley, G. 2000. *The Greek World after Alexander, 323–30 BC*. Routledge.

Sidebottom, H. 2005. "Roman Imperialism: The Changed Outward Trajectory of the Roman Empire." *Historia: Zeitschrift für Alte Geschichte* 54, no. 3: 315–30.

Silva, A. C. F. 1983–1984. "A cultura castreja no Noroeste de Portugal: habitat e cronologias." *Portugalia* 4/5: 121–29.

Simon, H. 1962. *Roms Kriege in Spanien, 154–133 v. Chr.* V. Klostermann.

Snodgrass, A. M. 1987. *An Archaeology of Greece: The Present State and Future Scope of the Discipline*. University of California Press..

Southern, P. 2006. *The Roman Army: A Social and Institutional History*. Oxford University Press.

Spivey, N., and M. Squire. 2004. *Panorama of the Classical World*. Thames & Hudson.

Steinbock, B. 2013. *Social Memory in Athenian Public Discourse: Uses and Meanings of the Past*. University of Michigan Press.

Stone, S. C. 2014. *Morgantina Studies*. Vol. VI: *The Hellenistic and Roman Fine Pottery*. Princeton University Press.

Straus, S. 2006. *The Order of Genocide: Race, Power, and War in Rwanda*. Cornell University Press.

———. 2010. "Political Science and Genocide." In *The Oxford Handbook of Genocide Studies*, edited by D. Bloxham and A. D. Moses, 163–81. Oxford University Press, 2010.

———. 2012. "'Destroy Them to Save Us': Theories of Genocide and the Logics of Political Violence." *Terrorism and Political Violence* 24, no. 4: 544–60.

———. 2015. "What Is Being Prevented? Genocide, Mass Atrocity, and Conceptual Ambiguity in the Anti-Atrocity Movement." In *Reconstructing Atrocity Prevention*, edited by S. P. Rosenberg, T. Galis, and A. Zucker, 17–30. Cambridge University Press.

Suha, M. 2011. "Further Observations on the Hellenistic Fortifications in the Kokytos Valley." In *Thesprotia Expedition II: Environment and Settlement Patterns*, edited by B. Forsén and E. Tikkala, 203–44. Finish Institute at Athens.

Sullivan, C. 2012. "Blood in the Village: A Local-Level Investigation of State Massacres." *Conflict Management and Peace Studies* 29, no. 4: 373–96.

Sumi, G. 2015. *Ceremony and Power: Performing Politics in Rome between Republic and Empire*. University of Michigan Press.

Taylor, E. R. 2006. *If We Must Die: Shipboard Insurrections in the Era of the Atlantic Slave Trade*. Louisiana State University.

Taylor, M. 2014. "Roman Infantry Tactics in the Middle Republic: A Reassessment." *Historia* 63: 301–22.

———. 2016. "The Battle Scene on Aemilius Paullus's Pydna Monument: A Reevaluation." *Hesperia* 85, no. 3: 559–76.

Thein, A. 2016. "Booty in the Sullan Civil War of 83–82 B.C." *Historia* 65, no. 4: 450–72.

Thill, E. W. 2011. "Depicting Barbarism on Fire: Architectural Destruction on the Columns of Trajan and Marcus Aurelius." *Journal of Roman Archaeology* 24: 283–312.

Thompson, F. H. 1993. "Iron Age and Roman Slave-Shackles." *Archaeological Journal* 150, no. 1: 57–168.

———. 2003. *The Archaeology of Greek and Roman Slavery*. Duckworth in association with Society of Antiquaries of London.
Thompson, H. A. 1980. "Stone, Tile and Timber: Commerce and Building Material in Classical Athens." *Expedition* 22, no. 3: 12–26.
Thompson, M. W. 1987. *The Decline of the Castle*. Cambridge University Press.
Thornton, J. 2006. "Terrore, terrorismo e imperialismo. Violenza e intimidazione nell'età della conquista romana." In *Terror et pavor. Violenza, intimidazione, clandestinità nel mondo antico*, edited by G. Urso, 157–96. Edizioni ETS.
Tomlinson, R. A. 2006. "Fortifications." In *Encyclopedia of Ancient Greece*, edited by N. Wilson, 79–81. Routledge.
Toner, J. 2013. *Roman Disasters*. Polity Press.
Tränkle, H. 2010. "Livy and Polybius." In *Livy*, edited by J. D. Chaplin and C. S. Kraus, 476–95. Oxford University Press.
Ulrich, R. B. 2007. *Roman Woodworking*. Yale University Press.
United States National Geospatial-Intelligence Agency. 2014. *Sailing Directions (enroute). Western Mediterranean*. National Geospatial-Intelligence Agency.
Valentino, B. 2004. *Final Solutions: Mass Killing and Genocide in the 20th Century*. Cornell University Press.
———. 2014. "Why We Kill: The Political Science of Violence against Civilians." *Annual Review of Political Science* 17: 89–103.
Valentino, B., P. Huth, and D. Balch-Lindsay. 2004. "'Draining the Sea': Mass Killing and Guerrilla Warfare." *International Organization* 58, no. 2: 375–407.
Van Wees, H. 2010. "Genocide in the Ancient World." In *The Oxford Handbook of Genocide Studies*, edited by D. Bloxham and A. Dirk Moses. Oxford University Press.
———. 2016. "Genocide in Archaic and Classical Greece." In *Our Ancient Wars: Rethinking War through the Classics*, edited by V. Caston and S.-M. Weineck. University of Michigan Press.
Verberne, L. 1993. Review of *Die Massenversklavungen der Einwohner eroberter Städte in der hellenistisch-römischen Zeit*, by H. Volkmann. *Mnemosyne* Fourth Series 46, no. 2: 276–78.
Walbank, F. W. 1957–1979. *A Historical Commentary on Polybius*. 3 vols. Clarendon Press.
———. 1971. "The Fourth and Fifth Decade." In *Livy*, edited by T. A. Dorey, 47–72. Routledge & Kegan Paul.
———. 1985a. "The Scipionic Legend." In *Selected Papers: Studies in Greek and Roman History and Historiography*, 120–37. Cambridge University Press.
———. 1985b. "Speeches in Greek Historians." In *Selected Papers: Studies in Greek and Roman History and Historiography*, 242–61. Cambridge University Press.
Waller, M. 2011. "Victory, Defeat and Electoral Success at Rome, 343–91 BC." *Latomus* 70: 18–38.
Walsh, P. G. 1982. "Livy and the Aims of 'Historia': An Analysis of the Third Decade." *ANRW* 2.30.2: 1058–74.
Walvin, J. 2013. *Crossings: Africa, the Americas and the Atlantic Slave Trade*. Reaktion Books.
Warrior, V. 1996. *Initiation of the Second Macedonian War: An Explication of Livy Book 31*. F. Steiner.
Waterfield, R. 2014. *Taken at the Flood: The Roman Conquest of Greece*. Oxford University Press.
Westington, M. M. 1938. "Atrocities in Roman Warfare to 133 B.C." PhD dissertation, University of Chicago.

Westlake, H. D. 1985. "The Sources for the Spartan Debacle at Haliartus." *Phoenix* 39, no. 2: 119–33.

Wette, W. 2006. *The Wehrmacht: History, Myth, Reality*. Harvard University Press.

Whatley, N. 1964. "On the Possibility of Reconstructing Marathon and Other Ancient Battles." *Journal of Hellenic Studies* 84: 119–39.

Whitby, M. 2007. "Reconstructing Ancient Warfare." In *The Cambridge History of Greek and Roman Warfare*. Vol. 1: *Greece, the Hellenistic World, and Rome*, edited by P. Sabin, H. Van Wees, and M. Whitby, 54–84. Cambridge University Press.

White, D. 1964. "Demeter's Sicilian Cult as a Political Instrument." *Greek, Roman and Byzantine Studies* 5: 261–79.

Wickham, J. 2014. "Mass Enslavement of War Captives by the Romans to 146 BC." PhD dissertation, University of Liverpool.

Williams, J. 2017. *The Archaeology of Roman Surveillance in the Central Alentejo, Portugal*. California Classical Studies.

Wiseman, J. 1979. "Corinth and Rome I: 228 B.C.–A.D. 267." *Aufstieg und Niedergang der römischen Welt* 2.7.1: 438–548.

Wood, E. J. 2006. "Variation in Sexual Violence during War." *Politics & Society* 34: 307–41.

———. 2014. "Conflict-Related Sexual Violence and the Policy Implications of Recent Research." *The International Review of the Red Cross* 96, no. 894: 457–78.

Wood, R. M. 2014. "From Loss to Looting? Battlefield Costs and Rebel Incentives for Violence." *International Organization* 68, no. 4: 979–99.

Yakobson, A. 2009. "Public Opinion, Foreign Policy and 'Just War' in the Late Republic." In *Diplomats and Diplomacy in the Roman World*, edited by C. Eilers, 45–72. Brill.

———. 2010. "Traditional Political Culture and the People's Role in the Roman Republic." *Historia* 30: 1–21.

Yang, S. 2017. "Letting the Troops Loose: Pillage, Massacres, and Enslavement in Early Tang Warfare." *Journal of Chinese Military History* 6: 1–52.

Yarrow, L. M. 2006. *Historiography at the End of the Republic: Provincial Perspectives on Roman Rule*. Oxford University Press.

———. 2012. "Decem Legati: A Flexible Institution, Rigidly Perceived." In *Imperialism, Cultural Politics, and Polybius*, edited by C. Smith and L. M. Yarrow, 168–83. Oxford University Press.

Zimmermann, K. 2011. "Roman Strategy and Aims in the Second Punic War." In *A Companion to the Punic Wars*, edited by D. Hoyos, 280–98. Wiley-Blackwell.

Zimmermann, M. 2006. "Violence in Antiquity Reconsidered." In *Violence in Late Antiquity*, edited by H. A. Drake, 343–58. Ashgate.

———. 2013. *Gewalt: Die Dunkle Seite der Antike*. Deutsche Verlags-Anstalt.

Ziolkowski, A. 1986. "The Plundering of Epirus in 167 B.C.: Economic Considerations." *Papers of the British School at Rome* 54: 69–80.

———. 1990. "Credibility of Numbers of Battle Captives in Livy, Books XXI–XLV." *Parola Del Passato* 45: 15–36.

———. 1993. "*Urbs Direpta*, or How the Romans Sacked Cities." In *War and Society in the Roman World*, edited by J. Rich and G. Shipley, 69–91. Routledge.

Index

Page references for figures are italicized.

Abdera, 144, 146–48, 155–56, 167–68, 206. *See also* Hortensius, Lucius
Acerrae, 57
Achaean League, Achaeans, 31, 134, 153, 157, 161–62
Achilles, 47, *48*, 49
Aegina, 151–52
Aequi, Aequians, 23
Aeschines, 99
Aetolian League, Aetolians, 68, 82, 134, 136–38, 148, 152–54, 157, 161–62, 167
Agathocles (tyrant of Syracuse, 317–289 BCE), 47, 210
Agrigentum, 73, 103, 110–11, 126, 204
Albinus, Lucius Postumius, 179, 183
Alexander the Great (Macedonian King, 336–323 BCE), 29, 33, 133, 149–51, 210
Alexandria (Egypt), 55
Ambracia, 82–84
Ammianus Marcellinus, 6–7, 50
Amphipolis, 161
andrapodize, andrapodizing, 13, 43, 163n
Andros, 44
Anicius, Lucius, 160

Anticyra, 152–53
Antigonid dynasty. *See* Macedonia, Macedonians
Antiochus III (Seleucid King, 222–187 BCE), 150
Antipater of Thessalonica, 45–46
Antipatrea, 33–34, 36, 39, 51, 57–60, 72
Antissa, 155n, 162, 167
Antium, 146
Antron, 143, 146
Apame, 139
Aperantia, 154
Appian of Alexandria, 3, 12, 34–35, 49, 53, 71, 73, 121, 123, 153, 157, 167, 177, 181, 183–187, 190, 192, 195, 207
Apulia, Apulians, 20, 95, 98
Apustius, Lucius, 33–34, 59–60, 72
Aquilonia, 51
Archimedes, 102–3, 109
Arevaci. *See* Celtiberians
Arpi, 98
arson. *See* urban destruction, methods
Artaxata, 72
Assyria, Assyrians, 13, 54, 55
Atella, 100–101

271

Athamania, 134
Athens, Athenians, 13, 55, 147–48, 149, 167
Atilius, Marcus (Lusitanian War, praetor), 181–82, 185, 189, 196
Atlanta, devastation of, 55–56
Atrax, 38
Atticus, Titus Pomponius, 68–69
Ausonians, 23

Babri Mosque, destruction of, 56
Baecula, Battle of, 120
Baeturia, 192
Belisarius, 53
Bergium, 43, 45
Boeotia, Boeotians, 139, 141–43, 145–46, 161, 167, 206
booty. *See* plunder, plundering
Bruttium, Bruttians, 20, 95
Brutus, Decimus Junius (Lusitanian War, consul), 195–96, 203–4

Caepio, Quintus Servilius (Lusitanian War, consul), 194–95. *See also* Viriathus, assassination of
Caesar, Gaius Julius, 38, 55, 73–74, 78, 79, 84
Calatia, 100–101
Cales, 99
Callaeci, 195
Callicinus, Battle of, 134, 136–38, 140, 144, 146, 148
Calpurnius Piso, Lucius (Lusitanian War, praetor), 176, 178
Campania, Campanians, 20, 95, 97–101, 111, 125–26, 201, 204–5, 208
Cannae, Battle of, 89–91, 93, 95, 96, 100, 113, 125, 158, 187, 206
Canusium, 90
Capsa, 71, 76
captives: methods of taking, 40–44, *41*; ransomed or otherwise spared, 44, 150–51. *See also* mass enslavement; mass killing; mass violence
Capua: besieged and captured by Rome, 98–101, 111, 205; leaders publicly executed, 99–100; joins Hannibal, 95, 97; punished by Rome, 6, 8, 75, 95, 100–101, 125–26, 201, 204. *See also* Second Punic War
Carmo, 184
Carnuns, 145
Carpetania, 194
Carystus. *See* Statellates
Carthage, Carthaginians, 2–4, 6–7, 23, 26–28, 31, 38, 40, 55–58, 71, 73–75, 78–79, 89–93, 96, 98–101, 103–4, 106, 108, 110–28, 138, 180–81, 187, 201, 203–6, 208–210. *See also* Carthage, destruction of; First Punic War; Hannibal Barca; Second Punic War; Third Punic War
Carthage, destruction of, 2–4, 6–7, 11, 23, 28, 57–58, 74–75, 100, 125, 201, 209; reactions to, 4
Casilinum, 98
Cassius Dio, 71, 191
Cassius Longinus, Gaius (Third Macedonian War, consul), 144–45, 147
Castulo, 121–22
Cato, Marcus Porcius "the Elder," 28, 43–44, 71, 74, 77, 166, 208; speeches against Galba, 187–88
Cato, Marcus Porcius "the Younger," 69
Cauca, 34–35, 39
Celtiberians, 74, 91, 177–79, 181–83, 189, 191
Ceremia, 145–46, 203
Cerro de la Cruz, 192–93
Cethegus, Marcus Cornelius (Second Punic War, praetor), 110
Cethegus, Publius Cornelius (Sullan commander), 49
Chaonians. *See* Epirus
Chalcis, 57, 72, 142, 144, 146–48, 156, 167–68, 203. *See also* Hortensius, Lucius; Lucretius, Gaius
Chiomara, 45
Cicero, Marcus Tullius, 2, 6–7, 46–47, 67–75, 77–78, 80, 83, 85–86, 100, 104, 175, 182, 192. *See also* Pindenissus
Cilicia, 67–70, 77
Cirrha, 53
Cisalpine Gaul, 22, 92
city-sacking, 5, 10, 13, 35, 45, 51, 59, 68–69, 73, 75–76, 79–86, 92, 102–3,

109, 126, 143, 147, 152–53, 163–64, 192, 201–3, 210; process of, 36, 39–40; as tool for commanders, 60, 85, 123, 145–46, 161, 166, 192, 203, 209. *See also* plunder, plundering; mass violence, economic motives
Claudius Pulcher, Appius (Second Punic War, consul), 98–99
Column of Marcus Aurelius, 40, *41*, 46–47, *48*, 49, 51, *52*
Cominium, 51
Conii, 184–85
Connobas, 192
consul. *See* Rome, government
contiones, 156
Corinth, destruction of, 1–2, 4, 6–7, 11, 23, 31, 43, 45–46, 57–58, 71, 73–75, 78, 100, 201, 209; reactions to, 2. *See also* Mummius, Lucius
Coronea, 141, 143–47, 156, 167–68
Corsica, 28, 31, 208
Cremona, 51
Croton, 95
Cynoscephalae, Battle of, 139, 151

Dardanians, 154
decem legati (ten commissioners), 161–62
deditio (ritual surrender to Rome), 80–85, 143, 185, 188
Delphi, 53, 140, *160*
Demetrius "the Besieger," 150
demolition. *See* urban destruction, methods
Didius, Titus, 186
Diodorus Siculus, 2, 4, 6–7, 12, 46–47, 51, 74, 77, 109, 177, 183–84, 209
Dionysius of Halicarnassus, 47
diripio, diripere. *See* city-sacking
Dodona, 164
Duillius, Gaius, 78
Dyme, 153
Dyrrachium, 144

Elea, 164, *165*
Enna: massacre at, 71, 104–106, 111, 125–26, 186, 202; strategic significance, 104. *See also* Pinarius, Lucius
Epicydes (pro-Carthage Syracusan leader), 96, 101–2, 108

Epirus, 134, 150, 160–61; defection to Perseus, 73, 148, 155, 195; devastation of, 73, 76, 162–64, *165*, 166–67, 182, 201, 203, 206; strategic significance of, 155, 167. *See also* Paullus, Lucius Aemilius (Third Macedonian War, consul); Third Macedonian War
Etruria, Etruscans, 20–22, 47, *48*, 49, 92, 95, 112
Euboea, 154
Eumenes II (King of Pergamum, 197–159 BCE), 134, 140, 157–58
Eusebius, 49–50

Fabius Maximus, Quintus (Second Punic War, consul and dictator), 92–93, 98, 111, 122
Fabius Maximus, Quintus (Second Punic War, son of the dictator), 98
Fabius Maximus Aemilianus, Quintus (Lusitanian War, consul), 192
Fabius Pictor, Quintus, 113n
Falerii, 7, 24, 208
Fidenae, 7
First Maccabees, 7, 210
First Macedonian War (214–205 BCE), 29, 95, 112, 138, 152–53
First Punic War (264–241 BCE), 3n, 27–28, 75, 91–92, 95, 185
First Roman-Jewish War (66–73 CE), 38, 78. *See also* Jerusalem, siege and destruction of
forced migration, 7n, 44, 82, 162
Flamininus, Titus Quinctius, 79, 84, 139, 155
Flavius Fimbria, Gaius, 81
François Tomb at Vulci, *48*, 49
Fregellae, 6, 58
Frontinus, 71, 192
Fulvius Flaccus, Quintus (Second Punic War, consul), 98–100. *See also* Capua

Gades, 120
Galatians, 45
Galba, Servius Sulpicius (Lusitanian War, praetor), 25, 190–92, 194; career prior to praetorship, 182; defeated by Lusitanians, 183–84; invades Lusitanian territory with Lucullus,

184–85; massacres Lusitanians, 71, 185–87, 189, 196, 202–3, 205–9; political controversy over massacre, 188; political pressure to defeat Lusitanians, 183, 189. *See also* Lusitanians, massacre of; Lusitanian War (155–139 BCE); Viriathus
Gaul, Gallic tribes, 20, 22, 23, 28, 38n, 73–74, 78–79, 89, 92, 101, 112, 134, 159
Genthius, 144, 154, 157, 160–61, 167
Gonnus, 141, 143
gladius (Roman sword), 27, 153, 159, 193, 246
Gracchus, Tiberius Sempronius, 77–78, 179–80, 183
Great Plains, Battle of the, 123
Greece, Greeks, 1–2, 5–6, 11, 21–22, 28–29, 53, 58, 73–74, 79, 82, 95, 109, 133–34, 136–157, 159, 161–68, 181, 205–7; reactions to Roman military conduct, 147–48, 152–53, 210. *See also* Hellenistic military norms; Third Macedonian War

Haliartus: defenses and strategic significance, 141, 145; dissolved as a community by Roman Senate, 168; Roman attack on and destruction of, 141–142, 145–46, 167, 206. *See also* Lucretius, Gaius
Hamilcar Barca, 91–92
Hannibal Barca, 28–29, 38n, 97–98, 100, 111–13, 120, 122–26, 178, 204–5; battlefield victories, 89–93, 112, 138, 158, 205–6; failure to protect Italian allies, 99, 101; invasion of Italy, 92–93; strategy, 93; use of mass violence, 210; wins allies in Italy and abroad, 95–96, 138. *See also* Cannae, Battle of; Second Punic War; Zama, Battle of
Hasdrubal (Carthaginian commander in Third Punic War), 3–4
Hasdrubal Barca, 112, 120
Hasdrubal "the Fair," 92, 114
Hellenistic military norms, 149–51; Roman violations of, 151–52. *See also* Greece, Greeks, reactions to Roman violence

Heraclea Pontica, 51
Herdonea, 112
Hiero II (tyrant of Syracuse, 270–215 BCE), 95–96
Hieronymus (tyrant of Syracuse, 215–214 BCE), 96
Himilco (Carthaginian commander), 103, 108
Hippocrates (pro-Carthage Syracusan leader), 96, 101–2, 108
Hortensius, Lucius (Third Macedonian War, praetor): censured by Roman Senate, 156; violence against Greeks, 144–47, 203. *See also* Abdera; Chalcis; Third Macedonian War
Hostilius, Aulus (Third Macedonian War, consul), 148, 154–56

Iberians, 28, 71, 79, 89, 91–92, 108, 113–14, 119–22, 126, 166–67, 178–79, 191, 208, 210
Ilipa, 120
Ilium, 53
Illyria, Illyrians, 73, 95, 138, 144–46, 154, 157, 160–62, 167, 203, 210
Ilurgia (Iliturgi), 34–35, 53, 121–22, 126
imperium, 24, 207–9, 244–45
Isaura, 78
Italian allies. *See socii*
Itucca, 192

Japha, 38
Jerusalem, siege and destruction of, 39, 42, 44, 55, 74
Josephus, 38–39, 55, 78

Laelius, Gaius (Lusitanian War, praetor), 192, 196
Laelius, Gaius (Second Punic War, commander), 113n, 114, 117, 124
Laenas, Marcus Popilius, 81–85, 188
Laevinus, Marcus Valerius (Second Punic War, consul), 73, 110–11
Lake Trasimene, Battle of, 92–93
Laodice, 139
Larisa Cremeste, 143, 146
Latins, Latin Colonies, 20–21, 23, 95, 112, 159
Latium, 20

legion. *See* Rome, armies
Leontini, 96, 101–2
Lepidus, Marcus Aemilius, 82–83. *See also* Ambracia
Leuven, burning of, 52
Lewes Priory, demolition of, 55
Libo, Lucius Scribonius, 187–88
Licinius Crassus, Publius (Third Macedonian War, consul), 134, 136–37, 140–41, 143–47, 155, 185, 206. *See also* Callicinus, Battle of; Coronea; Perseus; Pteleum; Third Macedonian War
Liguria, Ligurians, 28, 44, 81–82, 84, 112, 157–58, 188
Livy, 25, 33, 38, 40, 43, 51, 53, 55–59, 71–73, 79–80, 82–83, 91–92, 96–114, 116–18, 121, 124–25, 134, 136, 142–47, 153, 160, 162n, 163n, 164, 166, 186, 188; on mass violence, 11–12, 34–35
Locha, 123
Locri, 95
Longus, Tiberius Sempronius (Second Punic War, consul), 92
looting. *See* plunder, plundering
Lucania, Lucanians, 20, 95, 98
Lucretius Gallus, Gaius (Third Macedonian War, praetor): censured by Roman Senate and fined, 156; violence against Greeks, 141–47, 203. *See also* Chalcis; Haliartus; Third Macedonian War
Lucullus, Lucius Licinius (Lusitanian War, consul), 34–35, 185, 190
Lusitanians, 49, 71, 91, 175–97, 201–9 *See also* Galba, Servius Sulpicius; Lusitanian War; Viriathus
Lusitanian War (155–139 BCE), 175–97; Lusitanian raids, 175–77, 181, 183, 190; Lusitanian way of war, 183–84; origins, 178–80; Roman anxiety, frustration, 181–83, 189, 197, 203; Roman invasions of Lusitania, 180–81, 185, 192–96; Roman use of mass violence, 185–89, 192–96, 201–3, 205–9; Viriathus's war against Rome, 190–95. *See also* Galba, Servius Sulpicius; Lusitanians; Viriathus

Macedonia, Macedonians, 29, 31, 33, 38, 53, 56, 68, 74, 76, 95, 112, 133–34, 135–41, 143–54, 155n, 157–62, 166–68, 181–82, 203, 205–8, 210; army, 133–34, 151, 159. *See also* First Macedonian War; Second Macedonian War; Third Macedonian War
Macella, 78
Mago (Carthaginian garrison commander), 116–19
Mago Barca, 112
Malian Gulf, 141, 143, 145, 206
Malloea, 143, 146.
Manilius, Manlius (Lusitanian War, praetor), 175–76, 178
Mantinea, 151
Marcellus, Marcus Claudius (Second Punic War, consul), 104, 154; approves of massacre at Enna, 106; attacks, besieges Syracuse, 102–3, 106–8; campaigns against Hannibal, 111; campaigns against Italian rebels, 98; career and character, 101–2; destroys Megara, 73, 111, 125; killed in ambush, 112; sacks Leontini, 102; sacks Syracuse, 108–10. *See also* Enna; Second Punic War; Syracuse
Marcellus, Marcus Claudius (Lusitanian War, consul), 180–81, 185, 189
Marius, Gaius, 71, 76, 81
mass enslavement, 1, 6–8, *8*, 23, 40–44, 67–68, 71, 73, 78–79, 81–84, 98, 100, 109–111, 120, 124, 126, 142–45, 147, 151–53, 156, 164, 187–88, 192, 201–4, 206–211; methods, 44–46, 202; profits, 75–77, 150. *See also* mass enslavement; mass killing; Rome, attitudes towards war, violence; urban destruction
mass execution. *See* mass killing
mass killing, 1, 5–8, *8*, 23–24, 33–35, 44, 59–60, 71–73, 77–79, 91, 98–100, 102, 106, 110–11, 119–21, 125–27, 142, 144, 146, 149, 159, *160*, 161, 185–89, 192, 196, 201–2, 206–8, 210; during assaults on cities, 38–40; execution methods, 46–47, *48*, 49–50. *See also* city-sacking; mass enslavement;

mass violence; Rome, attitudes toward war, violence; urban destruction
mass violence: ancient sources, 9–12, 34–35; anger and, 102, 109, 124, 126, 203; characteristic feature of Roman warfare, 5–7, 201, 209–210; common in ancient warfare, 210–11; counterproductive, 98, 100, 102, 106, 111, 125–26, 147–48, 154–56, 205; definition, 7–8, *8*; economic motives, 75–76, 123, 145–47, 166, 203; effectiveness, 204–5; effort and organization in, 44, 49–52, 56–57, 59, 126, 145, 166, 187; exemplary punishment, 23–24, 72–75, 98–100, 103, 106, 109–111, 121–22, 125–26, 146–47, 166, 193–96, 203, 206–7; failure on the battlefield and, 91, 196, 205–6; historiography, 13–14; imperialism and, 207–9; instrumental strategy, 85–86, 202–4; leaders' role in, 59–60, 202; politically motivated, 77–79, 147, 197, 203–4; pragmatic motives, 71–72, 119, 124–26, 145–46, 166–67, 188–89, 194, 196, 203, 206; restraints on, 79–81, 84–85, 149–51, 155–57; theoretical approaches, 14–15. *See also* city-sacking; mass enslavement; mass killing; Rome, attitudes towards war, violence; urban destruction
massacre. *See* mass killing
Massilia, 60
Massinissa, 122–24
Megara Hyblaea, 73, 103, 111, 125
Memnon of Heraclea, 51
Metaurus, Battle of the, 122
Metellus, Lucius Caecilius, 25
Metellus, Marcus Caecilius, 90–91
Molossians. *See* Epirus
Montgomery Castle, demolition of, 56
Morgantina, 103–4, *105*, 106, 110
Mummius, Lucius, 1–2, 43, 78, 177–78, 180, 183–84, 189, 196. *See also* Corinth, destruction of
Mylae, 210

Nero, Gaius Claudius, 112
New Carthage: defenses and strategic significance, 114–17, *115*, *116*; division of captives, 43, 120; Romans assault, 38, 113n, 117–20; Roman massacre, 5, 60, 73, 119–21, 126, 204, 206. *See also* Scipio Africanus, Publius Cornelius; Mago (Carthaginian garrison commander); Second Punic War
Nobilior, Marcus Fulvius, 82–85
Nola, 98, 102, 125
Numantia, 6–7, 43, 74
Numidia, Numidians, 71, 89, 99, 101, 110, 122–23, 159

Onasander, 79
Oreum, 152
Orongis, 121
Orraon (Horreum), 164, *165*
Ovid, 196
Oxthracae, 181

Panormus, 44
patres. *See* Senate, Roman
Paullus, Lucius Aemilius (Second Punic War, consul), 89–90, 158
Paullus, Lucius Aemilius (Third Macedonian War, consul): campaign in Macedonia, 158–60, *160*; career and character, 158; conflict with soldiers over loot, 76, 182; devastates Epirus, 73, 76, 162–64, *165*, 166–67, 203, 206; settlement of Greece and Macedonia, 161–62, 168, 207. *See also* Perseus; Pydna, Battle of; Third Macedonian War
Perseus (Macedonian King, 179–168 BCE), 74, 76, 133–34, 138–46, 150, 156–62, 167–68, 203, 205–8; military successes against Rome, 136–37, 154, 205–6; revives Macedonia's fortunes, 139–140; use of mass violence, 141, 210; wins Greek support, 137–38, 141, 148, 154–55, 157–58, 161–62, 205. *See also* Callicinus, Battle of; Macedonia, army; Pydna, Battle of; Third Macedonian War

Persia, Persians, 29, 210
phalanx. *See* Macedonia, army
Phanote (Doliani), 155, 160, 164, 167
Philip II (Macedonian King, 359–336 BCE), 29
Philip V (Macedonian King, 238–179 BCE), 112, 149, 151; alliance with Hannibal, 95, 138; rebuilds kingdom after Second Macedonian War, 139. *See also* Cynoscephalae, Battle of; Second Macedonian War
Philippus, Quintus Marcius, 157–58
Phocaea, 40
Picenum, 112
Pinarius, Lucius, 71, 104–6, 111, 125, 202. *See also* Enna; Second Punic War
Pindenissus, 44, 67–70, 72–73, 75, 77–79, 83, 85
Piraeus, 55
plunder, plundering, 2, 4, 6, 10, 13, 15, 23, 26, 33–34, 39–40, 51–53, 54, 59–60, 68–69, 78, 83, 98, 106, 108–9, 120, 142–44, 147–48, 156, 163, 175–77, 179–80, 185, 187, 189, 191, 196, 210; sought by Roman commanders, soldiers, 69, 75–76, 108, 123, 146, 166, 182, 203, 209. *See also* city-sacking; mass violence, economic motives
Plutarch, 12, 43, 47, 53, 76–77, 109n, 110, 134, 150, 159–60, 163n, 166
Po River, 22, 24
Polyaenus, 106
Polybius, 2, 4, 27n, 31, 38, 40, 43, 45–46, 53, 56, 73–74, 77, 79–80, 102, 113n, 114–15, 118n, 119–20, 123–24, 137, 143, 148–52, 155, 158, 162n, 163n, 164, 181, 182, 185, 203, 207; on mass violence, 5, 7, 11–12, 210
Polystratus (poet), 2
Pontefract Castle, demolition of, 56
Praeneste, 49
praetor. *See* Rome, government
Prusias II (King of Bithynia, 182–149 BCE), 139
Pteleum, 143, 145–46
Ptolemaic Egypt, 29, 149
Punicus (Lusitanian war-leader), 175–77
Pydna (town), 158, 161
Pydna, Battle of, 158–160, *160*, 168

Pyrrhus of Epirus, 22, 29, 148n, 150, 195
quaestor, 43, 45, 90, 163, 245
Quinctius Crispinus, Titus, 112

ransom. *See* captives
Rhetorica ad Herennium, 6–7
Rhodes, 157, 167
Rome: alliances, 22–23; aristocracy, 24–25; armies, 27, 245–46, 246; attitudes towards war, violence, 25–26, 77–85, 137, 209; conquest of Italy, 20–24; conquest of the Mediterranean, 26–31; government, 24–25, 243–45; origins, 19–20; provinces, 31, 95, 178, 208; violence and imperialism, 207–9. *See also* mass violence; Senate, Roman; *socii*; Triumph, Roman
Rufus, Marcus Caelius, 68–69
Rwandan genocide, 39, 186

Saguntum, 92, 210
Salinator, Marcus Livius, 112
Sallust, 4, 71
Samnium, Samnites, 20, 22, 24, 95, 98, 208
Sardinia, 27, 31, 78, 208
Satricum, 23–24
Scipio Aemilianus, Publius Cornelius, 2–4, 7, 11, 40, 43, 125, 159. *See also* Carthage, destruction of; Numantia; Third Punic War
Scipio Africanus, Publius Cornelius (Second Punic War, consul): campaigns in North Africa, 122–126; campaigns in Spain, 113–122, 126; destroys Ilurgia, 121–22; failed armistice with Carthage, 123–24; massacre at New Carthage, 5, 60, 119–20; peace negotiations with Carthage, 123, 125; refuses surrender of Punic towns, 73, 124; reputation for courage, 90; stops conspirators after Cannae, 90–91; treatment of prisoners, 53, 79, 84, 120.
See also Ilurgia; New Carthage; Second Punic War; Zama, Battle of

Scipio, Gnaeus Cornelius, 112–14, 121
Scipio, Publius Cornelius (Second Punic War, consul, father of Scipio Africanus), 90, 92, 112–14, 121
Scupi, 46–47
Second Macedonian War (200–197 BCE), 29, 38, 72, 139, 146, 155
Second Punic War (218–201 BCE), 3n, 8, 28–29, 71, 73, 79, 89–126; in Africa, 122–25; in Italy, 97–101, 111–12; origins, 91–92; Roman strategy, 91, 97, 100–101, 111, 120–21, 125–26; Roman use of mass violence, 98–102, 106, 108–111, 119, 121, 123–26. in Sicily, 101–111; in Spain, 112–22; *See also* Cannae, Battle of; Hannibal Barca; Scipio Africanus, Publius Cornelius; Zama, Battle of
Seleucid Empire, 29, 31, 139, 149–50, 153
Seleucus IV (Seleucid King, 187–175 BCE), 139
Senate, Roman, 81, 84, 92, 97–100, 109–113, 122–25, 137–38, 140, 143–44, 147, 153, 161, 168, 178, 181–84, 189, 194, 197; censures violence of Roman commanders, 82, 156; debates whether to award triumphs, 26, 68–69, 76–77, 83, 177, 182; debates whether to destroy cities, peoples, 60, 100; orders destruction of cities, peoples, 2, 4, 73, 76, 162, 166; restrains conduct of Roman commanders, 155–57; role, function, 24, 244–45. *See also* Rome, government
Seneca, 15
Senones, 23, 73
Servilianus, Quintus Fabius Maximus (Lusitanian War, consul), 192–94, 196, 203, 206
Servilius, Marcus, 25, 182
Servilius Vatia, Publius, 78
sexual violence, 10, 13, 15, 19, 40, 45, 59, 143. *See also* andrapodize, andrapodizing
Sibylline Oracles, 6–7, 210
Sicily, 26–28, 31, 71, 73, 92, 95–96, 101–111, 125–26, 204, 208
siege warfare, 36

slave trade, Atlantic and American, 41
socii (Italian allies of Rome), 22–23, 46, 28, 89–91, 93, 95, 97, 99–101, 112, 125–26, 134, 159, 191, 206, 208, *246*
Sora, 23–24
Spain, 5, 12, 42–43, 46, 71, 80–81, 158, 167, 178–80, 210; Carthaginian expansion, 26, 28, 91–92; in the Lusitanian War, 175–78, 180–97, 203, 206; Roman expansion, provinces, 28–29, 31, 178–80, 208. in the Second Punic War, 28, 38, 46, 112–22, 126, 204, 208; *See also* Celtiberians; Iberians; Lusitanians; Lusitanian War
Spanish tribes. *See* Celtiberians; Iberians; Lusitanians
Sparta, 13, 149, 152
Statellates, 81–82
stipendium, 22
Strabo, 12, 73, 106, 142, 163n, 164, 183–84
strategic perspective (theoretical framework). *See* mass violence, theoretical approaches
Stratus, 154
Successors, wars of, 29, 149–50
Sulla, 49
Syracuse, 47, 51, 95–96, 101, *107*, 111, 126, 180n, 210; defenses and strategic significance, 102; joins Carthage, 102; Roman assault, siege, 102–3, 106–8; sack of, 108–110, *109*. *See also* Marcellus, Marcus Claudius; Second Punic War

Tabula Alcantarensis, 81. *See also deditio*
Tacitus, 39, 49–51, 72, 185
Tarentum, 22, 38n, 95, 98, 100, 111, 125
Tarraco, 113
Teanum, 99
terrorism. *See* mass violence, exemplary punishment
Thebes, 149
Thesprotians. *See* Epirus
Thessaly, Thessalians, 134, 140–41, 143, 154–55, 161
Third Macedonian War (171–167 BCE), 76, 133–68, 182; Epirotes join Perseus, 73, 148, 155, 195; Genthius

joins Perseus, 157; Greek reactions to Roman conduct, 147–48; Greek sympathies drift towards Perseus, 137–40, 161–62, 167–68, 205, 207; origins, 138–40; Perseus's military successes, 136–37, 154, 205–6; postwar settlement, 161–62; Roman campaigns in Greece, 134, 136–37, 140–47, 155, 167–68; Roman invasion of Macedonia, 157–61; Roman Senate censures violent commanders, 157; Roman Senate mollifies Greeks, 155–57. Roman use of mass violence, 73, 142–47, 163–68, 203, 205–8, 210; *See also* Callicinus, Battle of; Epirus; Greece, Greeks, reactions to Roman violence; Paullus, Lucius Aemilius (Third Macedonian War, consul); Perseus; Pydna, Battle of

Third Punic War (149–146 BCE), 3–4, 28. *See also* Carthage, destruction of

Thisbe, 141–43, 145–47, 156, 167–68

Thracians, 134, 136–37, 139, 159, 210

Thudydides, 55

Thurii, 98, 100, 125

Tiber River, 20

Ticinus River, Battle of, 90, 92

Titus (Roman emperor, 79–81 CE), 74, 78. *See also* Jerusalem, siege and destruction of

Trajan's Column, 42, 51, *52*, 53, *54*

Trebia, Battle of the, 92–93

Trebonius, Gaius, 60

Tribola, 190–91

tribune, military, 25, 90, 182

tribune of the plebs, 156, 187, 245

triumph, Roman, 26, 43, 69, 76–78, 83–84, 110, 133, 160, 177, 182–83, 204

Truceless War (241–237 BCE), 28, 92

Turdetania, 185, 190

urban combat, 38

urban destruction, 2–7, *8*, 11–12, 23–24, 28, 31, 33–35, 46, 59–60, 67, 69–70, 71–74, 77–79, 85, 100, 103, 110–11, 121, 124–25, 141–43, 145–46, 149–53, 162, 164, *165*, 166–68, 180–81, 192–93, 201, 203, 206–10; extent and limitations, 57–58; methods, 51–57, *52*, *54*, 202. *See also* city-sacking; mass enslavement; mass killing; mass violence; Rome, attitudes towards war, violence

urbs capta (literary motif), 10–11

Utica, 122

Uxellodunum, 74

Valens, 50

Valentia, 12, 46–47

Valerius Maximus, 12, 49, 77, 187

Varro, Gaius Terentius, 89

Veii, 7, 21–23

Veneti, 73–74,

Vetilius, Gaius, 190–91

Viriathus, 179, 193, 196, 204–8; assassination of, 195; background and character, 191; military successes against Rome, 190–92, 194, 205; use of mass violence, 210. *See also* Galba, Servius Sulpicius; Lusitanian War; Servilianus, Quintus Fabius Maximus

Vitruvius, 55, 58

Volsinii, 208

Washington, DC, burning of, 52

World War II, atrocities in, 50

Xenophon, 149

Zama, Battle of, 124, 126, 138

About the Author

Gabriel Baker received his PhD in history from the University of Iowa in 2016 and has taught in college and high school classrooms. He lives in the Chicago area with his family.

War and Society

Series Editors: Michael B. Barrett and Kyle Sinisi

The study of military history has evolved greatly over the past fifty years, and the "War and Society" series captures these changes with the publication of books on all aspects of war. The series examines not only traditional military history with its attention to battles and leaders, but it explores the broader impact of war upon the military and society. Affecting culture, politics, economies, and state power, wars have transformed societies since the ancient world. With books that cut across all time periods and geographical areas, this series reveals the history of both the conduct of war and its societal consequences.

Gabriel Baker, *Spare No One: Mass Violence in Roman Warfare*
Marc Gallicchio, *The Scramble for Asia: U.S. Military Power in the Aftermath of the Pacific War*
Brian D. Laslie, *Air Power's Lost Cause: The American Air Wars of Vietnam*
Geoffrey Megargee, *War of Annihilation: Combat and Genocide on the Eastern Front, 1941*
Lawrence Sondhaus, *German Submarine Warfare in World War I: The Onset of Total War at Sea*
Haruo Tohmatsu and H. P. Willmott, *A Gathering Darkness: The Coming of War to the Far East and the Pacific, 1921–1942*
Alan Warren, *Slaughter and Stalemate in 1917: British Offensives from Messines Ridge to Cambrai*
H. P. Willmott, *The War with Japan: The Period of Balance, May 1942–October 1943*
Thomas W. Zeiler, *Unconditional Defeat: Japan, America, and the End of World War II*

www.ingramcontent.com/pod-product-compliance
Lightning Source LLC
Chambersburg PA
CBHW072127290426
44111CB00012B/1806